Polymeric Nanofibers

ACS SYMPOSIUM SERIES **918**

Polymeric Nanofibers

Darrell H. Reneker, Editor
The University of Akron

Hao Fong, Editor
South Dakota School of Mines and Technology

Sponsored by the
ACS Division of Polymer Chemistry, Inc

American Chemical Society, Washington, DC

Library of Congress Cataloging-in-Publication Data

Polymer nanofibers / Darrell H. Reneker, editor ; Hao Fong, editor ; sponsored by the ACS Division of Polymer Chemistry, Inc.

    p. cm. — (ACS symposium series ; 918)

    "Developed from a symposium sponsored by the Division of Polymer Chemistry, Inc. at the 226th National Meeting of the American Chemical Society, New York, NY, September 7–11, 2004"—Pref.

    Includes bibliographical references and index.

    ISBN-13: 978–0–8412–3919–7 (alk. paper)

    1. Textile fibers, Synthetic—Congresses. 2. Polymers—Congresses. 3. Nanostructured materials—Congresses.

    I. Reneker, Darrell H., 1929- II. Fong, Hao, 1970- III. American Chemical Society. Division of Polymer Chemistry, Inc. IV. American Chemical Society. Meeting (226th : 2004 : New York, N.Y.). V. Series.

TS1548.5.P624   2005
677'.47—dc22

                           2005048311

The paper used in this publication meets the minimum requirements of American National Standard for Information Sciences—Permanence of Paper for Printed Library Materials, ANSI Z39.48–1984.

Distributed by Oxford University Press

ISBN 10: 0-8412-3919-3

PRINTED IN THE UNITED STATES OF AMERICA

Foreword

The ACS Symposium Series was first published in 1974 to provide a mechanism for publishing symposia quickly in book form. The purpose of the series is to publish timely, comprehensive books developed from ACS sponsored symposia based on current scientific research. Occasionally, books are developed from symposia sponsored by other organizations when the topic is of keen interest to the chemistry audience.

Before agreeing to publish a book, the proposed table of contents is reviewed for appropriate and comprehensive coverage and for interest to the audience. Some papers may be excluded to better focus the book; others may be added to provide comprehensiveness. When appropriate, overview or introductory chapters are added. Drafts of chapters are peer-reviewed prior to final acceptance or rejection, and manuscripts are prepared in camera-ready format.

As a rule, only original research papers and original review papers are included in the volumes. Verbatim reproductions of previously published papers are not accepted.

ACS Books Department

Contents

Indexes

Preface

This book is based upon the *Symposium on Polymeric Nanofibers* that was held in New York, New York at the 226th National Meeting of the American Chemical Society (ACS) September 7–11, 2003. The symposium had about 75 oral presentations and posters.

Although the use of electrical forces to produce fibers traces back into the 1800s, the development of the science and technology relevant to the production of polymeric fibers with nanometer-scale diameters never became an important part of the textile fiber industry, which came to rely upon fibers with much larger diameters.

Nonwoven sheets that contained short nanofibers were produced inexpensively from polyethylene. Expanded polytetrafluoroethylene sheets containing nanofibers were produced and used as a water barrier in clothing. Both served large and profitable markets but neither was widely recognized as containing nanofibers. In the last quarter of the 1900s, electrospun polymeric nanofibers were used in the filtration industry in the United States and for gas masks in the Soviet Union, but the processes were kept secret by the manufacturers.

As the broad field of nanotechnology gained widespread recognition in the 1990s, it was recognized that electrospinning provides a route to the creation of very long fibers with nanoscale diameters. The required apparatus is simple and operates well at the laboratory scale. The electro–hydro–dynamical science relevant to electrospinning is intellectually challenging. The growth of interest in that decade was rapid. At about the time this symposium was held, electrospinning and nanofibers were recognized, by analysis of publication rates and patterns of citations, as a fast moving front of materials science.

This book contains papers that were among the first to describe the many directions in which the science and technology of polymer nanofibers is now evolving. This symposium was cosponsored by the ACS Division of Polymer Science, Inc. and by the Fiber Society. Advice on the organization was provided by Frank Ko, Heidi Schreuder Gibson,

Gregory Rutledge, and Wayne Jones, who served as chairpersons for the four sessions in which oral presentations were made.

Darrell H. Reneker

Maurice Morton Institute of Polymer Science
The University of Akron
Akron, OH 44325–3909

Hao Fong

Department of Chemistry and Chemical Engineering
South Dakota School of Mines and Technology
501 East St. Joseph Street
Rapid City, SD 57701–3901

Polymeric Nanofibers

Chapter 1

Polymeric Nanofibers: Introduction

Darrell H. Reneker[1] and Hao Fong[2]

[1]Maurice Morton Institute of Polymer Science, The University of Akron, Akron, OH 44325–3909
[2]Department of Chemistry and Chemical Engineering, South Dakota School of Mines and Technology, 501 East Saint Joseph Street, Rapid City, SD 57701

Polymeric nanofibers are rapidly finding their place in nanomaterials technology. Electrospinning makes it easy to produce many kinds of nanofibers for many purposes.

1

A pioneering spirit pervades this book. Each chapter tells about the early successes of polymeric nanofibers, produced mostly by electrospinning, and reveals the promise of future efforts. Many topics and phenomena related to nanofibers and electrospinning do not yet have widely accepted names. No dogma was applied in the selection or editing of the chapter. Each is simply what the authors wrote. For each paper in this book, there are (up to the year of 2005) about 20 other papers now published on electrospinning. Several (1,2) recent reviews are available.

The creation of fibers, by electrifying a fluid, traces back into the last years of the 1800's. Electrostatic machines that generated potentials of hundreds of kilovolts were then available in many laboratories. The effect of electric fields on materials was of contemporary interest. Piezoelectric effects were discovered and characterized. Roentgen, prior to his discovery of X-rays, used an electrostatic machine to observe electrostriction in a sheet of rubber. An amusing report of this experiment is given, in English translation, in a paper by Ma and Reneker (3).

In 1917, John Zeleny published a paper (4) entitled "Instability of Electrified Liquid Surfaces", which described the observation of liquid jets created by electrical forces. Near the end of this paper, Zeleny wrote, "The long known experiment of threads being pulled from highly electrified sealing wax is doubtless an example of the action described in this paper". No other reports of the "long known experiment" have been found, but recent experiments show that this makes a dramatic demonstration with the red sealing wax that is presently available.

The observation of jets issuing from an electrified liquid droplet is much older. G. I. Taylor, in a paper entitled "Electrically Driven Jets" (5), wrote that in about 1600 William Gilbert described the observation that a spherical drop of water on a dry surface is drawn up into a cone when a piece of rubbed (and thereby electrically charged) amber is held a suitable distance above it. Taylor cited the book "de Magnete", (Book 2, Chapter 2, translated by P.F. Mottelay). Electrified jets of non-polymeric fluids usually break up into electrically charged liquid droplets. The process is often called electrospraying, and has been a subject of continuous scientific study and development.

The earliest substantial report of electrospinning to make fibers is in patents by A. Formhals, the first of which (6) was issued in 1934. This was after research in the 1920's clearly established that polymers are long linear molecules, and at the time when the synthesis of fiber forming molecules was making dramatic progress. Formhals described electrospinning as a process for making textile fibers, but as the textile fiber making industry developed, other methods for making fibers came to be used.

Electrospinning lived on and was occasionally examined as a curious way for making fibers thinner than the usual textile fibers, but neither the science nor the technology attracted widespread attention. Baumgarten (7), in 1971, working at Dupont, reported comprehensive experiments on the electrostatic spinning of acrylic microfibers, and published many excellent stop-motion photographs of electrospinning jets. A decade later a series of papers on electrospinning from polymer melts was published by Larrondo and Manley (8), working at McGill University. Any of these publications, or any of several electrospinning patents that issued from the 1930's to 1990, might have stimulated widespread interest in the scientific study and development of polymer nanofibers and electrospinning, but this did not happen. Although polymeric nanofibers were known in the textile industry, they are not yet been incorporated into that industry in substantial ways.

It has since become apparent that the usefulness of electrospun nanofibers in filtration technology was recognized by the Donaldson Company in Minneapolis, where many useful filters that contained polymer nanofibers were developed, but the technology was maintained as a trade secret. It is also reported that gas masks, for the protection of soldiers, were developed and used in the former Soviet Union. Perhaps these early, substantial, and successful efforts with polymer nanofibers will someday be reported in the open literature.

Only in the 1990's, as the current broad interest in nanomaterials and nanotechnology was growing, were the circumstances right for the rapidly growing effort on nanofiber applications and technology that the contributors to this book are now leading.

Many perceptions, facts, and circumstances became evident in the 1990's. The examples below, and many others, affect our view of the future:

- The production of nanofibers is a very effective way to create surface area on a polymeric solid. The geometrical surface area to mass ratio for electrospun nanofibers is around 300 square meters per gram. For fibers with a diameter of 3 nanometers, in which the molecules have a diameter of around half a nanometer and are extended along the fiber axis, simple geometry shows that half of the molecules in a cross section are at the surface of the fiber. Even thinner segments of fibers are occasionally observed, and there is a strong possibility that even single polymer molecules can be held in an extended form by the forces associated with the electrospinning process. Single polymer molecules, extended by the excess electrical charge they carry, may already be present in the electrostatic spray methods used for the injection of molecules into a mass spectrometer.
- The equipment required for electrospinning is simple, readily available, and inexpensive. The process is robust, and results are

reproducible, although the process controls to produce samples with high uniformity at a specified diameter are only now emerging. The collection of the fibers into some kinds of desirable structures also presents complex problems. The mechanical properties of electrospun nanofiber make them candidates for nanoscale support structures of many kinds, including scaffolds for the growth of artificial biological organs. The fact that electrospinning easily produces fibers with lengths of many kilometers offers many possibilities for their use in gravity free outer space where molten polymers can be subjected to very high electric fields without the occurrence of sparks.

- Electrically conducting polymers can be electrospun into nanofibers. Nanofibers made from electrically insulating polymers can be made conductive by coating with metals or carbon by chemical deposition, or be deposition from vapor.

- The internal structure of nanofibers can accommodate molecules, chemical reactions, separated phases, and even hold large particles.

- Polymeric molecules that contain metal atoms can be pyrolized to produce ceramic nanofibers which are stiff and retain their mechanical strength at high temperatures.

- Some kinds of organic polymers that can be made into nanofibers can be converted to carbon nanofibers. Hierarchical carbon structures can be produce by chemically growing even smaller carbon nanotubes tubes on the carbon nanofibers.

- The electrospinning process does not depend upon the mechanical formation of a jet issuing from a hole, but surface tension and electrical forces work against each other is such a way that a jet emanates from a liquid surface. It is often convenient to support a droplet of fluid on the outlet of a hole in the container, but even in such a case, surface tension often holds a small volume of the fluid between the outlet of the hole and the beginning of a simple jet.

- Most of the scanning electron micrographs of collected nanofibers show segments of the same long fiber. The variations in diameter and other morphological features in such micrographs occur at different places along the fiber. The variations may be gradual, such as a change in diameter, or the variation may be a consequence of the onset (or disappearance) of a distinctive feature, and referred to, in the language of fluid dynamical theories, as instabilities in the process.

- Nanofibers add to the geometrical base of nanomaterials in the form of particles or thin sheets.

Synopsis of this book

This book contains information about electrohydrodynamical models of jets, process control systems, and ways to measure the path of a jet. Observations were made of the diameter and velocity of segments of the jet. Bicomponent fibers were produced by creating a single jet that was fed by two closely spaced orifices that supplied different solutions. Factors that affect the final diameter are described. Electrode arrangements that converge and direct electrospun fibers toward desired locations were explored. Increased productivity, control of nanofiber diameter, and the orientation of nanofibers in membranes with engineered porosity were sought.

Electrostatic assembly of nanofibers, which were chemically modified to be polyelectrolytes, were used to construct composites with alternating layers of nanofibers and nanoparticles. The internal morphology of nanofibers made from a blend of two polymers was described. Co-continuous and core-sheath structures were made. Orientation development in electrospun liquid-crystalline polymer nanofibers was observed. Nanofibers were chemically modified during and after electrospinning by cross-linking and by coating. Chemical modifications were made by chemical reactions occurring during electrospinning.

Scaffolds for tissue engineering were made from biocompatible nanofibers, chosen to be compatible with essential physiological processes, and to support normal cell growth and differentiation while establishing the desired orientation and relative position of differentiated cells.

Processing parameters were explored to produce electrospun fibers for use in tissue engineering. Porous polystyrene and collagen nanofibers were produced. Fine porous protein nanofabrics with biological activity were made by an electrospray deposition method.

Morphological effects of clay nanosheets, in electrospun nanocomposite nanofibers, on the crystallization and morphology of a polymer matrix were characterized.

Electrospinning a mixture of carbon nanotubes into polymer nanofibers provided a route to macroscale composite structures. Conducting polymers, combined with carbon nanotubes and ionic salts, inside electrospun nanofibers were characterized as detectors for humidity sensors, glucose sensors, and complementary DNA sequences. Effects of carbon black on the structure of electrospun nanofibers and on the tensile properties of butyl rubber membranes prepared from electrospun nanofibers were characterized. Formation of carbon and graphite nanofibers from nanofibers of mesophase pitch is described.

The mechanical behavior of non-woven sheets of electrospun nanofibers and yarns was measured as a function of the degree of alignment of the nanofibers that was produced by collecting the fibers on a rapidly moving surface. Uniaxial alignment of electrospun nanofibers collected between conductive strips separated by an insulating gap was described. The optical transparency of nanofiber reinforced polymer matrix composites was characterized. Self-assembled fibrillar gels, composed of dumbbell shaped molecules with hydrophobic end groups, were formed as alcohol solutions cooled.

Summary

The chapters in this book provide "bench marks" that mark the field of electrospinning as it is presently known. These "bench marks" are scattered, and use many different polymers in a variety of ways. Science and technology based on nanofibers is in a position to expand dramatically in the next decades as even more uses are identified.

References

1. Huang, Z.-M., Zhang, Y.-Z., Kotaki, M. K., Ramakrishna, S., *Composites Science and Technology,* **2003,** *63,* 2223-2253.
2. Li, D., Xia, Y., *Advanced Materials* **2004,** *16,* 1151-1170.
3. Ma Y., Reneker D. H., *Rubber Chemistry and Technology*, **1996,** *69,* 674-685.
4. Zeleny, John, *Physical Review, Second Series*, **1917,** *10, 1*
5. Taylor, G. I., *Proceedings of the Royal Society, A,* **1969,** *313,* 453-475.
6. US patent 1,975,504 **1934**.
7. Baumgarten, P. K., *Journal of Colloid and Interface Science* **1971**, *36,* 71-79.
[8] Larrondo, L., Manley, R. St.J. *Journal of Polymer Science: Polymer Physics Editio,n* *1981,* *19,* 909-20.

Chapter 2

Nanofiber Manufacturing: Toward Better Process Control

Darrell H. Reneker[1], Alexander Yarin[2], Eyal Zussman[2], Sureeporn Koombhongse[1], and Woraphon Kataphinan[1]

[1]Maurice Morton Institute of Polymer Science, The University of Akron, Akron, OH 44325–3909
[2]Faculty of Mechanical Engineering, The Technion, Haifa 3200, Israel

This paper considers the development of a comprehensive model of the electrospinning process. All stages of the electrospinning process, from the preparation of the solution to the collection of the nanofibers are discussed together from the point of view of manufacturing. Experimental observations of the three-dimensional path of the jet can be compared with the path calculated from the model.

Introduction

As a fluid polymer solution is electrospun into a nanofiber, an electrified jet emerges from the electrified surface of the fluid. The cross sectional area of the jet decreases by a factor as large as a million. To conserve volume, the length of the emerging jet increases by a similar factor, so that a one-millimeter long segment near the beginning of the jet may become as long as one kilometer as the diameter decreases from 150 microns to 150 nanometers. Solvent evaporation decreases the elongation factor; but a velocity at the leading end of the order of $V_f = 3.5 \cdot 10^4 \, V_0$ would be required to keep the jet straight.

The jet velocity at the beginning, V_0, often is of the order of 10^{-1} m/s, which yields V_f of the order of 3.5 km/s. The energy required to increase the velocity to such a high value, which is higher than the speed of sound in most solids, is much larger than that required to create a nanofiber. The electrically driven elongation and bending creates a nanofiber many kilometers long, in a volume with linear dimensions of a fraction of a meter, by producing multiple coils with many turns. Bending and elongation, driven by the mutual repulsion of the electrical charge carried on the surface of the jet, continues and the fractal like bending instability occurs repeatedly at smaller scales. The multiply coiled path of the elongating jet defines a cone-shaped region called the envelope cone. Experimental observations show that the jet of polymer solution solidifies and produces a very long, thin nanofiber with a large surface area per unit mass, without accelerating any part of the jet to a high velocity (1,2,3,4).

The operational simplicity and low capital costs of laboratory scale electrospinning apparatus make it possible for researchers in almost any laboratory to produce nanofibers for their own purposes. Nanofibers, easily made from a variety of polymers, are finding many uses. Nanoscale combinations of materials are used in filtration, in wound dressings, and in many other applications. Carbon nanofibers (5) and ceramic nanofibers (6) can be produced by pyrolysis of nanofibers made from polymers that contain carbon or metal atoms.

Markets for nanofibers are emerging in areas that presently use hundreds of tons of conventional material per year. The scale-up of nanofiber production facilities is underway, even as important aspects of the electrospinning process still are being elucidated. The needs for engineering methods and performance data, production, process control, product characterization, and quality control are evident and growing. Compact sets of parameters, which characterize both the electrospinning process and the resulting nanofibers, are sought.

A comprehensive process model combines the solution properties and processing parameters to predict the parameters that characterize the polymer nanofibers. Correlation, of observations of the process and the predictions of models, is at the core of process control. A useful model of the jet connects the volumetric flow rate, the surface charge density, the surface tension, and the viscoelastic properties of the solution to observations of the jet path.

It is natural to ask what will happen to the diameter of the nanofiber if one process parameter, such as the viscosity, is varied. The modeling required to answer such a simple and obvious question is the subject of this paper. The answer that emerges is typically quite complex; particularly if the question is enlarged to ask what process variables must be observed and adjusted to ensure the diameter of the electrospun nanofiber will be maintained at a desired value.

This paper is illustrates possibilities for making the results of complex and powerful theoretical models available to the designers and operators of electrospinning apparatus. The models (7,8,9) developed and used by Yarin,

Reneker and their various colleagues, serves as a specific basis for examples of what can be done. Papers by Spivak and Dzenis (*10*), Hohman, Shin, Rutledge, and Brenner, (*11*), Fridriks (*12*) and Feng (*13*) describe other models.

Description of the Model

This paper describes a state of the art example of the connections between solution properties, observable process parameters, and nanofiber properties. It is intended to address the need for manufacturing nanofibers with predictable, consistent, and useful properties. The components of the electrospinning process are categorized into broad and interacting classes as shown in Table 1. Each of the categories are discussed separately.

Table I. Sequential categorization of the electrospinning process.

1. The polymer solution	2. Launching the jet	3. Elongation of the straight segment of the jet	4. The onset of bending, with continuing elongation	5. Solidification of the nanofiber	6. Collection of the nanofiber

The Yarin-Reneker papers (*7,8,9*) list about 25 dimensional parameters that may affect the electrospinning process. This is not a unique or complete set of parameters, but is inclusive enough to be useful, and small enough that most of the parameters can be either measured or estimated with reasonable confidence. This list of parameters was reduced to 12 non-dimensional parameters by the methods of dimensional analysis, and these parameters were used to calculate the first parts of the path of the electrified jet that is involved in electrospinning. Solidification by solvent evaporation, which leads to dry nanofibers, was included in the model. The mathematical model was used to create a computer model of the path of the jet. Experimentally observable information about the path of the jet was computed from the model. The computer model provides a practical way for predicting the effect of changes in the properties of the polymer fluids, or changes in other experimental parameters, and for comparing the predicted changes with experimental observations. A sophisticated and intricate mathematical model can be made available to the process designer and the production engineer as computer software. Fortunately, a perfect model of the electrospinning process is not a prerequisite to a practical process control system, and a realistic electrohydrodynamical model of the electrospinning jet can improve process control.

The Polymer Solution

The electrospinning process begins with a polymer solution. Many polymers, copolymers and mixtures of polymers that are soluble can be electrospun into nanofibers. The choice of suitable solvents is broad, including water, organic solvents, sulfuric acid, and paraffin oil. Solvents that do not evaporate can be electrospun into jets that are coagulated in a liquid. Some polymer melts can be electrospun. Properties such as surface tension, electrical conductivity and viscoelasticity are important, and the model can accommodate a wide range of values for these properties. The dependence, of the effective viscosity and elasticity of the solution on the extension and orientation of the polymer macromolecules in the solution, as the jet elongates and solvent evaporates, was included.

Conventional fluid characterization supplies values of surface tension, zero-shear viscosity, elasticity, and relaxation time (14). In electrospinning of mixtures and melts the rheological properties may be more complicated, and require additional parameters to be incorporated into a process model. Examples include the addition of a small concentration of co-solvent that can affect the electrospinning process by altering surface tension or solidification behavior. Where such considerations are important, a good process model should be expanded to include the added parameters.

Launching the Jet

The electrical and fluid flow conditions change dramatically during the few milliseconds required to form a jet and to adjust the shape of the region from which the jet emerges. The solution is usually delivered through an orifice. The pressure inside the orifice was essentially constant, and was supplied by a hydrostatic "head" that was a few centimeters high.

The details of the creation of the jet often involve the formation of a Taylor cone and the emergence of the jet from the tip of the cone. A constant displacement pump can be used to force the solution through an orifice at a flow rate equal to the flow rate at which the electrified jet carries the solution away. Dripping and oscillatory effects involving the shape and size of droplet from which the jet emerges occur, but are not described here. Jets can be created by mechanically pulling a jet out of a pool of fluid held in the palm of the hand in the presence of an electric field, and in many other ways. From the viewpoint of a practical process for making nanofibers, the creation of the jet is a "start-up" phenomenon that happens within a few milliseconds after the electric potential is applied (2,15).

Elongation of the Straight Segment of the Jet

While surface tension tends to make liquid surfaces contract in area, thus leading toward a spherical shape, electrical charge tends to make the fluid change its shape so that the surface area increases. In electrospinning, the interplay of these two sets of forces favors the electrical forces. Electrification of a droplet that is held together by surface tension causes a thin, electrically charged, and cylindrical jet to emanate from the surface of the droplet, and to extend in the direction of the electric field.

The model describes the evolution in time and space, of a cylindrical fluid jet with an excess electrical charge on its surface. The diameter and velocity near the beginning of the jet, and the current carried by the jet can be measured and used to determine the electrical charge per unit surface area, assuming that the charge is distributed on the surface and moving with the surface. The axial components of the self-repulsion forces of the electrical charges that are effectively attached to the surface cause the jet to elongate. The jet tapers, becoming smaller in diameter as the jet elongates. The charge per unit surface area decreases as the surface area per unit mass of the jet increases, since the flow of additional charge through the jet is small.

In the straight segment described above, each element of the fluid moves in a direction parallel to the axis of the jet. The words segment and element are used in the following way. A segment of a jet has a contour length that is longer than the diameter of the jet. A segment may be straight, bent, coiled, tapered, or have other attributes. An element of a jet is a coin shaped volume, that is, a cylinder that is thinner than its diameter. The axis of an element is always in the direction of the path at the centroid of the element. The axis of all the segments defines the path of the jet. In a bent segment that is elongating, an element may move in a direction perpendicular to the path.

Evaporation of solvent makes increasingly significant reductions in volume as the diameter of the jet decreases. The electrically driven elongation of the jet, combined with the viscoelastic rheological properties of polymer solutions, tends to prevent the capillary instability, which otherwise causes liquid jets to break up into separate droplets and participates in the formation of beads on electrospun nanofibers. Neutralization of the charge on the surface of the jet, by airborne counter-ions from adventitious corona discharges, allows the capillary instability to produce beads on the electrospun fibers (16). Under some conditions, such as the use of a very high potential, or a high volumetric flow rate, the straight segment may be very short. Then, bending may start immediately after the jet appears.

Process Control Measurements in the Straight Segment

The length of the straight segment between the origin of the jet and the onset of the bending instability is usually easy to observe and measure when the fluid is delivered to the tip at a constant pressure. The length of the straight segment is a useful parameter for monitoring the process.

The electrical current carried by the jet, usually a few microamperes, is conveniently measured with a meter that is sensitive to changes of 10 nanoamperes. An ammeter that allows both its terminals to be at a high voltage is very helpful in assessing leakage and corona currents.

The diameter of the jet at any position along the straight segment can be measured optically, either by using a telescope to produce an enlarged image of a segment of the jet, by observing the interference patterns of a monochromatic laser beam, or by observing the intensity of light scattered from an element of known length of a jet with a diameter smaller than the wavelength of the light. Eyes and video cameras are sensitive to the interference colors, which are associated with particular jet diameters, as was shown by Xu (*17*). Variations of the position of a distinctive color with time can be recorded or followed with the eye. With video cameras, changes in the diameter and taper can be perceived and recorded in one millisecond or less. Observations by Xu of the straight segment of a flowing jet during electrospinning indicated that the birefringence is small until the jet becomes thinner than 80 nm. The relatively high orientation of the molecules observed in dry nanofibers is developed and "locked in" only during the last part of the electrospinning process.

The velocity of the fluid in the straight segment can be measured by laser velocimeter (*18*), by observing the motion of tracer particles or other features that move with the jet, and by using videographic methods. This information can be used to calculate the diameter, volume, and flow rate at points along the straight segment. Since the electrical current in the axial direction is constant in the absence of airborne ions, the charge per unit area on the surface of the jet can be determined wherever the diameter and velocity are known. Predictions of the model can be compared with these observations.

The Onset of Bending, with Continuing Elongation

The tapered, elongating cylinder of fluid continues to flow along its axis until an electrically driven bending instability occurs. The end of the straight segment and the start of the bending instability can be defined as the point at which the jet bends by an amount equal to the projected diameter of the straight segment. A bending perturbation of the axis of the straight jet creates a new set of electrical forces, with components perpendicular to the jet axis, which tend to make the bending perturbation grow. The jet quickly bends through an angle of approximately 90 degrees and the path develops into a series of loops, generating a coil around the original direction of the axis. See Figure 1. Illuminating the jet with a short and intense flash of diffuse light makes the instantaneous path of the jet observable by eye or with a camera. The loops grow larger in diameter as the electrically driven elongation process lengthens the perimeter of each loop. The entire electrified coil is pulled in the direction of the axial electric field established by the electrical potential difference between the origin of a jet and the surface toward which the jet is attracted.

The charges carried by each element of the jet all continue to interact with each other by Coulomb's law. At the onset of bending, these off-axis forces generate a radial component of velocity that is added vectorially to the downward velocity. The elongation of the bent segment along its own curving local path continues, generating a helical coil with many turns. The geometry of the off-axis electrostatic interactions becomes very complicated when the path of the jet is multiply coiled (7).

The bending and elongation of the jet makes it essential to distinguish between the instantaneous path established by the local direction of the axis of the jet and the trajectory of an element of the jet. The trajectory of each short element of the bending jet is predominantly perpendicular to the axis of that element. The trajectories of some elements of the coiled jet produce moving "glints" which are specular reflections of an intense beam of light reflected off a segment of the jet. These moving glints are observed in electrospinning experiments, both by eye and by a video camera with an exposure time of several milliseconds. The moving glints are often seen as streaks, which can cause confusion if an observer attributes such a streak to the path of the jet. The motion of a glint depends on the position of the camera, the light source, the jet, the curvature of the jet, and the occurrence of some rotations of the segment of the jet. Two components of the velocity of a glint are easily determined from video images, but all the above-mentioned factors must be considered to relate the images of the streaks to the velocities of the elements of the jet.

Figure 1 shows a stereographic image made with a bright flash, about 100 microseconds long, that illuminated the coiling loops when more than three bending instabilities were present. A solution of polyethylene oxide in a mixture of water and alcohol was used. The straight segment was not illuminated.

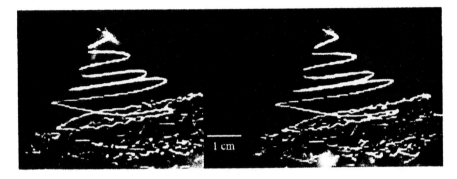

*Figure 1. Sterographic image of the instantaneous position of the
coils of a bending jet superimposed on the integrated image of
glints from the passage of many such coils.*

The onset of the second bending instability is clearly shown, as it developed
on the largest diameter loop of the first instability. The bottom part of the figure
shows small glints from the more complicated path that resulted from higher
order bending instabilities, but many details of the path are too small to be
resolved by the camera. Right- and left-handed coils are observed with about the
same frequency of occurrence. The handedness of the coils in the running jet
switched randomly, since no deliberate control was provided. The instantaneous
image of the jet path was superimposed upon a weak, longer exposure time
image of streaks from moving glints that were illuminated by a steady beam of
light. The stereographic view of the diffuse lines produced by the passage of
many glints shows that these diffuse lines do not lie in the plane perpendicular to
the view direction because the beam of light that illuminated the moving glints
was not in a plane that contained the view direction and the projected axis of the
straight segment. The cone angle of these diffuse lines is less than the cone
angle of the extremities of the path of the first bending instability.

Figure 2 illustrates concepts and terms used in this paper to describe the
path of an electrospinning jet. The straight segment extends from the source of
the jet to the region near A. The direction of the axis of each element of the jet
turns by almost 90° as that element leaves the straight segment and becomes
involved in the bending instability. The onset of the bending instability is
evident near point A when the path is deflected by a distance equal to the local
diameter of the jet. The line AD is the projection of the path of the straight
segment of the jet. After the onset of bending, the path of the jet lies within a
volume that is nearly conical in this diagram, but may have other interesting
shapes. The surface of this volume is called the envelope of the path of the jet,
or, when appropriate, the envelope cone. The lines AB and AC are at the

intersection of the envelope and a plane that contains the axis (AD) of the straight segment of the jet.

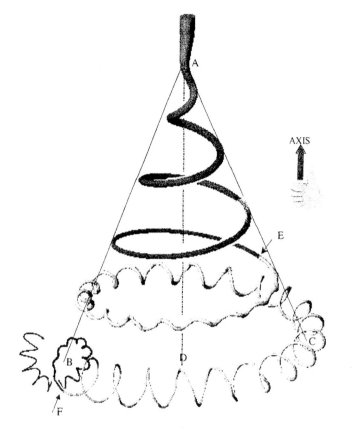

Figure 2. A diagram showing the end of the straight segment, followed by the onset and development of three successive generations of right handed bending instabilities. The convention used for handedness is that if the right hand is rotated in the direction of the fingers, the hand will advance along a right-handed helix in the direction of the thumb.
(See page 1 of color inserts.)

After the first 360° turn of the coil generated by the bending instability, the trajectory of the centroid of a particular segment nearly follows a line such as AC. The segment bends, elongates, and becomes smaller in diameter to maintain the continuity of the path. The velocity of the centroid of the element surrounding point A, which was originally parallel to the axis of the straight segment, acquires a radial component of similar magnitude. The element elongates into a thinner segment along an arc, and forms a coil. Additional bending instabilities often occur. In Figure 2, the second bending instability begins near point E and the third bending instability begins near point F.

The envelope of the coil formed by the first bending instability in this model is a simple cone, with a well-defined half angle at its vertex. The envelope is commonly observed by eye or with an ordinary video camera. Most real jets depart from the idealized shape shown in the Figure 2, often in interesting ways. There are several phenomena that may cause the shape of the envelope to be more complicated. Evaporation of the solvent reduces the volume of fluid in the path, and at the same time changes the viscoelastic parameters, making the elongation more difficult, with the result that the successive turns of the coil grow to smaller diameters and do not reach the cone BAC. If several generation of bending instabilities are present, the length of coils formed by the higher order instabilities may grow larger by increasing the pitch of the coil without the necessity for increasing its contour length, so that the envelope can sometimes continue to grow larger even after the fluid solidifies. Many of these possibilities can be sorted out and evaluated by observations of the material inside the envelope cones.

The model predicts the path, in the straight segment and in the set of coils after the first bending instability, but it would require a much larger calculation to predict the path through the successively smaller coils, which form after every subsequent bending instability. If the solvent is volatile enough, as is often the case, the thin jet solidifies in flight.

Process Control Measurements in the Bending Segments

The average shape of the envelope cone can be observed visually or with a camera that has an exposure time of 10 ms or longer. Even after the jet has become too thin and complicated to resolve its path in stop-motion images, the overall shape of the envelope is still observable. The envelope cone often has a well-defined half angle near the onset of the bending, where the influence of solvent evaporation is small and the diameter of the coils grows uniformly. If the fluid solidifies before the second bending instability occurs, the diameter of the coils of the first bending instability no longer increases, so the envelope cone becomes an "envelope cylinder". The shape of the envelope thus contains information about the solidification of the jet and the shape of coils created successively by all the bending instabilities that occur.

Where the diameter of the fluid jet is commensurate with the wavelength of visible light, blue light is scattered so strongly that it often appears to the eye or to a video camera as a white, or overexposed area of the image. The position at which this intense scattering occurs provides an important marker in the electrospinning process. The interference colors, which are more diffuse but otherwise similar to those seen in the tapered straight segment of the jet, provide additional information about the diameter of the moving jet inside the envelope cone.

Solidification of the Nanofiber

Influences, such as drying rate, solidification rate of a melt, or coagulation rate if a liquid is used to remove the solvent, can be brought into a useful relation to the electrohydrodynamical model, as described (7, 8, 9). If the electrospun fiber is still somewhat fluid when it is collected, the crossing points may merge together with enough flow in the immediate vicinity of the crossing point to distinguish the collection of a fluid jet from the collection of a solid jet. The insertion of probes disrupts the path of the jet, but the diameter of the extracted segments, their coiled path and their fluidity as revealed by the extent of merging at crossing points, provides useful information about the process.

Other instabilities lead to nanofiber based structures

Morphological examination of nanofibers on the collector or of larger diameter fibers collected on a probe inserted into the straight segment or the envelope cone provides evidence for several kinds of instabilities, which are sometimes present. When the capillary instability occurs during the electrospinning of a polymeric fluid, beaded nanofibers form (16). Sometimes the jet path, after the onset of the bending instability, follows a more complicated path and successive loops of the coil touch in flight and form permanent connections. The resulting yarn-like network is called a garland (19). Under some conditions the radial electric field surrounding a charged jet is large enough to cause radial branches to form on the primary jet (20). With some solutions, a skin forms on a polymeric jet, and the skin collapses into a ribbon as the solvent evaporates (21). All of these phenomena affect the shape and morphology of the dry nanofibers and provide recognizable signals about the electrospinning process that may be helpful in particular cases. Other influences, such as air currents, gravity, humidity, and corona discharges may become important as new ranges of electrospinning parameters are explored.

Polymer crystallites, solid particles or molecules added to the fluid, chemical reactions, phase separation of multiple polymeric components, and many morphological features can be observed with transmission electron microscopy, in combination with appropriate sample preparation techniques. Much remains to be done in these areas.

Collection of the Nanofiber

The collection of a series of loops into a non-woven sheet on a stationary or moving collector is straightforward, although the details are complicated by the complex coiled morphology of the arriving dry nanofibers. The smaller coils may be collected as coils with many turns, or pulled out into waves or more nearly straight segments. Zussman et al.(22) achieved conditions where the winder surface speed and the structure of the electric field were sufficient to alter the shape of the bottom region of the envelope cone. By using a sharp edge on the surface of a rapidly rotating collector wheel, they were able to collect yarns in which the nanofibers were highly parallel. They further showed that the electrical charge on the solidified fibers could, under appropriate conditions, create evenly spaced arrays of parallel fibers, and related these conditions to the magnitude of the charge carried on the fiber. They also made nanofiber crossbars, and nanofiber ropes using this method (23).

Post collection treatments such as annealing and pyrolysis are required for the manufacture of some kinds of polymers, such as carbon nanofibers or ceramic nanofibers. Optical diffraction patterns of light scattered by oriented arrays of nanofibers provide a quantitative way of characterizing the average alignment of nanofibers. Various ways of managing electric fields at the collector, such as arrays of wires maintained at particular potentials also produce useful alignments.

Calculation of Observable Quantities from a Model

Seventeen input quantities, shown in Table 2, are called for in the computer model, which utilizes the continuum model (7,8,9). When all these parameters are specified, the program calculates a predicted path of the jet. This computer model, written in FORTRAN, is available from the authors.

Table 2. Parameters used for calculation

Solution concentration	Relaxation time
Charge density on jet	Temperature
Perturbation frequency	Recording interval
Elongational viscosity	Initial radius of jet
Solution density	Distance from tip to collector
Electrical potential difference	Surface tension
Relative humidity	Relaxation time vs. concentration
Viscosity/concentration (1st coeff.)	Viscosity/concentration (2nd coeff.)
Cutoff polymer concentration	

Examples of the observable and experimentally interesting items that may be calculated from the model include: 1) A three-dimensional representation of the path evolving in time; 2) The viscosity of the fluid as a function of position along the jet, as the solvent evaporates. The effects of orientation of the molecules in the jet are included in the model via the Maxwell rheological constitutive equation; 3) The concentration of polymer in the jet as a function of position along the path of the jet; 4) The axial tensile stress in the jet; and 5) The radius of the jet along its length. It is possible to compute other things from this model, including production oriented quantities, such as feed rates, solvent evaporation rates along the path, mass production rate of nanofibers per unit time, and production rate of surface area per unit mass of polymer.

Summary and Conclusions

Mathematical models for the behavior of an electrospinning jet, combined with practical observations of all accessible parts of the process, and measurements of ancillary properties of the solution and nanofibers, will provide valuable information for monitoring and controlling the electrospinning at every stage of the process. Many steps toward improved process control are described in this paper.

Utilization of sophisticated experimental observations to produce better, more uniform nanofibers of more kinds of materials is an interesting scientific and technological problem. This is an area in which companies can work together to improve models at the same time they compete in the market place. All have an interest in access to the best information about the general features of the process.

Acknowledgments

We acknowledge support from the National Science Foundation, Grant DMII-0100354, from the Coalescence Nanomaterials Consortium which is supported by Donaldson, Ahlstrom, Parker Filtration, Hollingsworth and Vose, and Fleetguard, and from the Israel Science Foundation, the Israel Academy of Science, grants 287/00 and 26/03.

References

1. Doshi, J., Srinivasan, G., Reneker, D. H., *Polymer News*, **1995**, *20*, 206-207.
2. *Structure formation in polymeric fibers;* Salem, D. R.; Sussman, M. V.; Eds.; Hanser, Munich, **2000:** pp 225-246.
3. Li, D., Xia, Y., *Advanced Materials* **2004,** *16*, 1151-1170.
4. Huang, Z.-M., Zhang, Y.-Z., Kotaki, M. K., Ramakrishna, S., *Composites Science and Technology,* **2003,** *63*, 2223-2253
5. Hou, H. and Reneker, D. H., *Advanced Materials*, **2004**, *16*, pp 69-73.

6. Teye-Mensah, R., Tomer, V., Kataphinan, W., Tokash, J. C., Stojilovic, N., Chase, G. C., Evans, E., A., Ramsier, R. D., Smith, D. J., Reneker, D. H., *Journal of Physics: Condensed Matter,* **2004,** *16*, 7557-7564.

7. Reneker, D. H., Yarin, A. L., Fong, H., Koombhongse, S., *Journal of Applied Physics,* **2000,** *87*, 4531-4547.

8. Yarin, A. L., Koombhongse, S., Reneker, D. H., *Journal of Applied Physics,* **2001,** *89*, 3018-3026.

9. Yarin, A. L., Koombhongse, S., and Reneker, D. H., *Journal of Applied Physics,* **2001** *90*, 4836-4846.

10. Spivak, A. F., and Dzenis, Y. A., *J. Appl. Mech.* **1999,** *66*, 1026.

11. Hohman, M.M., Shin, M., Rutledge, G., and Brenner, M. P., *Physics of Fluids,* **2001,** *36*, 2201-20.

12. Fridrikh, S.V., Yu, J. H., Brenner, M. P., Rutledge, G. C. *Phys. Rev. Lett.* **2003,** *90*, 144502. See also, in this book, Chapter by Fridrikh.

13. Feng, J. J., *Physics of Fluids,* **2002,** *14*, 3912.

14. Theron, S. A., Zussman E., Yarin, A. L., *Polymer,* **2004,** *45*, 2017-2030.

15. Reznik, S. N., Yarin, A. L., Theron, S. A., Zussman, E., *J. Fluid Mech.,* **2004,** *516*, 349-377.

16. Fong, H., Chun, I., Reneker, D. H., *Polymer,* **1999,** 4582-4592.

17. Xu, H. Dissertation, Submitted to Graduate Faculty, University of Akron, **2003.**

18. Warner, S. B., Buer, A., Grimler, M., Ugbolue, S. C., Rutledge, G. C., Shin, M. Y., **1999** *Annual Report(M98-D01), National Textile Center*, Available: http://heavenly.mit.edu/~rutledge/PDFs/NTCannual99.pdf.

19. Reneker, D. H., Kataphinan W., Theron, A., Zussman, E., Yarin, A. L., *Polymer,* **2002,** *43*, 6785-6794.

20. Yarin, A. L., Kataphinan, W., Reneker, D. H. *submitted to Journal of Applied Physics,* fall 2004.

21. Koombhongse, S., Liu, W., Reneker, D. H., *J. Polym. Science Part B: Polymer Physics,* **2001,** *39*, 2598-2606.

22. Theron, A., Zussman, E., Yarin, A. L., *Nanotechnology,* **2001,** *12*, 384-390.

23. Zussman, E., Theron, A., Yarin, A. L., *Appl. Phys. Lett.,* **2003,** *82*, 973-975.

Chapter 3

Characterization of Electrospinning Jets Using Interference Color Technique

Han Xu[1,2] and Darrell H. Reneker[1]

[1]Maurice Morton Institute of Polymer Science, The University of Akron, Akron, OH 44325–3909
[2]Current address: Proctor & Gamble Company, 6280 Center Hill Avenue, Cincinnati, OH 45224

A comprehensive model combining electromagnetic theory of light and quantitative science of color perception was developed and experimentally validated to characterize the diameter of a fluid electrospinning jet from the interference colors shown on the jet. One light source and one camera provide real-time micron scale information about the jet diameter. The jet segments may be moving during the measurements. This method provides continuous measurements of jets with diameters from 15 to 0.5 microns. Observation of color patterns also provides information about the taper of the jet.

The electrospinning jet is a continuous tapering cylinder of fluid created from the surface of a polymer solution or melt when the electric force overcomes the surface tension of the fluid. Since the modeling of many electrospinning parameters, such as the fluid flow rate, strain, strain rate and stress, all depend on diameter information as input, it is important to know the jet diameter at many places along its path during electrospinning.

Optical imaging was used to observe the development of jets with a diameter of only a few microns during electrospinning (1). As the jet diameter approaches the wavelength of light, optical imaging becomes more difficult. Observation of a micron diameter jet requires very high magnification, short exposure times and high intensity illumination. Images of segments of the jet that are moving require intense, short flashes of light.

When the jet diameter is around 10 microns and tapering as it elongates, interference colors can be observed. Each of the interference colors is associated with a particular diameter of the jet. For a fixed position of the camera and the light source, the interference color of a segment does not depend upon the orientation or motion of a segment, although the intensity may. A laser diffraction method was used to measure the diameter of an electrospinning jet at selected points along the straight segment at the beginning of the jet. The cylindrical jet was treated as a double slit (2).

Faint colors were occasionally observed, on the spinning jets illuminated with white light, in Reneker's laboratory since 1997. Pastel red, yellow, green and blue bands were observed in the envelope cone area, where the bending instability takes place. The following chapter presents a systematic investigation of the relationship between the observed colors and corresponding jet diameters, as well as validation and applications of this method (3).

The Origin of the Interference Color on a Cylindrical Jet

When a monochromatic light impinges upon a cylindrical jet in a direction perpendicular to the jet axis, the light waves scattered from the jet create intensity maxima, which can be captured on a screen. At angles away from the incident direction, if the path-length difference of light coming from the two sides of the jet is equal to the wavelength of the light, a series of constructive and destructive intensities is produced. Higher orders of interference, due to more than one integer difference in path-length, lie further away from the incident direction. The interference pattern is symmetric with respect to the incident direction. The position of the constructive interference is determined by the diameter of the jet, and the wavelength of the incident light. The diffracted light intensities from a cylindrical jet at different observation angles are calculated in a following section using Lord Rayleigh's electromagnetic theory of light (4).

If the incident beam is polychromatic, as shown in Figure 1, a cylindrical jet will diffract the incoming light to different scattering angles according to the light wavelengths. These light waves produce constructive interference peaks at different positions on the screen. A camera placed at a certain observation angle from the incident direction will be able to catch those colors that enter its aperture. The captured colors will change with the diameter of the cylindrical jet as well as the observation angle. If we fix the observation angle, by placing the camera at a particular position, each mixture of color recorded by the camera is correlated to a certain diameter of the jet. The mixture of colors seen in a particular range of angles is complex, since the observed colors are the sum of the incident white light added to the maxima and minima of the interferences. This complication was explained in detail for soap films by Boys (5). Here we provide a similar explanation for the case of cylinders with diameters up to about ten times the wavelength of visible light.

In order to quantitatively correlate the observed color with the jet diameter, we first need to explore the diffraction behavior of cylindrical jet at all visible wavelengths. The actual color observed will be the recombination of all the diffracted light at that specific observation angle including light in that spectral range from the white light source. Secondly, understanding the color response of human eyes to various combinations of visible light is needed to determine the jet diameter from observations of the color. Visual interpretation of mixed color is not accurate at every diameter, but for a typical combination of the white light source and the angular position of a camera, several distinctive pastel colors can be followed by an observer as the diameter of the jet changes. In practice it is best to use laser diffraction to measure the diameters associated with distinctive colors, and then observe the motion of distinctive color bands along the jet to monitor the diameter and taper of a jet. Distinctive colors can be observed by eye even after the first bending instability causes the segments to move in complicated ways.

Experimental Setup

Poly (ethylene oxide) (PEO), with a molecular weight of 400,000 g/mole, was dissolved in water at a concentration of 6%. A metal measuring spoon with a 1 mm diameter hole drilled on the bottom was used to hold the polymer solution. The pressure in the solution, at the hole, was nearly constant and amounted to a hydrostatic head pressure of a few millimeters of water. A power source manufactured by Gamma High Voltage Research supplied a voltage around 9 kv to the spoon. A grounded copper plate was set 18.5 cm below the spoon. The electrospinning station was mounted on an elevator that provided controlled up and down adjustments of the spinneret and the collector with respect to the optical system. The experimental setup is shown in Figure 1.

24

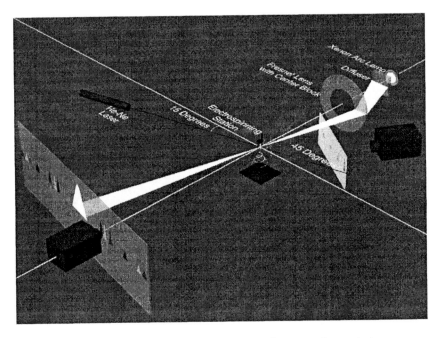

Figure 1. Experimental setup of the interference color technique
(See page 2 of color inserts.)

A xenon arc lamp with a color temperature of 6500 °K was used as a light source. The light emitted from the xenon arc source first passed through a ground glass diffuser to produce homogeneous white illumination. It was then converged at the electrospinning jet by a Fresnel lens with the central region black. Note that the incident illumination was a converging hollow cone at the sample and in the absence of a jet, it created a ring of intensity on a screen placed as shown at the lower left corner of Figure 1. The presence of a jet created interference colors. In this paper, vertically aligned jet segments are discussed in greatest detail. A color camera, which captured the interference colors, was stationed to avoid the high intensity ring of unscattered light. Since the straight segment of the jet is a slowly tapering circular cone with a vertical orientation, only the rays shown in white in Figure 1 produce colored interference peaks that enter a camera placed at the position shown. This camera was sometimes slightly defocused to prevent saturation of pixels by the bright, sharply focused image of the jet, in order to get optimal color effects. The angle between the camera axis and incident beam direction is defined as the observation angle. An optimum observation angle was determined by observing the colored images. The method used to select the optimum observation angle is described later. Laser diffraction was used to measure the diameter of the distinctively colored segments observed in the image produced by the camera. A 3.32mW He-Ne laser (λ = 632.8nm) with a spot size of 0.65mm in diameter, a second screen, and a second camera shown in the right side of Figure 1, were used in this experiment.

Spectral Power Distribution of the Light Diffracted by a Tapered Jet

Lord Rayleigh's work on electromagnetic theory of the light (4) provides a way to calculate the light intensity scattered by an infinitely long cylinder. For a cylindrical jet with a radius of a and refractive index of n_{jet}, if the incident beam has a wavelength of λ, the diffracted light intensity at scattering angle χ will be I, which is expressed in Equation 1. The refractive indexes of common polymers range from 1.30 to 1.70 and can be found in polymer handbooks (6). The refractive index of the solution can be calculated by the sum of contributions from solvent and polymer at the known concentration of the solution.

$$I = [\Delta K^{-1} \frac{\pi a}{k \cos \frac{1}{2}\chi} J_1(2ka \cos \frac{1}{2}\chi)]^2 \tag{1}$$

Where, a : Jet radius

$$k = \frac{2\pi}{\lambda}$$

χ : Observation angle

(The angle between the camera axis and the direction of the incident light, which is determined by the Fresnel lens, as shown in the diagram in Figure 1)

$$\Delta K^{-1} = \frac{1}{n_{jet}^2} - \frac{1}{n_{air}^2} \quad \text{(n : Refractive index)}$$

J_1 : Bessel's Function as $J_1(z) = \frac{z}{2}(1 - \frac{z^2}{2 \times 4} + \frac{z^4}{2 \times 4^2 \times 6} - \frac{z^6}{2 \times 4^2 \times 6^2 \times 8} + ...)$

Human Eye Response to the Spectral Power Distribution

According to the trichromatic theory of color vision, for an unknown color, it is always possible to mix certain amount of three primaries to create the same visual effect. The color matching equations (Eq. 2) were utilized to obtain the exact amount of red, green and blue to achieve color match. The quantities $\bar{x}(\lambda)$, $\bar{y}(\lambda)$ and $\bar{z}(\lambda)$ describe a color model based on human perception developed by the CIE (Commission Internationale de l'Eclairage) (7,8). Details of the calculation are described by Xu (9). Non-uniform spectral distribution of the light source was taken into account by adding $S(\lambda)$, the spectral power distribution of the light source, which can be measured by spectrometer or obtained from the light manufacturer.

$$X = k \int_{380}^{780} I(\lambda)S(\lambda)\bar{x}(\lambda)d\lambda \tag{2}$$

$$Y = k \int_{380}^{780} I(\lambda)S(\lambda)\bar{y}(\lambda)d\lambda$$

$$Z = k \int_{380}^{780} I(\lambda)S(\lambda)\bar{z}(\lambda)d\lambda$$

$$where, k = 100/ \int_{380}^{780} S(\lambda)\bar{y}(\lambda)d\lambda$$

The spectral power distribution of the diffracted light $I(\lambda)$ from a jet with a certain diameter was calculated from Equation 2. An integration of the product of $I(\lambda)$ and $\bar{x}(\lambda)$ over a range of visible wavelengths represents the amount of red in $I(\lambda)$. The corresponding amounts of green and blue are represented by Y and Z in Equation 2. A coefficient k was used to adjust the full strength of Y, which is the intensity of green, to 100, so that colors with different intensities are comparable with each other.

We calculated the perceived color associated with the spectral distribution of scattered light at each angle. This led to a choice for setting the camera angle that optimized the observation of distinctive colors. The details of this optimization process are described Xu. (9) A practical way to use the interference colors for monitoring the jet diameter and taper is to identify two distinctive color bands with known diameters and determine the distance between them.

Comparison of the Interference Color and Light Diffraction Results

To calibrate the jet diameter associated with each distinctive color, light diffraction experiments were carried out at the same time using the setup shown in Figure 2. Electrospinning parameters were maintained at constant values. The electrospinning station was moved upward or downward so that the laser illuminated a distinctive color band of the straight segment. The diffraction pattern was projected on a semi-transparent screen and was recorded by a camera behind the screen. A second camera recorded the laser beam position on the jet when the primary white source was off, and also recorded the interference color at the position when the xenon light was on and the He-Ne laser was off. The constant position of the interference colors on the jet confirmed that the spinning was stable and there were only minimal jet diameter fluctuations during the experiment.

The diameters at seven different distinctively colored positions along the tapered jet were measured by diffraction of laser light. The measured diameters and the observed interference colors are correlated in Figure 2. The interference colors were recorded at an observation angle of 12 degrees. Figure 2 also shows the calculated interference colors generated by a tapered jet with diameters from 2 to 17 microns at an observation angle of 12 degrees. Comparison of the observed color of a segment with the color calculated from the diameter measured by laser diffraction produced satisfactory, but not quantitative agreement.

Since the interference colors can be directly observed with a simple illumination system, familiarity with the diameters associated with a particular distinctive color allows an operator to observe fluctuations and other features of the electrospinning jet and, for example, to adjust the voltage to move a color band with a known diameter to a particular point along the straight segment of the jet.

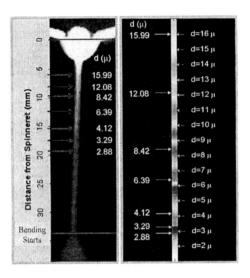

Figure 2. Comparison between interference colors and jet diameters measured at points along the jet, using light diffraction. The diagram at the right shows the range of colors predicted from the trichromatic theory of color vision. (See page 2 of color inserts.)

Application of Interference Color Technique

The interference color method provides detailed diameter information on segments along the axis of fluid jets with diameters from about half a micron to over 10 microns. Distinctive colors are observed even when the jet path is moving. The interference color method is a non-invasive method to study the flow of electrified jets of polymeric fluids. After proper illumination and viewing conditions are established, the diameter and taper of the jet can be estimated usefully and continuously by visual observation or video photography.

Behavior of Single Jet

Figure 3a shows the how the interference colors of a jet change as the voltage is increased. A particular distinctive color band on the top part of the jet was always shorter than the same color band when it was moved further down the jet by the increased applied voltage. Here, the taper rate is defined as da/dl, where a is the jet diameter and l is the position along the axis of the jet.

Figure 3 also shows that increasing the electrospinning voltage caused the portion of every segment that has a particular color to elongate and shift downward. This downward shift shows that jet diameter increased with an increase of spinning voltage for this constant pressure feed system. The elongation of a color band is caused by a taper rate that decreases when spinning voltage is increased. Increasing the spinning voltage increases fluid flow. In this electrospinning jet, decreased spinning voltage increased the taper rate and reduced the diameter of a segment at a given distance from the spinneret. Figure 3b supports similar conclusions for two jets flowing from the same droplet.

Laser diffraction measurements of jet diameter confirmed the above result. Figure 4 shows the diameter at different positions (L) along the axis, measured using laser diffraction. For a constant value of L, higher voltages produced a thicker jet and, for the larger jet diameters, decreased the magnitude of the taper rate, as shown by comparing the slopes of the three curves. These point by point diameter measurements yield the same result as the visual observation of color bands, in Figure 3, where the distinctive colors elongated and moved downward when the voltage was increased.

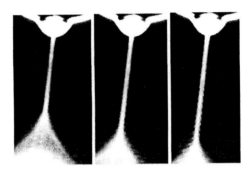

Figure 3a. Interference colors on a single jet during electrospinning
(Increasing voltage from left to right)
(See page 3 of color inserts.)

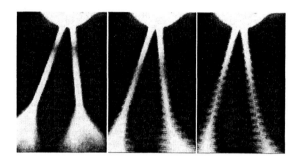

Figure 3b.   Interference colors on a twin jet initiated from a single droplet
(Increasing voltage from left to right)
(See page 3 of color inserts.)

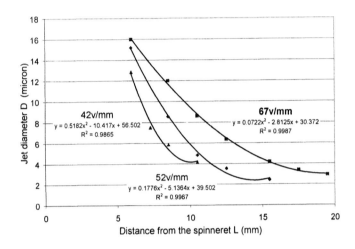

*Figure 4. Laser diffraction measurement of electrospinning jet diameter meas-
ured at different voltages. Polynomials fitted to the data are shown.
(See page 4 of color inserts.)*

Behavior of Multiple Jets

Two or more jets can flow from the same droplet. These jets are quasi-stable. Within limits, an increase in the flow rate of one jet of a pair is associated with a decrease in the flow rate of the other jet. After a longer time, one jet grew steadily larger as the other jet disappeared, for reasons that are presently unknown. These changes are easily observed in the motion of the distinctive interference colors. The behavior of such multiple jets, as observed with interference colors, may lead to better process control and improved productivity. Interference color observations of jets are helpful for exploring how to increase electrospinning throughput and how to improve the uniformity of fiber diameters.

Figure 5 shows a competition, of 3 quasi-stable jets from one drop, in which two jets die. During the competition, which was initiated by a modest decrease in the applied voltage, the same color bands as shown in Figure 3 shifted downward and become larger in the leftmost jet, showing that it grew in diameter (contrary to the usual effect of lowering the voltage on a single jet). The other two jets exhibited upward movement and narrowing of color bands, as would be expected for a decreased voltage. The flow rate in the leftmost jet increased by approximately the same amount as the flow rate of the other two jets decreased.

The taper rate near the spinneret of the other two jets increased to a high value and after a hundred milliseconds, both of these jets stopped completely. The lower portion of two right jets showed a bluish white color. When jet diameter becomes close to the wavelength of the light, the diffracted intensity of different wavelengths tends toward equality in all radial directions. Then, summing the intensity of all the visible wavelengths produces blue which become whiter and more intense, often to the point of saturating the camera. This bright spot is easily recognized visually, even when it occurs after the onset of bending instabilities as shown in Figure 3a. The dense color band on the top of the thin segment and the long bluish white tail are characteristics of a jet with a submicron range of diameters. Generating a thin jet by "starving" a jet provides a route to ultra thin electrospun fibers, but effects of solvent vaporization rate become important (1) as the jet becomes thin.

In earlier, independent experiments, Koombhongse (10) observed that when several jets were present, the total electrical current did not change during the disappearance of a jet, indicating that electrical current increased in the surviving jets by compensating amounts as an adjacent jets died.

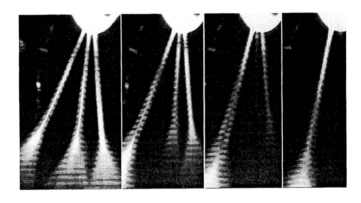

*Figure 5. Interference colors on a triple jet initiated from a single droplet
(From left to right: 34ms, 68ms, 102ms and 136ms after voltage was dropped)
(See page 4 of color inserts.)*

Conclusions

Real-time characterization of the fluid jet by observation of interference colors can be used to monitor and control fiber formation in electrospinning and other fiber spinning processes. This method is also suitable for characterizing diameters, and diameter distributions of some solid fibers in the 0.5 to 15 micron diameter range, even when the fiber is moving.

Acknowledgement

Financial support from the National Science Foundation Project DMII-9813098 is acknowledged.

References

1. Reneker, D. H.; Yarin, A. L.; Fong, H.; Koombhongse, S. *J. Apply Phys.* **2000**, 87(9), 4531-4547.
2. Doshi, J. N. Ph.D. thesis, University of Akron, Akron, OH, 1994.
3. Xu, H.; Galehouse, D. C.; Reneker, D. H. *Polymeric Materials Science and Engineering*, **2003**, *88*; 37.
4. Boy, C. V. *Soap bubbles: Their colors and the forces which mold them,* Dover, 1959.
5. Rayleigh, J. W. S. *The collected optics papers of Lord Rayleigh;* Optical Society of America: Washington, D.C., 1994; pp 527-528.
6. *Physical Properties of Polymers Handbook;* Mark, J. E., Ed.; AIP Press: Woodbury, NY., 1996; pp 535-542.
7. Wyszecki, G.; Stiles, W.S. *Color Science: Concepts and Methods, Quantitative Data and Formulae*, 2nd Ed., John Wiley & Sons: New York, NY, 1982; pp 131-143.
8. URL: http://www.cs.bham.ac.uk/~mer/colour/cietorgb.html
9. Xu, H. *Ph.D. Dissertation, University of Akron,* Akron, Ohio, 2003.
10. Koombhongse, S., *Ph. D. Dissertation, University of Akron*, Akron, Ohio 2001.

Chapter 4

Nonlinear Whipping Behavior of Electrified Fluid Jets

S. V. Fridrikh[1], J. H. Yu[1], M. P. Brenner[2], and G. C. Rutledge[1]

[1]Department of Chemical Engineering, Massachusetts Institute
of Technology, 77 Massachusetts Avenue, Cambridge, MA 02139
[2]Division of Engineering and Applied Sciences, Harvard University,
29 Oxford Street, Cambridge, MA 02138

Electrostatic production of fibers (electrospinning) has
recently become an object of intensive experimental research
due to potentially broad markets for the fibers as small as
100nm (nanofibers). The understanding of the physics of the
process and the fundamental model thereof are still lacking.
We present the theory of electrospinning accounting for non-
linear effects such as the thinning of the jet and strain
hardening. The model predicts the amplitude of the whipping
instability causing the formation of nanofibers to saturate and
the jet to reach a terminal diameter controlled by the surface
tension, surface charge repulsion and solvent evaporation.

Introduction

Electrostatic production of micro- and nano-fibers (electrospinning) from polymer solutions and melts has been known for a long time (*1*), but only recently has it acquired importance as a method of production of non-woven fabrics. The technology has attracted much attention (*2-12*) due to the ease with which nanometer diameter fibers can be produced from either natural (*2*) or synthetic (*4, 5*) polymers. Such small fibers have numerous and diverse potential applications including filtration (*6*) and composite materials (*7, 8*). Their high surface area makes nanofibers attractive as catalyst supports (*9*), and in drug delivery (*10*). Electrospun nonwoven fabrics are being developed for tissue engineering (*11*). Electrospun conducting polymers have been used to fabricate nanowires (*12*).

During electrospinning a charged fluid jet emanating from the nozzle is accelerated and stretched by an external electric filed acting on the charges located on the jet surface (*1*). Though the experimental assembly for electrospinning is rather straightforward, understanding of the physics of the phenomenon and control of the properties of the fabrics produced are less trivial. It has been demonstrated experimentally (*14, 13*) that electrified jets experience a whipping instability, causing the jet to bend and stretch; this instability has been attributed to the surface charge repulsion (*17, 20*) and eventually leads to the formation of nanofibers. Depending on the operating and material parameters, Rayleigh and conducting varicose instabilities also may develop during electrospinning (*20*), thus preventing formation of uniform fibers and causing breakage of the jet into droplets or appearance of bead-on-a-string structures.

The basic principles for dealing with electrified fluids have been developed in a series of papers by G. I. Taylor (*18,19*). He proposed the model of a "leaky dielectric", treating the fluid neither as a perfect dielectric nor as a perfect conductor. It is assumed in the model that the fluid is dielectric enough for the volume charge to be ignored and at the same time conducting enough for the charges to build up at the surfaces in the presence of the electric field. It has later been shown that this approach is adequate for the range of fluids and operating conditions encountered in electrospinning (*20*).

The behavior of the jet before the onset of the instabilities has been addressed both theoretically (*21-24*) and experimentally (*21, 22*). The one-dimensional model treating the jet as a slender object and based on Taylor's "leaky dielectric" approach gives numerical prediction for the shape of the stable jet away from the nozzle in good agreement with experiments (*22*). The asymptotic solution of the model away from the nozzle also agrees well with experiments (*21*). The one-dimensional model, however, breaks down in the vicinity of the nozzle. A more realistic two-dimensional model, accounting for the details of the experimental geometry and allowing for the treatment of the jets near the nozzle, describes the shape of the jet in good agreement with the experimental data (*24*). Both models, however, produce reliable results for low conductivity fluids only.

The major challenge in the physics of electrically driven jets is to understand the evolution of the unstable jet from the onset of the instability up to the moment the jet hits the collector. The linear stability analysis represents the first natural step in this direction. This analysis for the uncharged finite conductivity jet was performed by Saville in the early 1970's (25, 26). The stability of the uniformly charged jet with infinite conductivity and zero tangential electric field has also been analyzed (27, 28). This analysis of linear stability captured important features of the dynamics of the jets in electric fields, most importantly the presence of an axisymmetric instability and an oscillatory "whipping" instability. However, quantitative characteristics of these instabilities disagree with experimental data. Hohman et al. performed the stability analysis of the slender charged fluid jet with the "leaky dielectric" properties in a tangential electric field (20). The long wavelength approximation and the assumption of the slenderness of the jet allowed for substantial simplification of the mathematical formalism. The model predicts the already known Rayleigh and whipping instabilities and a "conducting" axisymmetric mode missing in Saville's theory. In the limit of zero charge density and for the long instability wavelength the model reproduces the results of Saville. Significantly, the Rayleigh mode competes at low electric field, while the conducting varicose mode competes at high electric field. For the often dilute solutions employed in electrospinning, the competition between varicose and whipping instabilities determines whether one obtain an electrospun fiber or an electrospray.

However, linear instability analysis covers only the early stages of the evolution of the instability. A model accounting for large amplitude stretching of the whipping jet and non-Newtonian properties of the fluid is needed to predict the shape of the envelope and the radius of the whipping jet as a function of the process and material parameters. A model capturing some of these effects has been proposed by Yarin and co-authors (16). In that model, bending is assumed to be driven by charge repulsion and resisted by fluid elasticity; expressions for the wavelength and growth rate of the bending instability were obtained by comparison to previous solutions for aerodynamically driven Newtonian jets (29). We present here a fluid dynamics model of the whipping behavior of an electrically charged jet beyond the linear regime and compare the model predictions with experimental data. The model accounts for non-linear effects such as stretching of the jet and strain hardening and describes the evolution of the whipping jet from the onset of the instability to the moment it hits the collector. The model predicts that the whipping amplitude saturates, its final value being set by the balance between surface charge repulsion and surface tension, as described previously. An asymptotic analysis of the whipping behavior of the slender charged jet was performed to derive a simple expression predicting the existence and magnitude of a "terminal diameter" that the whipping jet can reach when the whipping amplitude saturates (30). This prediction is in a good agreement with a broad range of experimental data, attesting to the validity of the model (*30*). It provides one with a simple tool for estimating the size of the fibers obtained by electrospinning. The effect of solvent evaporation can also be incorporated into the model.

The Model

During electrospinning, a reservoir of polymer fluid is contacted with a large electric potential, and the fluid is delivered to the tip of a small capillary. The electrical charge that develops at the fluid's free surface interacts with the external electric field, resulting in the formation of the Taylor cone and the emission of a steady fluid jet that thins as it is being accelerated by the electric field. In most operations of interest, the jet experiences a whipping instability, leading to bending and stretching of the jet, observed as loops of increasing size as the instability grows. The jet thins dramatically, by as much as 3-4 orders of magnitude, while traveling the short distance between the electrodes (up to 30 cm). The presence of polymer in solution leads to the formation of fine solid fibers as the solvent evaporates.

The model presented here describes the dynamics of the jet as a function of material properties: conductivity (K), dielectric permittivity (ε), dynamic viscosity (η), surface tension (γ), and density (ρ); as well as operating characteristics: flow rate (Q), applied electric field (E_∞), and electric current (I). The current is determined dynamically by the material properties, flow rate and external field. However, prediction of the current requires modeling of the jet as whole, including the nozzle domain. Such a calculation is beyond the scope of this paper, and thus we consider electric current here as an independent operating parameter which should be measured experimentally.

The whipping jet is treated as a slender viscous object. We assume that the radius of curvature R of the loops of the whipping jet is much larger than the radius of the jet h. This assumption allows for a dramatic simplification in description of the dynamics of the jet and makes the model effectively one-dimensional. This approach has been used in analysis of various slender objects such as viscous jets (*13, 31*), catenaries (*32, 33*), elastic rods (*34*) and 2-D fluid sheets (including lava flows, processes of candy making and blowing of glass and polymer vessels) (*35-37*), and we employ here some useful analogies with these systems.

Let us consider a slender, charged, viscous axisymmetric jet in external electric field E_∞ and gravitational field, both parallel to the centerline of the originally straight jet. We restrict the centerline deformations of the jet to the *xz* -plane of an orthogonal right-handed coordinate system $\hat{x}, \hat{y}, \hat{z}$, with z being parallel to the electric field. We define $\hat{t}$ as a unit vector tangential to the centerline and $\hat{\xi}$ as a unit vector in the xz -plane normal to the centerline. Locally, the shape of the (initially straight) jet can be parameterized by the arc length s measured along the centerline and the coordinate X normal to the centerline measured along $\hat{\xi}$. The kinematic variables of interest in the curvilinear coordinates are the angle of the centerline with the z-axis $\theta(s, t)$, and the cross-sectional area of the jet $A(s, t) = \pi h^2$, where $h(s,t)$ is the jet radius. If the instability is assumed to have sinusoidal shape, the problem can be simplified by switching from curvilinear to Cartesian coordinates. In Cartesian coordinates $\theta(s, t)$, can be replaced by the amplitude of the whipping $H(z, t)$ measured along x–axis; and $A(s, t)$ can be replaced by $A(z, t)$ (see Figure 1).

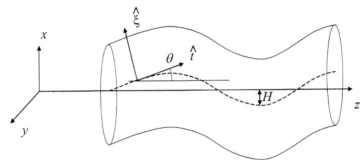

Figure 1. Coordinates used to describe the motion of the whipping jet. Dashed line is the centerline of the jet. See the text for details.

The equations for balance of forces and torques for this system can be written down following the derivation by Love for a thin elastic rod (*34*) and by Teichman & Mahadevan for a viscous catenary (*33*):

$$T_s + N\theta_s + K_{\parallel} = \rho A \ddot{s} \qquad (1)$$

$$N_s + T\theta_s + K_{\perp} = \rho A \ddot{X} \qquad (2)$$

$$M_s + N = 0, \qquad (3)$$

where $(a)_b \equiv \partial a/\partial b$; $T(s, t)$, $N(s, t)$, $M(s, t)$ and $K(s, t)$ are the resultants of the tangential stresses (tension), normal stresses (shear), torques and external body forces acting on the cross-section of the jet, correspondingly. The external body forces are due to gravity and the external electric field. In deriving equations (1)-(3) we assumed that the whipping behavior is dominated by stretching and neglected the inertial terms related to rotational motion. To close the system we need to add a continuity relation accounting for the extensibility of the jet (formulated for a short segment δs of the jet) and expressions for T, M and K:

$$M = M_{int.} + (M_{ext.})_s \tag{4}$$

$$M_{int.} = \frac{3h^4\mu}{4\pi}\theta_{st} - \frac{\bar{\varepsilon}h^4 E_\infty^2}{16}\theta_s \cos(\theta)^2$$

$$+ \frac{\pi h^5}{8}(\beta+1)\sigma_0 E_\infty \ln(\chi)\theta_{ss}\cos(\theta) \tag{5}$$

$$(M_{ext.})_s = \frac{\pi^2 h^4 \sigma^2}{\bar{\varepsilon}}(1-\ln(\chi))\theta_{ss} + \frac{\bar{\varepsilon}h^2 E_\infty^2}{2}\theta_{ss}\cos(\theta)\sin(\theta)$$

$$+ 16\pi h^3(\beta+1)\sigma_0 E_\infty \ln(\chi)\,\theta_s + \frac{\bar{\varepsilon}\beta h^4 E_\infty}{2}\theta_{ss}\sin(\theta)^2 \tag{6}$$

$$T = -\pi h\gamma + 3\mu A_t + \frac{2\pi^2 h^2 \sigma_0^2}{\bar{\varepsilon}}$$

$$+ \frac{\bar{\varepsilon}h^2 E_\infty^2}{4}\sin(\theta)^2 + \frac{\bar{\varepsilon}(2\beta+1)h^4 E_\infty}{2}\cos(\theta)^2 \tag{7}$$

$$K_\perp = \frac{4\pi^2\sigma_0^2}{\bar{\varepsilon}}(1-\ln(\chi))\theta_s - 2\pi\sigma_0 E_\infty \sin(\theta)$$

$$+ \frac{\bar{\varepsilon}(\beta-3)h^2 E_\infty^2}{4}\theta_s\sin(\theta)^2 - \frac{\bar{\varepsilon}\beta h^2 E_\infty^2}{8}\theta_s\cos(\theta)^2 \tag{8}$$

$$- \frac{\beta\pi h^3\sigma_0 E_\infty}{4}(1-\ln(\chi))\theta_{ss}\cos(\theta) + 2\pi h\gamma\theta_s - \rho g\sin(\theta)$$

$$K_\parallel = 2\pi\sigma_0 E_\infty \cos(\theta) + \rho g\cos(\theta) \tag{9}$$

$$(A\delta s)_t = 0 \tag{10}$$

where σ_0 is the surface charge density, $\chi=\lambda/h$ is the aspect ratio of the jet.

In derivation of the above equations we assumed that the terms with tangential gradients are negligible compared to the normal ones and that the jet is uniform in diameter at scales of the order of the wavelength of the whipping instability.

From the full set of equations given by (5)-(10), we now develop a simplified version of the model sufficient to capture the major physical effects responsible for the stretching of the jet. The electrostatic terms in the dynamic equations originate from the discontinuity of the electric field energy density at the surface of the jet ($\sim \bar{\varepsilon}\, E_\infty^2$), the electric field pulling on the surface charges ($\sim \bar{\varepsilon}\, \sigma_0 E_\infty$) and the surface charge repulsion ($\sim \sigma_0^2 / \bar{\varepsilon}$). We can estimate the ratio of the energy density term to the surface repulsion term for values of the external field, flow rate and current for experiments typical of aqueous solutions of polyethylene oxide ($E_\infty = 10^5$ V/m, $Q = 10^{-8}$ m^3/min, $I = 10^{-7}$A). The mass conservation law implies $\sigma_0 = hI/2Q$, from which we obtain that this ratio is $\sim \bar{\varepsilon}^2 E_\infty^2 Q^2 /(h^2 I^2) \sim 10^{-16} h^{-2}$ (where h is measured in meters) and that it is small for any realistic radius of the jet ($h > 10nm$). Thus the electrostatic energy density term can be ignored compared to the surface charge repulsion. This assumption corresponds to the case of strongly charged whipping jets.

We further assume that the jet is advected downstream at a constant velocity v_0 and ignore the effects of gravity. Thus the behavior of the jet is controlled by surface charge repulsion, viscosity, surface tension and also by inertia. The jet is assumed to be thin enough for most of the charges to be located at the surface. With these simplifications, continuity and closure relations become:

$$M_{int.} = \frac{3h^4 \mu}{4\pi}\theta_{st} \tag{11}$$

$$(M_{ext.})_s = \frac{\pi^2 h^4 \sigma^2}{\varepsilon}(1 - \ln(\chi))\theta_{ss} \tag{12}$$

$$T = -\pi h\gamma + 3\mu A_t + \frac{2\pi^2 h^2 \sigma_0^{\,2}}{\varepsilon} \tag{13}$$

$$K_\perp = \frac{4\pi^2 \sigma_0^{\,2}}{\varepsilon}(1 - \ln(\chi))\theta_s + 2\pi h\gamma\theta_s \tag{14}$$

$$K_\| = 0 \tag{15}$$

At the early stages of the whipping instability the length and the diameter of the jet may be considered constant, and linear instability analysis may be preformed. For small periodic perturbations, the equations for normal and tangential components decouple (to leading order in the amplitude of the perturbation) and thus $X \sim H$, $s \approx z$, $\partial/\partial s \approx \partial/\partial z$ and $\theta \approx \partial X/\partial s$, where H is the perturbation amplitude. The surface tension at this stage may be ignored and the dynamic equation for the normal component becomes:

$$-\frac{3h^4\mu}{4\pi}\theta_{ssst} + \frac{2\pi^2 h^2 \sigma_0^{\ 2}}{\bar{\varepsilon}}(3 - 2\ln(\chi))\theta_s = \rho A\ddot{X} \qquad (16)$$

For a small periodic perturbation of the centerline of the form

$$r(z,t) = z\hat{\mathbf{z}} + rH\exp(\omega t + ikz)\hat{\mathbf{x}}, \qquad (17)$$

where $k=2\pi/\lambda$ is the wave number of the instability and ω is the instability growth rate, the dispersion relation is:

$$-\omega k^4 \frac{3h^4\mu}{4\pi} + k^2 \frac{2\pi^2 h^2 \sigma_0^{\ 2}}{\bar{\varepsilon}}(3 - 2\ln(\chi)) = \omega^2 \rho A. \qquad (18)$$

The high charge density favors the short wave length perturbations while the viscosity tends to damp them. The balance of these two forces results in the maximum growth rate ω_{max} characterizing the fastest growing mode:

$$\omega_{max} = \left(\frac{9\pi^4 \sigma_0^{\ 4}}{8h^2 \rho\mu\bar{\varepsilon}^2}(2\ln(\chi) - 3)^2\right)^{1/3} \qquad (19)$$

and the corresponding wave number is:

$$k = \left(\frac{\pi^2 \sigma_0}{h^2 \mu}\right)^{1/3} \left(\frac{2\pi\rho}{\bar{\varepsilon}}(2\ln(\chi) - 3)\right)^{1/6}. \qquad (20)$$

The wavelength $\lambda = 2\pi/k$ estimated from (20) for 2% aqueous solution of PEO (M = $2 \cdot 10^6$, μ = 1.6 Pa·s, jet diameter h at the onset of the varicose instability is approximately 50 μm, I = 100 nA, Q = 0.02 ml/min) is about 0.1mm, which is more than an order of magnitude less than the experimentally observed values. One possible source of this discrepancy is an overestimation of the surface charge. We have assumed that the asymptotic formula, $\sigma=Ih/2Q$, holds close to the nozzle; previous work (20,21) shows that this greatly overestimates the surface charge. Elastic tension in the jet, for example due to the stretching of polymeric chains while the steady jet is in flight, may also be a source of discrepancy; highly elastic jets have been shown to yield slightly longer waves than Newtonian jets in capillary breakup, and elastic tension has a significant stabilizing effect on the draining filament-bead configuration (38).

Similar expressions for the growth wave length and growth rate of the whipping mode have been derived by Yarin et al. (their eq 21, 23 and 24) (16). In that work, the nonlogarithmic contribution due to charge repulsion arising from the electrostatic pressure due to the surface charge has not been accounted for. Both works ultimately drop the contribution due to surface tension, which is minor at the onset of the whipping instability. The results of Yarin et al. are presented in dimensionless form. Both models are based on the assumption that in the linear approximation the wave length and the growth rate of the instability are set by viscosity, inertia and surface charge repulsion. The comparison of growth rates for varicose and whipping instabilities (eqn 25 of Ref 16) is obtained by analogy to the linear instability analysis for uncharged Newtonian jets destabilized by air drag (29) and allows to conclude that the whipping instability grows faster than the Rayleigh varicose instability when the fluid is highly viscous ($\mu = 10^3$ Pa·s). Hohman et al. have analyzed the competition between whipping and varicose instabilities for a range of operating conditions and fluid properties (20, 22).

The above equations (1-3) and (11-15) are quite general and can be applied to describe the non-linear behavior of the charged jet. Recently we have performed an asymptotic analysis of these equations and have found that at the late stages of the whipping instability the behavior of the jet is controlled by surface charge repulsion and the surface tension (30). As the jet thins during the whipping due to repulsion of the surface charges, the surface charge density decreases and is eventually balanced by the surface tension. At that moment the jet reaches its terminal diameter and becomes stable. These results apply to both Newtonian and non-Newtonian fluids. The model predictions for the terminal diameter have been tested extensively for various polymeric fluids and have been found to be in a good agreement with experimental data (30). Though successfully explaining the few factors setting the fiber diameter, the above mentioned analysis rather gives a lower bound for this value and does not account for important effects such as solvent evaporation and non-Newtonian viscosity. Besides, the evolution of the envelope of the whipping jet was not described; any material that arrives at the collector prior to attainment of saturation of the whipping instability was not considered.

To address this shortcoming, we now further advance this model to account properly for non-linear effects such as stretching of the jet and strain hardening and to describe the evolution of the whipping jet from the onset of the instability to the moment it hits the collector.

We assume that due to non-Newtonian effects the jet may be considered uniform at the scales comparable to the instability wavelength λ. This assumption is confirmed by the uniformity of the resultant fibers. We also assume that the unstable whipping jet is advected downstream at a constant speed. These approximations allow us to draw clear parallels between the whipping jet and viscous filament sagging under the influence of gravity (viscous catenary) (*33*). Figure 2 shows the schematic of the model jet. Surface charge repulsion bends and stretches the jet in the direction normal to the centerline, while viscosity and surface tension try to stabilize it. Unlike the case of the viscous catenary where the material is "nailed" at the ends, we assumed that the $\lambda/2$ segment of the jet is "hinged" at the ends and can freely rotate around the points $z = 0$ and $z = \lambda/2$. There are other important differences between the two systems. The viscous catenary described in Ref. *33* is composed of a Newtonian fluid that thins in the middle faster than at the ends and eventually breaks under the influence of gravity. The late stage whipping jet considered here consists of a non-Newtonian fluid with strong elastic tension that suppresses varicose instability and keeps the jet uniform. This jet does not break, but rather eventually reaches its terminal diameter set by the surface tension and surface charge repulsion. The analysis presented below can also be applied to highly viscous Newtonian fluids where varicose instability is suppressed long enough for the jet to reach the terminal diameter.

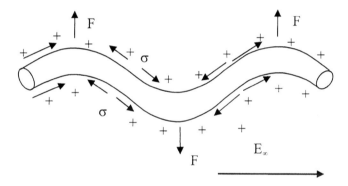

Figure 2. The schematic of whipping instability driven by the surface charge repulsion. F is the normal component of the charge repulsion forces.

The uniformity of the jet segment under consideration suggests that $T_s = 0$ and equations for normal and tangential components can then be decoupled. From now on we consider only equation for the normal component. After substituting (13) into (2), we obtain the dynamic equation for the component x:

$$x_{zz}\left(-3\eta\, A_t - \frac{2\,A^2\,\Sigma^2}{\varepsilon}(2\ln(\chi)-3)+\gamma\,\pi^{1/2}A^{1/2}\right)=\rho\,A\,\ddot{x}, \quad (21)$$

46

where $\Sigma = I/Q$ is the volume charge density. When the charge is located at the surface of the jet, then this represents that charge associated with the surface of a volume element of fluid. We assume that $x(z,t) = H(t) \sin(k z)$ (ignoring higher order harmonics) and obtain the equation for the evolution of the amplitude of whipping instability H:

$$H k^2 \left(-3\eta A_t - \frac{2 A^2 \Sigma^2}{\bar{\varepsilon}} (2\ln(\chi) - 3) + \gamma \pi^{1/2} A^{1/2} \right) = \rho A \ddot{H}, \quad (22)$$

and the mass conservation law:

$$A_t = -A \frac{H_t}{H} \left(1 - \frac{1}{1 + k^2 H^2} \frac{K(a)}{E(a)} \right), \quad (23)$$

where $a = H^2 k^2/(1 + H^2 k^2)$, $K(a)$ and $E(a)$ are elliptic integrals of the first and the second kind correspondingly. In eq. (22) the three terms in the brackets on the left are due to extensional viscosity, surface charge repulsion and surface tension correspondingly. On the right hand side is the inertial term. The shear viscosity term is important only at the very early stages of the whipping instability and has been dropped.

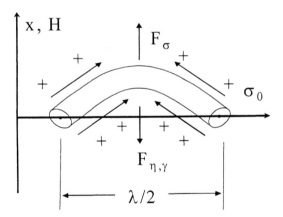

Figure 3. A $\lambda/2$ segment of the whipping jet presented as a "sagging charged viscous filament" hinged at the ends.

Results and Discussion

We expect that stretching of the segment of the jet will first be driven by surface charge repulsion and resisted by inertia and viscosity. Depending on the interplay between the inertia and extensional viscosity, the jet either stretches monotonically and approaches the terminal jet diameter prescribed by the balance of surface tension and charge repulsion (viscosity dominated behavior) or "overshoots" and oscillates around the terminal amplitude and terminal jet diameter of the whipping mode (inertia dominated behavior). Figure 4 shows that if realistic numbers for process and material parameters are chosen (PEO, Mw = 2 × 10^6 million, 2% in water, μ=1.6 Pa s, γ = 0.06 N/m, Σ=10 C/m^3) and the fluid is assumed to be Newtonian, the inertia dominated scenario is realized. The jet reaches its terminal whipping amplitude and diameter after a few strongly damped oscillations. This contradicts the experimental observations, which show no signs of such oscillations, and suggests that some effects have not been taken into account. In addition, the predicted amplitude of whipping (> 1m) is much greater than anything observed experimentally.

One of the important effects not included in the model up to this point is strain hardening. The experimental times of electrospinning are of the order of $10^{-2} - 10^{-3}$ s. The molecular (Rouse) times for our model PEO system are ~ 0.1 s. Thus we have Deborah numbers well above one ($De > 1$) and, under the conditions of the extensional strain being the primary mode of deformation, strain hardening can not be ignored in the intermediate stage of whipping. To account for the strain hardening we employ a phenomenological model used by Feng (20) that describes the growth of the transient extensional viscosity $\overline{\mu^+}$ as a function of strain:

$$\frac{\overline{\mu^+}}{\mu_0} = \begin{cases} \exp\left\{ p\left[1 - \cos\left(\frac{\varepsilon^2}{\varepsilon_s^2} \pi \right) \right] \right\}, & if \ \varepsilon \le \varepsilon_s \\ \exp(2\,p), & if \ \varepsilon > \varepsilon_s \end{cases} \tag{24}$$

where μ_0 is zero shear rate viscosity, ε is the Hencky strain and ε_s is the strain at which the steady state extensional viscosity is attained. At high strains the Trouton ratio of the model fluid reaches a value of $exp(2p)$, where p is a phenomenological model parameter. We solve the dynamic equation for the whipping amplitude numerically (eq. 22) using the above expression for the extensional viscosity. Figure 5 shows changes in stretching behavior as a function of p for $\varepsilon_s=6$. For values of p between 2 and 3, the oscillations are completely suppressed. This corresponds to a rather strong strain hardening;

transient viscosity is about two orders of magnitude above the viscosity at zero strain rate.

Further analysis and the complete non-linear description of the whipping jet will require incorporation of drying effects into the model and a simple model of fluid elasticity with model parameters that can be independently verified experimentally. Nevertheless, we have found that an estimate for the diameter that can be reached by the charged jet can be made based on asymptotic analysis of eqs. (1,2) (30) or can be derived directly from eq. 22. We take the second route here for the sake of clarity.

If we assume that at the late stages of whipping instability the stretching slows down and the viscosity term can be dropped from the dynamic equation, the stability of the jet with respect to the whipping mode is controlled by a competition between surface tension and surface charge repulsion. The dynamic equation for the whipping amplitude then takes the following form:

$$Hk^2 h \left(-\frac{2\pi^2 h^3 \Sigma^2}{\varepsilon} (2\ln(\chi) - 3) + \gamma \pi \right) = \rho \pi h^2 \ddot{H}, \qquad (25)$$

The two terms in the brackets originate from the surface tension (stabilizing term) and surface charge repulsion (destabilizing term).

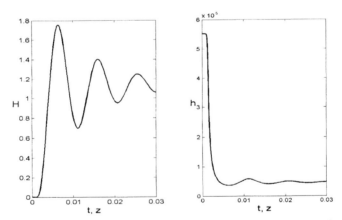

Figure 4. Amplitude of whipping H (m) and jet diameter h (m × 10⁻⁶) as a function of time (s × 0.1) or, equivalently, vertical coordinate z(m) for a Newtonian fluid (see text for details). The velocity is constant at 10 m/s.

One can see that, at the late stages of whipping, the surface charge repulsion term decreases as h^3 until it exactly balances the surface tension term. At this point, the dramatic stretching of the jet due to the whipping instability ceases and the terminal diameter of the jet is reached. The model assumes that the stretching after the saturation of the whipping mode is minimal, and the jet is advected to the collector with essentially constant diameter. From the balance of charge repulsion and surface tension, we obtain the following relation for the terminal jet radius h_t:

$$h_t = \gamma^{1/3} \Sigma^{2/3} \left(\frac{2\bar{\varepsilon}}{\pi(2\ln\chi - 3)} \right)^{1/3}. \tag{26}$$

Equation (26) predicts that the terminal radius of the whipping jet is controlled by the volume charge density and the surface tension of the fluid.

To account for the solvent evaporation and to calculate the fiber radius we employ the assumption made by Reneker et al. (17) that the jet is first stretched to a particular draw ratio, in this case the limiting draw ratio, and that drying takes place only after the limiting jet diameter is reached. The dry fiber then is deposited at the collector. It is reasonable to assume that during drying the wet fiber shrinks only in the directions normal to the contour length. Thus the fiber radius h_f can be calculated from $h_f = h_t\, c^{1/2}$, where c is the polymer concentration in the solution.

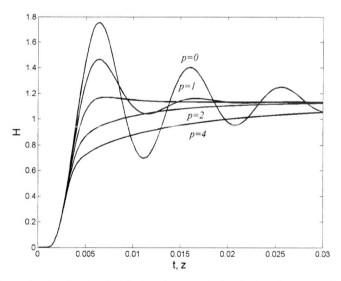

Figure 5. Amplitude of whipping for a strain hardening fluid assuming various values of the parameter p.

The above prediction for the fiber diameter has been extensively tested experimentally. In Figure 6 we present the correlation between the terminal jet diameter and the volume charge density for the solutions of polycaprolactone (PCL) with the concentrations ranging from 8 to 12%. The fiber diameters for PCL and other polymers have been measured by scanning electron microscopy (SEM). To calculate the terminal jet diameters, fiber diameters measured experimentally were divided by the square root of the polymer concentration in the solution as described above. All the data sets form a master curve with a slope of 0.64, in excellent agreement with the model prediction of 2/3. Several other polymer fluids have been electrospun to test the model predictions. Though

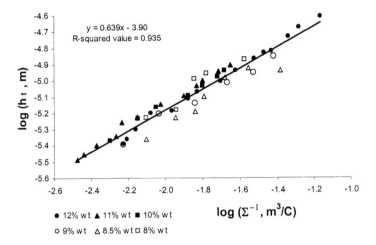

Figure 6. Terminal jet diameter for PCL solutions as a function of the inverse charge density. Master plot. (Reproduced with permission from reference 30. Copyright 2003 American Physical Society.)

none of them provided us with data covering a range of fiber diameter broad enough to check for $h_f \sim \Sigma^{-2/3}$ scaling, all the results are consistent with this scaling and agree with the model prediction for absolute value of fiber diameter within a 10-30% error margin (Figure 7). Only PCL fibers appear to be about twice larger than the model predicts. The charge carrier in PCL solution is the relatively volatile methanol. We conjecture that charged methanol molecules may leave the jet before stretching is complete and still hit the collector, thus causing overestimated jet current readings and consequently overestimated jet stretching.

The model outlined above predicts that the fiber diameter is independent of the zero shear rate viscosity. This assumes of course that the whipping jet has had enough time to reach the terminal diameter before impacting the collector.

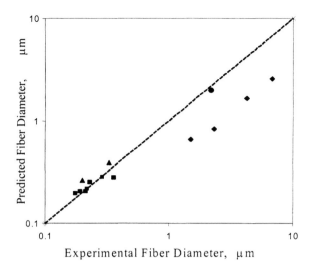

Figure 7. Parity plot for the diameters of the electrospun fibers. ♦ - *PCL (Mw = 150k, c = 11%),* ■ - *PEO (Mw = 600-2000k, c = 2%),* ▲ - *PAN (Mw = 250 k, c = 8%),* ● – *Boger fluid (37% of poly(ethylene glycol) Mw = 10 k and 0.25% of PEO Mw = 1500 k in water).*

This insensitivity of jet diameter to viscosity has been tested for PEO solutions of different molecular weights and concentrations and a sample of Boger fluid. Boger fluids are elastic fluids with constant shear viscosity (*39, 40*). We use this system to separate non-Newtonian effects in shear and extensional flows and toinvestigate the role of strain hardening in controlling the whipping behavior of the charged jet during the electrospinning. Samples with zero shear rate viscosities ranging from 0.2 to 20.75 Pa.s have been electrospun under similar operating conditions (Q =0.01 ml/min, E = 300-400 V/cm, I = 40-80×10^{-9} A). In agreement with the model, an order of magnitude change in zero shear rate viscosity causes less that a factor of two change of the diameter of PEO fibers. These small diameter variations can be attributed to variations in operating conditions and concentrations (Figure 8). (The left-most point in Figure 8 is special; it corresponds to the Boger fluid composed of 0.1 wt% PEO (Mw = 1500 k) and 37wt% polyethylene glycol (Mw=10 k) in water, so the bigger fibers in this case are a consequence of the much more concentrated solution than in case of PEO.) The experimental values for the fiber diameters presented in Figure 8 have been normalized by the corresponding model predictions and re-plotted in Figure 9. One can see that all the experimental points but one (corresponding to μ=20.75 Pa.s) agree with the predicted values

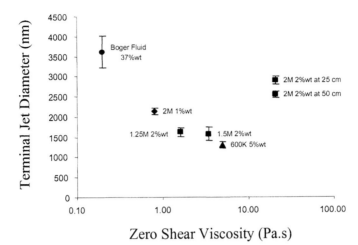

Figure 8. Terminal jet diameter calculated from the fiber diameter measured by SEM as a function of the zero shear rate viscosity fo PEO solutions and a sample of Boger fluid electrospun at similar operating conditions.

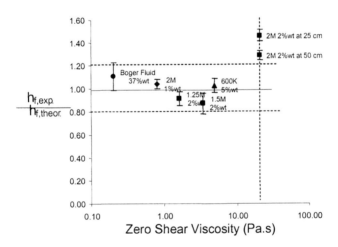

Figure 9. The ratios of experimentally measured fiber diameters to the ones predicted theoretically for PEO solutions and the Boger fluid.

within an error margin of 20%. This agreement deteriorates slightly only for the most viscous fluid, which is 10 times more viscous than glycerol. The two experimental points corresponding to this highly viscous fluid have been taken at different inter-plate distances of 25 and 50 cm. The voltage was varied to keep the applied field approximately the same between the plate electrodes of the parallel-plate configuration (15). The change of the inter-plate distance from 25 to 50 cm is accompanied by roughly a 15% drop in the fiber diameter. This suggests that, if given sufficient time to stretch, even highly viscous fluids are able to approach their terminal jet diameter. The attempts to collect a sample at distances over 50 cm have not been successful, with most of the material being scattered and never reaching the collector.

Conclusions

A simple model of a slender electrified fluid jet has been proposed to describe non-linear whipping behavior of non-Newtonian fluids during electrospinning. At the early stages of the instability, the instability wavelength and growth rate are set by inertia, charge repulsion and shear viscosity of the fluid. As the instability grows, the deformation of the jet is dominated by strain, and strain hardening causes the transient extensional viscosity to increase and to bind the inertial forces. This strong non-Newtonian effect accounts for dissipation of the kinetic energy and slowing down of the stretching. Ultimately, if the time spent by the jet in the air is sufficiently long, the jet will reach the terminal diameter in agreement with the theoretical prediction reported earlier (30). In this limit, the fiber diameter is shown to be independent of the zero shear rate viscosity of the fluid, and dependent on the extensional viscosity only in the rate at which the terminal diameter is approached.

Acknowledgements. This research was supported by the National Textile Center (project MD01-D22) and the U.S. Army through the Institute for Soldier Nanotechnologies, under Contract DAAD-19-02-D0002 with the U.S. Army Research Office.

References:

1. Formhals, A. U.S. Patent 2,077,373, 1934.
2. Demir, M. M.; Yilgor, I.; Yilgor, E.; Erman, B. *Polymer* **2002**, *43*, 3303-3309.

3. Norris, I. D.; Shaker, M. M.; Ko, F. K.; MacDiarmid, A. G. *Synthetic Met.* **2000**, *114*, 109-114.
4. Matthews, J. A.; Wnek, G. E.; Simpson, D. G.; Bowlin, G. L. *Biomacromolecules* **2002**, *3*, 232-238.
5. Jin, H.-J.; Fridrikh, S. V.; Rutledge, G. C.; Kaplan, D. L. *Biomacromolecules* **2002**, *3*, 1233-1239.
6. Tsai, P. P.; Schreuder-Gibson, H.; Gibson, P. *J. Electrostat.* **2002**, *54*, 333-341.
7. Kim, J.-S.; Reneker, D. H. *Polym. Eng. Sci.* **1999**, *39*, 849-854.
8. Ali, A., Ph.D. thesis, Drexel University, Philadelphia, PA, 2003.
9. Jia, H.; Zhu, G.; Vugrinovich, B. *Biotechnol. Progr.* **2002**, *18*, 1027-1032.
10. Kenawy, E.-R.; Bowlin, G. L.; Mansfield, K.; Layman, J.; Simpson, D. G.; Sanders, E. H.; Wnek, G. E. *J. Control. Release* **2002**, *81*, 57-64.
11. Li, W.-J.; Laurencin, C. T.; Caterson, E. J. *J. Biomed. Mater. Res.* **2002**, *60*, 613-621.
12. MacDiarmid, A. G.; Jones Jr., W. E.; Norris, I. D.; Gao, J.; JohnsonJr., A. T.; Pinto, N. J.; Hone, J.; Han, B.; Ko, F. K.; Okuzaki, H.; Llaguno, M. *Synthetic Met.* **2001**, *119*, 27-30.
13. Taylor G. I., *Proc. Roy. Soc. Lond. Ser. A* **1969**, 453-475.
14. Baumgarten P. K., *J. Colloid Interf. Sci.* **1971**, 36, 71-79.
15. Shin, Y. M.; Hohman, M. M.; Brenner, M. P.; Rutledge, G. C. *Polymer* **2001**, *42*, 9955-9967.
16. Yarin, A. L.; Koombhongse, S.; Reneker, D. H. *J. Appl. Phys.* **2001**, *89*, 3018-3026.
17. Reneker, D. H.; Yarin, A. L.; Fong, H.; Koombhongse, S. *J. Appl. Phys.* **2000**, 87, 4531-4547.
18. Taylor, G. I. *Proc. Roy. Soc. Lond. Ser. A* **1964**, 383-397.
19. Taylor, G. I. *Proc. Roy. Soc. Lond. Ser. A* **1965**, 145-158.
20. Hohman, M. M.; Shin, Y. M.; Rutledge, G. C.; Brenner, M. P. *Phys. Fluids* **2001**, *13*, 2201-2220.
21. Kirichenko, V. N.; Petrianov-Sokolov, I. V.; Suprun, N. N.; Shutov, A. A. *Dokl. Akad. Nauk SSSR* **1986**, *289*, 817-820.
22. Hohman, M. M.; Shin, Y. M.; Rutledge, G. C.; Brenner, M. P. *Phys. Fluids* **2001**, *13*, 2221-2236.
23. Feng, J. J. *J. Non-Newt. Fluid* **2003**, *116*, 55-70.
24. Yan, F.; Farouk, B.; Ko, F. *J. Aerosol Sci.* **2003**, *34*, 99-116.
25. Saville, D. A. *J. Fluid Mech.* **1971**, *48*, 815-827.
26. Saville, D. A. *Phys. Fluids* **1970**, *13*, 2987-2994.
27. Saville, D. A. *Phys. Fluids* **1971**, *14*, 1095-1099.
28. Huebner, A. L.; Chu, H. N. *J. Fluid Mech.* **1970**, *49*, 361.
29. Entov, V. M.; Yarin, A. L. *J. Fluid Mech.* **1984**, 140, 91-111.

30. Fridrikh, S. V.; Yu, J. H.; Brenner, M. P.; Rutledge, G. C. *Phys. Rev. Lett.* **2003**, *90*, 144502.

31. Tomotica, S. *Proc. Roy. Soc. Lond. Ser. A* **1935**, 150, 322.

32. Mahadevan, L.; Ryu, W. S.; Samuel, A. D. T. *Nature* **1998**, *392*, 140-140.

33. Teichman, J.; Mahadevan, L. *J. Fluid Mech.* **2003**, *478*, 71-80.

34. Love, A. E. H. *A treatise on the mathematical theory of elasticity*; Dover publications: New York, NY, 1944.

35. Ribe, N. M. *J. Fluid Mech.* **2001**, *433*, 135-160.

36. Ribe, N. M. *J. Fluid Mech.* **2002**, *457*, 255-283.

37. Ribe, N. M. *Phys. Rev. E* **2003**, *68*.

38. Chang, H. C.; Demekhin, E. A.; Kalaidin, E. *Phys. Fluids* **1999**, 11, 1717-1737.

39. Boger, D. V. *J. Non-Newt. Fluid* **1977**, *3*, 87-91.

40. Dontula, P.; Macosko, C. W.; Scriven, L. E. *AICHE J.* **1998**, *44*, 1247-1255.

Chapter 5

Key Parameters Influencing the Onset and Maintenance of the Electrospinning Jet

J. M. Deitzel[1], C. Krauthauser[2], D. Harris[3], C. Pergantis[3],
and J. Kleinmeyer[4]

[1]Center for Composite Materials, University of Delaware,
Newark, DE 19702
[2]U.S. Army Research Laboratory, Terminal Effects Division,
Armor Mechanics Branch, Aberdeen, MD 21003
[3]U.S. Army Research Laboratory, Materials Division, Multifunctional
Materials Branch, Aberdeen, MD 21003
[4]XIOtech Corporation, MD 21043

Electrospinning is a process by which submicron polymer fibers can be generated through use of an electrostatically driven jet of polymer solution. Use of electrospun nanofibers in electronic, biomedical, and protective clothing applications often involves the incorporation of some sort of functionalized particulate (i.e. carbon nanotubes, activated carbon, clay silicates, etc.). The current work uses conventional and high speed imaging techniques to study the motion of particles in the linear portion of electrostatically driven jets of Polyethylene oxide/water solutions. Observation of the motion of carbon particles using conventional video reveals the presence of eddy currents in the meniscus from which the jet originates. High-speed video of the motion of urethane particles in the liquid jet has been used to measure jet velocities, which range from 1-2 meters/second depending on initial processing conditions. The effect of solution viscosity, field strength, and flow rate on jet velocity and shear rate have implications for the scale up of the electrospinning process and are discussed in detail.

Introduction

Although the phenomena of electrospinning has been observed and studied for more than a century [1-21], it is only in the last five years that interest [2-11] in the technology has reached a critical mass in the academic community, spurred on by the on going interest in materials engineering on the nanoscale, and by biotechnology. The recognition that electrospun fibers can be assembled into structures that mimic morphologies seen in biology, and that the surface area associated with electrospun textiles can be a platform from which nanoscale chemical and biological processes can operate has driven the explosion of scholarly interest and publications. There are now clear cut applications in biotechnology and microelectronics which provide the economic incentive to fund the research needed to make the technology commercially viable, while maintaining the level of control demonstrated on the lab scale.

Electrospinning is a process, which uses an electrostatically driven jet of polymer fluid (solution or melt) to generate submicron diameter polymer fibers. Typically, a voltage is applied to the fluid that is pumped through a capillary orifice. This can be a glass pipette, a syringe needle, or simply a hole in a conductive plate. Flow rates are very slow, and range from 0.5 to a few tens of ml/hour. Prior to application of the voltage, a drop of fluid will form at the exit of the capillary. Once the voltage is applied, the drop deforms into the shape of a cone in the presence of the resulting electric field. When the electric field strength at the tip of this cone exceeds a critical value, a jet of fluid will erupt from the apex of the cone and proceed to the nearest target that is at a lower electrical potential, usually a collection plate or drum that is electrically grounded. Initially, the solution jet follows a linear trajectory, but at some critical distance from the capillary orifice, the jet begins to whip about in a chaotic fashion. This is commonly referred to in the literature as a bending instability. [2, 3, 4, 5] At the onset of this instability, the jet follows a diverging helical path. As the jet spirals toward the collection mechanism higher order instabilities manifest themselves [5] resulting in a completely chaotic trajectory. For this reason, electrospun materials are usually collected in quantity in the form of a randomly oriented non-woven mat.

The process described here is typical of a laboratory scale system and the rate of fiber production is very slow by commercial fiber spinning standards. The objective of the current research is to use real time imaging techniques to gain needed insight into mechanisms that govern the stability of the electrospinning jet. It is a general goal of this work to provide an introduction to the technology for those entering the field and to summarize the research we have conducted over the last decade. In spite of what has gone before, our understanding of the underlying mechanisms fundamental to the electrospinning process is still in its infancy. The authors feel that visualization of the electrospinning process is key to understanding how very subtle changes in processing conditions can affect the final product. It is our hope that our observations will serve to inspire the reader in their quest to advance the state of the art and adapt this fascinating technology to new applications.

Experimental

Model polymer solutions of Polyethylene oxide (PEO) and water are used in the experiments reported here. The solution concentrations were 7% and 10% of PEO in water by weight. The basic experimental setup is depicted in figure 1. The electrospinning apparatus consisted of a 30 kV high voltage power supply from Gamma High Voltage Research and a syringe pump from Harvard Apparatus. Polymer solution is pumped from a syringe through an 18 Gauge stainless steel syringe needle. The electrospinning voltage was applied directly to the syringe needle. Solution feed rates ranged from 1 to 8 ml/hour. Standard video of the electrospinning process was collected with a Coho CCD camera using a telephoto lens. The resolution of the CCD sensor was 512 X 512 pixels. High-speed video of the electrospinning process was obtained using a Kodak HG 1000 video camera with an Infinity K2 video lens. 70 micron carbon black and 50 micron diameter urethane spheres were used as tracer particles to observe flow behavior. The electrospinning jet was illuminated from several angles in order to bring out the appropriate level of contrast needed to make effective observations.

Results and Discussion

Video Observation

One of the driving forces behind the explosion of interest in the electrospinning process is possibility of incorporating meso- and nanoscale particles into a continuous network. Uses for such a composite system include a variety of applications such as sensors (chemical, biological, electrical, strain), Nanoelectronic devices, wound repair, tissue engineering, and protective barriers for chemical and biological defense. Although electrospun fibers have been doped with numerous types of nanoparticles [7-11], very little is understood regarding their effect on the spinning characteristics of the polymer solution. To investigate this, a 7% solution of polyethylene oxide (400k MW) in water was made with approximately 0.5% by wt. of carbon black particles (average diameter~50 microns) were mixed into the solution using a hand held electronic mixer. Activated carbon is commonly used in both liquid and gas filtration devices and the mixture of PEO and carbon black was meant to simulate the incorporation of activated carbon directly into an electrospun fiber mat.

Because of the small size of the electrospinning jet, and the chaotic motion it undergoes during its journey to a collection apparatus, observing flow behavior directly can be a very difficult challenge. However, in the vicinity of the capillary opening, both the Taylor cone and the initial part of the jet are usually stable enough and large enough to observe with standard video equipment. Video data of the electrospinning experiment was acquired through standard video rates at 30 frames/second. Upon inspection it is immediately apparent that the carbon particles have formed agglomerations in the PEO/water solution. It is probable that these agglomerations or aggregates are the result of insufficient mechanical agitation during mixing or possibly due to unfavorable chemical interactions between the particle and the PEO/Water solution. The key point of this discussion, however, is that these larger aggregates are easily observed and can be used as "tracers" that provide information about fluid flow behavior in the Taylor cone and the initial part of the electrospinning jet.

In figure 2, four large agglomerations can be observed, one near the needle, one in the middle of the meniscus, and two more at the apex of the cone, from which the electrospinning jet originates. The agglomerated particles are much larger than the initial diameter of the jet, and do not pass through. As more and more particles collect at the apex of the cone, the spinning process slows down, but never stops, and the particles can be seen to spin about the apex of the cone. Eventually (~10 minutes) the mass of the collected carbon particles becomes too great and the entire meniscus/drop separates from the needle and falls to the target below. This process effectively filters out all particles larger than the initial diameter of the electrospinning jet. Figures 3a and 3b are optical micrographs of the fiber mats obtained from the electrospinning experiment described above. Figure 3a is a micrograph taken of the portion of the mat situated directly under the needle. Clearly seen is a pile of carbon that has not been incorporated into fibers, but has simply fallen from the needle in drops. Figure 3b is a micrograph of the outer edge of the fiber mat, and is essentially free of carbon particles.

It is clear from this data that a key factor in successful incorporation of nanoparticles into electrospun fibers is good particle dispersion in the polymer solution. Large particles or aggregates will be effectively filtered from the electrospinning jet. However, it should be pointed out that size exclusion is only

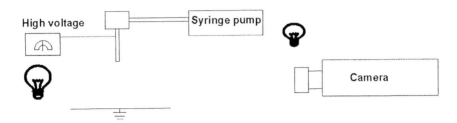

Figure 1: Basic electrospinning set up.

*Figure 2: Electrospinning of PEO/Water solution doped with carbon black
(Reproduced with permission from* Polym Rrepr. **2003**, 44(2), 57–58.
Copyright 2003 J. M. Deitzel.)

Figure 3a: Area of fiber mat directly under the syringe tip
(Reproduced with permission from Polym Rrepr. 2003, 44(2), 57–58.
Copyright 2003 J. M. Deitzel.)

Figure 3b: Area at the edge of the fiber mat. (Reproduced with permission from
Polym Rrepr. *2003, 44(2), 57-58. Copyright 2003 J. M. Deitzel.)*

one possible mechanism affecting the distribution of particles in an electrospun mat. It is likely that particle and biological molecules carrying an electric charge will be greatly affected by both local (field induced by local charge in fluid) and global (field induced by applied voltage) electric field strengths and shapes.

The fact that the electrospinning process continues even when multiple particles are trapped at the cone apex suggests that the electrospinning jet is being feed primarily from the surface of the droplet/cone. This is further illustrated in figure 4, which shows a series of images taken at 30 frames per second. At t=0 seconds two particles, A and B can be clearly seen near the center of the droplet, while particle C is visible at the edge of the droplet. As time progresses, particle A travels in a direction away from the electrospinning jet and toward the needle. At time =24 seconds, a third particle, C has moved from the lower right hand corner to the center of the droplet and travels up the centerline of the drop toward the needle outlet (time = 30 seconds). This phenomenon was first reported by Hayati [12, 13], et al who observed similar motion of fluorescent tracer particles in viscous, low molecular weight oils. He provided the following explanation, which is illustrated in figure 5. When a potential is applied to a drop of a semi-conducting fluid, (often referred to as a "leaky dielectric")[12, 13], a potential difference exists between the base of the resulting cone and its tip. As a result, there is a net shear force tangent to the liquid surface that is described by the following equation:

$$T = \varepsilon_0 E_n E_t \qquad (1)$$

Where, T is the shear force, ε_0 is the permittivity of free space; E_n is the electric field strength normal to the liquid surface and E_t is the electric field tangential to the liquid surface. This shear stress moves fluid at the surface toward the apex of the cone.

The observations described above clearly illustrate the complexity of the fluid flow at the point of jet initiation. Particles, such as clay silicates and carbon nanotubes must be well dispersed in a solution and must not exceed a critical size (i.e. the initial diameter of the jet) if they are to be evenly distributed through out the polymer nanofibers. The observations are also consistent with the theory that the free charge aggregates at the liquid surface [2, 3, 14, 15, 16, 17], providing the motive force needed to initiate and stabilize the electrospinning jet. This has interesting implications for nanofibers spun from solutions that have been doped with conductive materials or antimicrobial compounds which are often ionic in nature. These materials will have a tendency to aggregate at the fiber surface as a result of electrostatic forces. Additionally, an excess of these charge-carrying additives can greatly impair the formation and stability of an electrospinning jet [5].

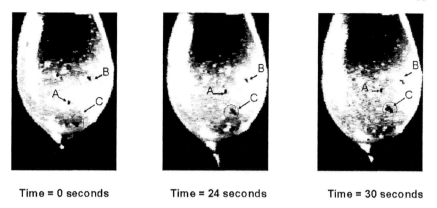

Time = 0 seconds   Time = 24 seconds   Time = 30 seconds

Figure 4: Time resolved images of particulate flow in the cone

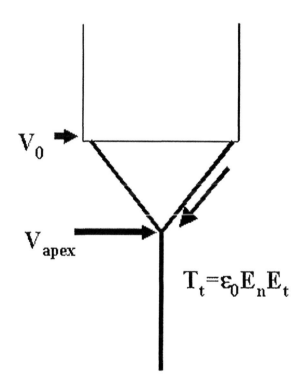

Figure 5: Diagram illustrating fluid flow at the surface in the cone region of the electrospinning process

High-Speed Imaging

As illustrated in the previous discussion, a great deal can be learned about electrospinning fluid dynamics from observing the motion of particles in the electrospinning jet. For polymer solutions, in the region of the Taylor cone the fluid flow is relatively slow and can be observed at standard video speeds (30 frames per second). However, once the fluid reaches the apex of the Taylor cone and passes into the straight portion of the electrospinning jet, the fluid undergoes rapid acceleration and high-speed cameras are required to follow tracer particles. High-speed video (1000-2000 fps, 70msec exposure time) has been used to measure the velocity of the fluid in the linear portion of the electrospinning jet for a variety of experimental conditions. A 10 % solution of 400,000 molecular weight PEO, in water, was used in these experiments. The particles used as tracers were ~ 50 micron diameter polyurethane spheres that are typically used as pigment in water dispersible coatings. High-speed video was obtained using a Kodak HG 100 high-speed video camera with an Infinity K2 video microscope lens. Individual frame resolution was 720 by 480 pixels. Distance calibrations for high magnification images were obtained from images of graph paper with known line spacing. A Redlake MotionScope^{tm} motion analysis software was used to measure particle velocities from the high-speed video.

Figure 6a shows an image of the straight portion of the electrospinning jet just below the apex of the Taylor cone. The diameter of the electrospinning jet decreases from 740 microns to 270 microns within a micron of the initiation point. The motion of particle A, seen at the apex of the cone, was tracked in successive video frames and the velocity of the particle was calculated and plotted as a function of time (see Figure 6b). The velocity profile shows that the particle reaches a maximum speed of 0.15 m/s at a distance of ~1.5 mm below the apex of the Taylor cone.

At lower magnifications the particles can be tracked for longer distances along the length of the linear portion of the electrospinning jet (Figure 7a). The full length of the electrospinning jet depicted in figure 7a is 7 mm. From the plot in figure 7b we see that particle B has an initial measured velocity of 0.1 m/s and reaches a maximum velocity of 2.5 m/s before it passes out of the field of view of the camera. This change in velocity occurs over a period of 8.5 milliseconds and corresponds to an average acceleration of 320 m/s^2. It should be noted that the electrospinning jet in figure 7a deviates somewhat from what would be considered the central axis of the electrospinning set up. The reason for this deviation is that the charged jet is being preferentially pulled toward the ground lead attached to the collection target. Because the ground lead was positioned perpendicular to both the needle and the line of sight of the camera, the observed portion of the electrospinning jet remained steady and in focus. As a result, the

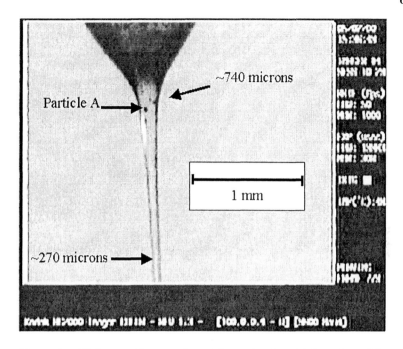

Figure 6a: High-speed image of particle motion in initial portion of the electrospinning jet.

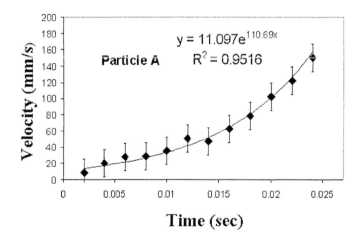

Figure 6b: Velocity profile of particle A in the initial portion of the electrospinning jet.

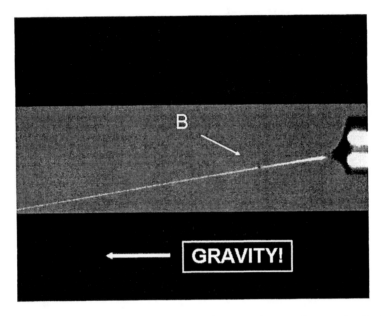

Figure 7a: Low magnification high-speed image of particulate motion in the electrospinning jet.

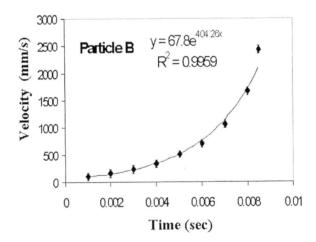

Figure 7b: Velocity profile of particle B in figure 7a.

authors are confident in the accuracy of our velocity measurements, although some small degree of uncertainty cannot be ruled out.

Jet Velocity

One factor that is important to the commercial scale up of the electrospinning process is the rate at which polymer nanofiber is produced. In lab scale experiments, fiber collection rates are on the order of a few tens to hundreds of milligrams per hour. While adequate for most proof of concept experiments, these spinning rates are much too slow for commercial viability. The speed of fiber deposition is ultimately controlled by the solution feed rate to the capillary and the electrospinning voltage. As has been extensively discussed in the literature, the rate of solution to the capillary must be comparable to the rate at which the solution is removed from the system through the electrostatically driven jet in order to obtain nanofibers of uniform diameter and morphology [6]. The most obvious way to increase fiber output would seem to be to simply increase the solution flow rate and electrospinning voltage. A series of experiments have been run to study the effect of flow rate and spinning voltage on fluid velocity and strain rates in the linear portion of the electrospinning jet. Figure 8 is a plot of particle velocity as a function time for 3 different processing conditions (Case I: 5 kV-2.5 ml/hr, Case II: 6 kV- 3.0 ml/hr, and Case III: 7 kV-6.0 ml/hr). All other conditions, including solution concentration and target distance were held constant.

For all three conditions, we see that particle velocity increases exponentially as a function of time; however, the rate at which the particle velocity increases (i.e. acceleration) is clearly dependent on the spinning conditions. Case I conditions represent the lowest combination of spinning voltage and solution feed rate for which it was possible to establish a stable jet. In Case II, both the spinning voltage and solution feed rate have been increased, and as a result the maximum particle velocity and average acceleration also increase slightly, as one would expect. In Case III, the spinning voltage and solution feed rate have again been increased, however, the maximum particle velocity and average acceleration both decrease with respect to the Case I condition.

Why does the fluid velocity and shear rate decrease for Case III, when both the electrospinning voltage and fluid feed rate have both been substantially increased compared to Case I conditions? It has been shown [4, 12,13] in the case of low molecular wt. fluids that the rate at which the radius of an electrostatically driven jet decays is very much dependent on the initial feed rate of the fluid. If the fluid feed rate is relatively slow, the radius of the jet decays very rapidly [16]. If the fluid feed rate is very fast, the radius of the jet will

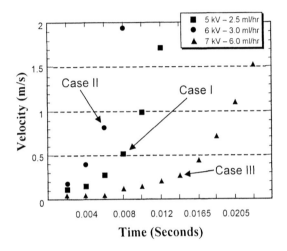

Figure 8: Velocity profiles of electrospinning jets for 3 different spinning conditions.

decrease much more slowly and in extreme cases, not at all [4, 13]. If a simple cylindrical geometry is used to represent a volume element of the electrospinning jet, then the ratio of surface area to volume (specific surface area) is proportional to the inverse of the jet radius (equation 2). This means that an increase in jet radius results in a corresponding decrease in specific surface area associated with that specific volume element. Assuming that the density of the polymer fluid is constant and that the surface charge density is constant, then it follows that the charge to mass ratio will decrease with increasing jet radius. Since the acceleration of the volume element is directly proportional to the ratio of charge to mass (see equation 3), we can expect that an increase in the initial jet radius will result in a decrease in the acceleration of the fluid. This effect can be mitigated to some extent by increasing the electrospinning field strength; however there are practical upper limits to this approach imposed by the dielectric breakdown threshold of the atmosphere.

$$A/V = 2/r \qquad (2)$$

Where: A is surface area of the cylindrical volume element
                  V is the volume
                  r is the radius of the jet

$$a = E*(q/m) \qquad (3)$$

Where: a is acceleration
                  E is the electric field strength
                  q is the available charge for a given volume element
                  m is the mass of a given volume element

This simplified model of a cylinder is meant only to provide a gross illustration of the interaction between solution feed rate, electrospinning voltage, and the charge to mass ratio, and is in qualitative agreement with our experimental observations. Such an illustration is felt to provide useful insights that can aid researchers new to the field in the design and scale up of electrospinning experiments. Most importantly, it suggests that for a given polymer solution there exists a finite processing window in terms of solution feed rate and electrospinning voltage. This idea is consistent with results reported by Baumgarten for solutions of Acrylic polymer and Dimethyl Formamide [14] where it was shown that the collected fiber diameter initially decreased, reached a minimum diameter, and then increased, when both spinning voltage and solution feed rate were increased. However, it should be understood that a more precise description of fluid flow in the electrospinning jet must take into account numerous other variables such as viscoelastic response, charge relaxation times, and solvent evaporation rate, all of which further complicate the interplay of mechanical and electrostatic forces. For a more detailed discussion of these issues readers are referred to the works of Hohman [3, 4], Yarin [2], and Feng [18, 19].

Jet Shear Rate

For each spinning condition (Case I-III), the particle velocity has also been plotted as a function of displacement along the length of the jet as shown in Figure 9. In each case, the initial part of the curve is linear, indicating a constant shear rate. The initial shear rate for Case I was determined to be 214 sec^{-1}. Case II exhibited a shear rate of 252 sec^{-1}, while case III had a shear rate of 135 sec-1. As the fluid travels farther away from the point of initiation, the increase in particle velocity becomes nonlinear, indicating that the rate of shear of the fluid increases. These results are similar to those reported by Larrondo and Manley[20-22] for Polyethylene spun from a molten state. In the case of electrospun polymer melts, this increase in shear rate has been attributed to a shift from extensional flow to a more chaotic flow behavior [20-22]. It has been suggested that as the jet moves away from the point of initiation, rotational components of shear manifest themselves due to the interaction of changing electrostatic and viscoelastic forces [2, 3, 20, 21]. It is possible that the observed shift in fluid flow behavior signals the onset of rotational shear components in the fluid jet that are ultimately responsible for the initiation of the much discussed [2, 3, 4, 5] bending instability.

Conclusion

The motion of particles in polymer solution in the linear portion of an electrostatically driven jet has been observed using high-speed imaging and standard video techniques. The motion of particles in the meniscus, or "cone" region just below the capillary orifice, suggests the presences of a shear force originating at the surface of the fluid. Particle motion in the center of the cone region was observed to run counter to the overall direction of flow. Particles larger than the initial diameter of the electrospinning jet were effectively filtered from the stream, and were not incorporated in the collected fibers, demonstrating the need for effective dispersion of particles in the spinning. Velocity profiles and shear rates were measured for a variety of spinning conditions. The results indicates that there exists an optimal spinning rate for single spinnerets that provides maximum jet stability and a minimum fiber diameter. This implys that there are limits to the degree that the rate of nanofiber production from a single

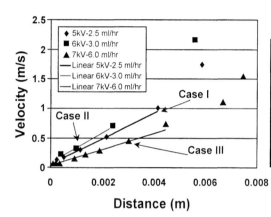

Figure 9: Plot of jet velocity as a function of distance from the point of jet initiation.

jet can be sped up without loosing control of the fiber morphology. Future efforts aimed at commercializing the electrospinning process will need to take this into account when designing the scaled up process.

Acknowledgements

The authors gratefully thank the Aberdeen Test Center for their help in obtaining the high-speed images of the electrospinning process.

References

1. Zeleny J.; *Physical Reviews*; **1914,3,69**
2. Yarin, A. L. ; Koombhongse, S. ; and Reneker, D. H. ; *J. Appl. Phys.*, **2001**, *89*, 3018
3. Hohman, M.M. ; Shin, M.; Rutledge, G.C. ;Brenner, M. P; *Phys. Fluids*, **2001**, *13*, 2201
4. Hohman, M.M. ; Shin, M.; Rutledge, G.C. ; Brenner, M. P; *Phys. Fluids* **2001**, *13*, 2221
5. Reneker, D. H. ; Yarin, A. L. ; Fong, H. ; Koombhongse, S. ; *J. Appl. Phys.*, **2000**, *87*, 4531
6. Deitzel, J.M. ; Kleinmeyer, J.; Harris, D. ; Tan, N. C. B.; *Polymer*; 2001, *42*, 261
7. Salalha W; Dror Y; Khalfin RL; Cohen Y; Yarin AL; Zussman E; *Langmuir* **2004**, *20*, 9852-9855
8. Guan HY; Shao CL; Wen SB; Chen B; Gong J; Yang XH; *Inorganic Chemistry Communications* **2003**, *6*, 1302-1303
9. Fong H; Liu WD; Wang CS; Vaia RA; *Polymer* **2002**, *43*, 775-780
10. Drew C; Wang XY; Samuelson LA; Kumar J; *Journal Of Macromolecular Science-Pure And Applied Chemistry A* **2003**, *40*, 1415-1422
11. Jiang HL; Fang DF; Hsiao BJ; Chu BJ; Chen WL; *Journal Of Biomaterials Science-Polymer Edition* **2004**, *15*, 279-296
12. Hayati, I.; Bailey, A.; Tadros, TH. F.; *Journal of Colloid and Interface Science* **1987**, *117*, 205
13. Hayati, I.; Bailey, A.; Tadros, TH. F.; *Journal of Colloid and Interface Science* 1987, *117*, 222
14. Baumgarten, P.K.; *Journal of Colloid and Interface Science*; **1971**, *36*, 71
15. *The Fundamental Laws of Electrolytic Conduction: Memoirs By Faraday, Hittorf, and F. Kohlrausch*; Goodwin, H.M., Ed; Harper&Brothers, New York, NY, 1899
16. Monk, P.; *Fundamentals of Electroanalytical Chemistry*; Wiley, Chichester, NY, 2001

17. Spivak, A. F. ; Dzenis, Y. A. ; *J. Appl. Mech.* **1999**, *66*, 1026
18. Feng, J.J.; *Physics of Fluids* **2002**, *14,* 3912
19. Feng, J.J.; *J. Non-Newtonian Fluid Mech.* **2003**, *116*; 55;
20. Larrondo, L.; St. John Manley, R.; *Journal of Polymer Science, Polymer Physics Edition* **1981**, *19*, 909
21. Larrondo, L.; St. John Manley, R.; *Journal of Polymer Science, Polymer Physics Edition* **1981**, *19*, 921
22. Larrondo,L.; St. John Manley, R.; *Journal of Polymer Science, Polymer Physics Edition* **1981**, *19*, 933

Chapter 6

Simultaneous Electrospinning of Two Polymer Solutions in a Side-by-Side Approach to Produce Bicomponent Fibers

Pankaj Gupta and Garth L. Wilkes[*]

Department of Chemical Engineering, Virginia Polytechnic Institute and State University, Blacksburg, VA 24061

Bicomponent fibers, in the range of 100 nm to a few microns, of miscible poly(vinylchloride)/segmented polyurethane (PVC/Estane®) and immiscible poly(vinyl chloride)/poly(vinylidienefluoride) (PVC/PVDF) were produced respectively by electrospinning two polymer solutions in a side-by-side approach. For each of the pairs investigated, PVC/Estane® and PVC/PVDF, energy dispersive spectroscopy was utilized to identify the respective components by detecting the signal corresponding to chlorine, oxygen and fluorine from PVC, Estane® and PVDF respectively. The ratio of the peak intensities of Cl to O in PVC/Estane® and Cl to F in PVC/PVDF were found to vary along the length of the fibers. The ratio of the peak intensities corresponding to Cl and O in the miscible PVC/Estane® was calibrated to the actual wt% of Estane®. The strength of this methodology is to effectively electrospun immiscible and miscible polymer pairs to yield submicron bicomponent fibers that are expected to exhibit a combination of properties from each of its constituent components.

Introduction

Electrospinning is a unique process to produce submicron polymeric fibers in the average diameter range of 100 nm-5 μm (1-4). Fibers produced by this approach have a diameter that is at least one or two orders of magnitude smaller than those produced by conventional fiber production methods like melt or solution spinning (5). In a typical electrospinning process, a jet is ejected from the surface of a charged polymer solution when the applied electric field strength (and consequently the electrostatic repulsion on the surface of the fluid) overcomes the surface tension. At a critical point, defined as the equilibrium between the electrostatic repulsion and the surface tension of the fluid, the free surface of the fluid changes to a cone, also commonly referred to as the Taylor cone (6). A jet ejected from the surface of this Taylor cone rapidly travels to the collector target located at some distance from the charged polymer solution under the influence of the electric field. Solidified polymer filaments are collected on the target as the solvent evaporates. It is well established that the jet undergoes a series of electrically driven bending instabilities (7-9) that gives rise to a multitude of looping and spiraling motions beginning in a region close to ejection of the jet. As the jet travels to the target, it elongates to minimize the instability caused by the repulsive electrostatic charges thereby causing the jet to undergo large amounts of plastic stretching that consequently leads to a significant reduction in its diameter. The large amount of plastic stretching is expected to cause orientation of the chains along the fiber axis although chain relaxation could also take place depending on molecular weight, molecular architecture and also the rate of solvent evaporation. These extremely small diameter electrospun fibers possess a high aspect ratio that lead to a larger specific surface. As a result, they have potential applications ranging from optical (10) and chemo-sensor materials (11), nanocomposite materials (12), nanofibers with specific surface chemistry (13), tissue scaffolds, wound dressings, drug delivery systems (14-16), filtration and protective clothing (17, 18).

The effects of several process parameters, such as the applied electric field strength, flow rate, concentration, distance between the capillary and the target have been explored in great detail for different polymer materials *(2, 4, 5, 19-21)*. Primarily, most of the systems that have been investigated have utilized electrospinning from a single polymer solution or melt. Recently, a few systems where blends of polymers (in the same solvent) and blends of polymer solutions (a four component system) have been electrospun. Blends of polyaniline, a conducting polymer, with poly(ethylene oxide), PEO, in chloroform were electrospun to produce filaments in the range of 4 – 20 nm *(22, 23)* that were investigated for their magnetic susceptibility behavior. More specifically, PEO was blended with polyaniline to increase the viscosity of the solution to achieve stable electrospinning. In another study, M13 viruses suspended in 1,1,1,3,3,3-hexafluoro-2-propanol were blended with a highly water soluble polymer, polyvinyl pyrrolidone (PVP) and later electrospun into continuous uniform virus-blended PVP nanofibers. The resulting electrospun mats of virus-PVP nanofibers maintained their ability to infect bacterial hosts after being resuspended in a buffer solution *(24)*. Blending of two polymeric components has also been performed to achieve certain specific chain conformations in polymeric biomaterials. For instance, blending of regenerated silk with polyethylene oxide (PEO) was performed to avoid the development of insoluble and brittle β-sheets of silk fibroin *(25)*. Furthermore, blending of silk fibroin with a water soluble and biocompatible polymer, PEO, enhanced the utility of the resulting electrospun mats in *in vitro* and *in vivo* conditions. Another biomacromolecule, dextran, was blended with biodegradable poly(D,L-lactide-*co*-glycolide) (PLGA) to prepare electrospun membranes for biomedical applications *(26)*. Due to the high water solubility of dextran and PLGA, the water solubility of the resultant electrospun mats was controlled by a post-spinning UV crosslinking process that involved irradiation of the methacrylate-substituted dextran in the presence of a photoinitiator. In another study, PLGA was blended with biocompatible poly(D, L-lactide) (PLA) to control the physical and biological properties of electrospun scaffolds, viz. degradation rate, hydrophilicity, mechanical properties and in vivo shrinkage *(27)*.

A few aspects need to be considered when electrospinning is performed from blends of polymer solutions. For a blend of two polymers (in the same solvent or different solvents), the mixture should be homogenous so that the resultant mat possesses a uniform spatial composition. In addition to being thermodynamically miscible, the interactions between the polymer and the solvent of the opposing pair are of critical importance in a four-component system (dissolution of polymers in different solvents). Hence, the thermodynamic and kinetic aspects of *mixing* need to be considered when utilizing blends for electrospinning.

Another way to produce electrospun mats comprising of two polymeric components, whether miscible or immiscible, is to electrospin two polymers simultaneously in a side-by-side fashion. Very recently, researchers have also been able to electrospin hollow silica *(28)* and ceramic *(29)* fibers by cospinning two solutions in a sheath-core fashion. As will be discussed later, in our approach the two polymer solutions do not come in physical contact until they reach the end of the spinneret where the process of fiber formation begins.

We have been able to design an electrospinning device where two polymer solutions have been electrospun simultaneously in a side-by-side fashion *(30)*. This allows having a bicomponent electrospun mat that possesses properties from each of the polymeric components. For instance, one of the polymers could contribute to the mechanical strength while the other could enhance the wettability of the resulting non-woven web. This could be useful for a protective clothing application. In fact by suitably choosing the constituent components based on their respective properties, the potential of these bicomponent fibers to be utilized in various applications becomes enhanced. These applications could include biomedical, protective, structural, sensing and so forth. The primary purpose of the present study, however, was to demonstrate the feasibility of this new methodology to produce submicron bicomponent fibers via electrospinning. As will be seen later, the strength of this methodology is to effectively electrospun immiscible and miscible polymer pairs to yield submicron bicomponent fibers.

In the following sections, the new bicomponent electrospinning device will be described. Preliminary results on poly(vinyl chloride)/segmented polyurethane (PVC/Estane®) and poly(vinyl chloride)/poly(vinylidiene fluoride) (PVC/PVDF) bicomponent fibers will be presented. It is important to note here that PVC/Estane® is a miscible system whereas PVC/PVDF is an immiscible system. PVC has a glass transition temperature of ca. 85 °C and is therefore a glassy and stiff material at room temperature. The mechanical properties of PVC, especially toughness can be enhanced by suitable plasticization *(31)*. When blended with a thermoplastic urethane-based polymer, Estane®, it is expected that the resulting mechanical toughness would be improved depending on the composition ratio. PVDF on the other hand is semi-crystalline and also displays piezoelectric behavior *(32)*. By incorporating a glassy and stiff polymer like PVC with PVDF, it is expected that the resulting system will possess characteristics of both the components. However, the choice of these specific systems was undertaken primarily to identify the two components easily in each polymer pair by means of energy dispersive spectroscopy (EDS), thereby demonstrating the feasibility of electrospinning a bicomponent fiber. The EDS detector, which was a part of the scanning electron microscope (SEM) utilized in this study to investigate the morphology of the fibers, had a minimum resolution of 1 μm x 1 μm x 1 μm. This allowed characterization the local composition of the bicomponent fibers at the micro level.

Experimental

Bicomponent Electrospinning Device

The schematic of the new device is shown in Figure 1, where the two plastic syringes each containing a polymer solution lie in a side-by-side fashion. A common syringe pump (K D Scientific, model 100) controlled the flow rate of the two polymer solutions. The platinum electrodes dipped in each of these solutions were connected in parallel to the high voltage DC supply (Spellman CZE 1000R). The free ends of the Teflon needles attached to the syringes were adhered together. The internal diameter of the Teflon needle was 0.7 mm with a wall thickness of ca. 0.2 mm and the length of the Teflon needles was ca. 6 cm. The grounded target used for collecting the solidified polymer filaments was a steel wire (diameter ~ 0.5mm) mesh of count 20x20 (20 steel wires per 1" each in the horizontal and vertical axes). It is also possible to use other kinds of substrates as well, e.g. the grounded target can be in the form of a cylindrical mesh that can be rotated to obtain filaments oriented in the extrusion direction. The fibers can also be collected on a wide array of substrates including wax paper, Teflon, thin polymer films and so forth. Choice of substrate can often facilitate easy collection and isolation of the fibers from the target.

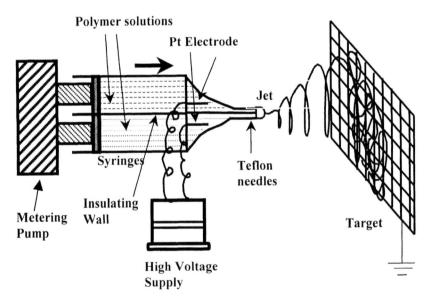

Figure 1. Schematic representation of the side-by-side bicomponent fiber electrospinning device

Materials

PVC, weight average molecular weight, M_w, of 135,900 g/mol in the form of a fine powder, PVDF, weight average molecular weight, M_w, of 250,000 g/mol in the form of pellets and Estane® 5750, a polyether based segmented polyurethane supplied by Noveon Inc., also in the form of pellets was utilized for this study. Molecular weight data for Estane® 5750 could not be obtained, as the information was deemed proprietary. All the three polymers were dissolved separately in N,N-dimethylacetamide (DMAc) at different weight concentrations ranging from 20-25-wt%. Polymer films were also cast from blends of PVC and Estane® in N,N-dimethylformamide (DMF) at different compositions (0, 25, 50, 75 and 100 wt% of Estane®). EDS was performed on solvent cast films of miscible PVC/Estane® blends to calibrate the peak intensities of oxygen (O) and chlorine (Cl) with the actual composition in the blend. Prior to SEM analysis all fiber and cast film samples were dried at 60 °C in vacuum for 8 h to eliminate any residual solvent. All solvents utilized in this study were purchased from Sigma Aldrich.

Measurements and Characterization

The viscosities of the polymer solutions were measured with an AR-1000 Rheometer, TA Instruments Inc. The measurement was done in the continuous ramp mode at room temperature (25 °C) using the cone and plate geometry. The sample was placed between the fixed Peltier plate and a rotating cone (diameter: 4 cm, vertex angle: 2°) attached to the driving motor spindle. The changes in viscosity and shear stress with change in shear rate were measured. A computer interfaced to the rheometer recorded the resulting shear stress vs. shear rate data. For the three solutions investigated in the present study, the shear stress vs. shear rate curve was linear within the range of shear rates investigated, thereby indicating Newtonian behavior. The slope of the linear shear stress-shear rate relationships gave the Newtonian or zero shear rate viscosities, η_0. The viscosity of the polymer solutions is of importance as it influences the final diameter of the electrospun fibers (7, 33, 34). It is important to note, however, that the extensional viscosity of the jet while in flight to the target is undoubtedly very influential in governing the stretching induced in the jet. However, a thorough study of the effect of extensional viscosity on fiber formation in electrospinning has not been reported to date. For the present study, the zero shear rate viscosities are reported. The η_0 of the polymer solutions were 8.7, 4.9 and 5.6 Pa·s respectively for PVC (25wt% in DMAc), Estane® (20wt% in DMAc) and PVDF (20wt% in DMAc) solutions.

An Oakton® conductivity tester, model TDStestr 20 was utilized to measure the conductivity of the polymer solutions. Prior to its use, the conductivity tester was calibrated by standard solutions procured from VWR Scientific®. The conductivities of the three polymer solutions were 7, 38 and 2 μS/cm respectively for PVC (25wt% in DMAc), Estane® (20wt% in DMAc) and PVDF (20wt% in DMAc).

A Leo® 1550 Field Emission Scanning Electron Microscope (FESEM) was utilized to visualize the morphology of the bicomponent polymer filaments. All the images were taken in the back-scattered mode, as the back-scattered detector is more sensitive to the electron density differences arising due to the presence of different chemical moieties, viz. Cl, O and F in PVC, Estane® and PVDF respectively. The samples were sputter-coated by a Cressington® 208HR to form a 5-10 nm conducting layer of Pt/Au layer on the surface of the fibers and films. This was done to reduce the charging of the non-conducting polymeric surfaces, when exposed to the electron beam in SEM. EDS in conjunction with SEM was utilized to investigate the morphology and local composition of the bicomponent fibers.

Results and Discussion

As shown in Figure 1, where the schematic of the bicomponent device is depicted, the two polymer solutions come in contact only at the tip of the Teflon needles. Even though the two polymer solutions are charged to the same polarity, some amount of mixing of the two components is expected to take place as the two solutions reach the end of the Teflon needle tips. Under stable electrospinning conditions, a fluctuating jet was observed for PVC/Estane® and PVC/PVDF at 14 and 15 kV respectively at a target distance of 20 cm and total flow rate of 3 ml/h. The corresponding average electric fields can be expressed as 0.7 kV/cm and 0.75 kV/cm respectively. Interestingly, when the distance between the Teflon needle-tips and the target was ca. 9 cm or larger, a single common Taylor cone was observed. From the surface of this Taylor cone, a fluctuating/pulsating jet was ejected. The position of ejection of the jet on the surface of the Taylor cone changed very rapidly with time and led to a somewhat non-steady flow of the polymer solution. These fluctuations/pulsations likely influence the extent of mixing of the two charged solutions when they come in contact at the tip of the Teflon needles. At distances larger than 25 cm, the jet was not continuous and the Taylor cone dripped due to weak field strength that did not convey the jet to the grounded target. At distances less than ca. 9 cm, two Taylor cones were observed to emanate from each of the two Teflon needles. As a result, two jets were observed to eject form each Taylor cone under these conditions. At such low distances (< 9 cm), the field strength was relatively strong, thereby inducing a strong electrostatic repulsion between the two polymer solutions emanating from each Teflon needle. This lead to the formation of two Taylor cones and subsequently two separate but identically charged (in terms of polarity) jets causing the formation of two zones of fiber collection on the target, each corresponding to only one of the two respective polymer components. For the systems investigated in this study, bicomponent electrospinning was conducted at a target distance of 10-25 cm from the needle tips. Recall that under these conditions, a single Taylor cone and consequently a single jet was observed to form.

The electrospinning conditions utilized for the miscible PVC/Estane® are summarized as: 14kV, 3ml/h, 15cm and 25wt% PVC with 20wt% Estane. For the immiscible PVC/PVDF, the conditions were: 15kV, 3ml/h, 20cm and 25wt% PVC with 20wt% PVDF. The back-scattered FESEM electron image of the dried PVC/Estane® electrospun web can be seen in Fig. 2a. EDS was performed on several spatial positions within the mat, but of particular importance are the two regions marked as 'A' and 'B'. Region 'A' exhibited an intense peak of chlorine (Cl), indicating the local composition to be principally comprised of PVC (Figure 2b), whereas, region 'B' exhibited an intense peak of oxygen (O), indicating the local dominance of Estane® (Figure 2c). It is important to note the presence of a smaller peak corresponding to oxygen in Figure 2b. This peak was relatively weak but it does indicate the presence of Estane® in the predominantly PVC rich fiber at region 'A'.

In Figure 3a, the back-scattered SEM on bicomponent PVC/PVDF electrospun web is shown. Similar to what was described above, EDS was performed on several spatial positions within the electrospun mat, but of particular importance are the two regions marked as 'A' and 'B'. Region 'A' (Fig. 4b) was observed to be predominantly composed of PVC, as indicated by a strong peak of chlorine in Figure 3b, whereas region 'B' was primarily composed of PVDF, as indicated by the strong peak of fluorine in Figure 3c. It is important to note that smaller peaks corresponding to trace amounts of fluorine and chlorine respectively can be observed as well in Figures 3b and 3c. In both the cases, it can be concluded that although fibers composed primarily of either component were observed to form, the presence of trace amounts of the other component indicates some level of physical mixing in the two solutions. To obtain a better understanding of this phenomenon, EDS was performed along the length of a given fiber to study the changes in local composition, if any. Bicomponent fibers chosen for these investigations were typically a micron or so in diameter as the minimum resolution of the EDS detector was 1µm x 1µm x 1µm. In fact, the diameter for these bicomponent fibers of PVC/Estane® and PVC/PVDF ranged from 100nm to a few microns.

When EDS was conducted on different regions along a 15 µm length of a 'PVC-rich' fiber in the PVC/Estane® mat, it was found that the ratio of the peak

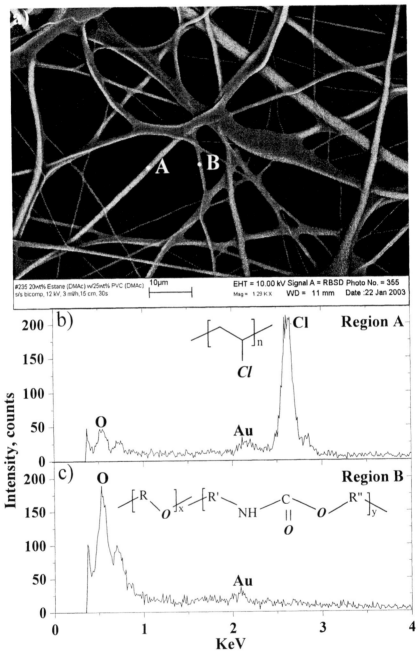

Figure 2 a) SEM of PVC/Estane® electrospun mat. EDS results conducted on regions 'A' and 'B' indicate the local composition of the fibers in b) and c) respectively. The chemical structures of PVC and Estane® are shown along the intense peaks arising due to chlorine (Cl) and oxygen (O) in b) and c) respectively.

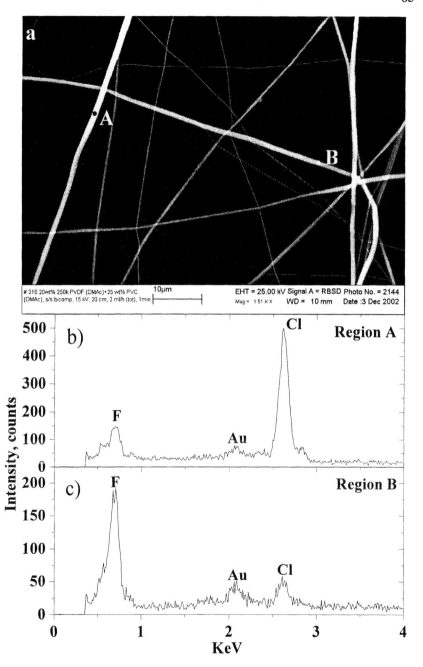

Figure 3 a) SEM of PVC/PVDF electrospun mat. EDS results conducted on regions 'A' and 'B' indicate the local composition of the fibers in b) and c) respectively. The chemical structures of PVC and PVDF are shown along the intense peaks arising due to chlorine (Cl) and fluorine (F) in b) and c) respectively.

intensity corresponding to chlorine to that of oxygen varied from 4.6 to 7.7 (Figure 4a). Similar measurements performed on regions along a 25 μm length of a 'PVDF-rich' fiber in the PVC/PVDF mat, indicated that the ratio of the peak intensity corresponding to fluorine with that of chlorine varied from 1.8 to 2.5 (Figure 4b). These results indicated that even though fibers comprised predominantly of either component were observed, the relative amount of a given component varied significantly along the length of the fiber. Restated, the mixing of the two components changes with time within the time frame of electrospinning process that promotes variations in the composition along the length of the fiber. These variations in the composition of the bicomponent fibers are attributed to the fluctuations of the common jet on the surface of the Taylor cone. Chain diffusion and relaxation can also enhance the mixing of the two components. It is useful to consider both the jets emanating from the Teflon needle tips carry the same electrostatic charge and therefore experience mutual electrostatic repulsion. This is expected to hinder the mixing of the two solutions to some degree. In addition, PVC can develop a low degree of crystallinity while PVDF distinctly crystallizes to a higher extent when solidified from solution. Hence, the amount of crystallinity induced in these fibers as the solvent evaporates during the flight of the fibers to the grounded target will also influence the extent of mixing of the two components. Therefore the mixing of the two charged solutions in bicomponent electrospinning is a competing phenomena between several effects - the fluctuations of the jet, electrostatic repulsion between the like-charged jets, diffusion of polymeric chains of one component in the other, chain relaxation, evaporation rate of the solvent and solvent induced crystallization in semi-crystalline polymers.

To estimate the composition of the PVC/Estane® bicomponent fibers in terms of actual wt% of either component, EDS analysis was conducted on solvent cast (non-spun) films of miscible PVC/Estane®. The peak intensity ratios of Cl to O as observed by EDS in these films of miscible PVC/Estane® were calibrated to the actual wt% of Estane®. Five films of PVC/Estane® were solvent cast from DMF at 0, 25, 50, 75 and 100 wt% of Estane® that were ca. 125 μm in thickness. As mentioned previously, prior to SEM these films were oven dried in vacuum at 60 °C for 8 h to eliminate any residual solvent. SEM results indicated uniform surface morphology, as expected. EDS analysis was performed on each film at different spatial positions to ascertain the local composition. The average peak intensities of Cl and O were plotted after correcting for the background in Figure 5a. It can be seen that the corrected peak intensities of Cl and O decrease and increase linearly with wt% of Estane® respectively. The ratio of corrected peak intensities of Cl to O was plotted as a function of wt% of Estane® and is shown in Figure 5b. The four data points (at 0 wt% Estane®, the Cl:O approaches infinity, and is thus not plotted) can be fitted empirically to a good approximation as an exponential decay. It is noted that the curve fitting of the

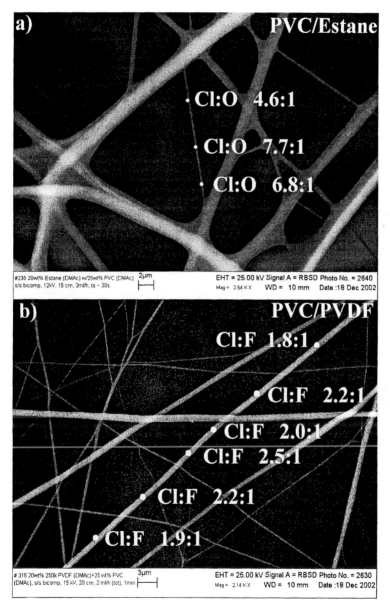

Figure 4. Variation in the peak intensity ratio of a) chlorine (Cl) and oxygen (O) in bicomponent PVC/Estane and b) chlorine (Cl) and fluorine (F) in bicomponent PVC/PVDF electrospun webs.

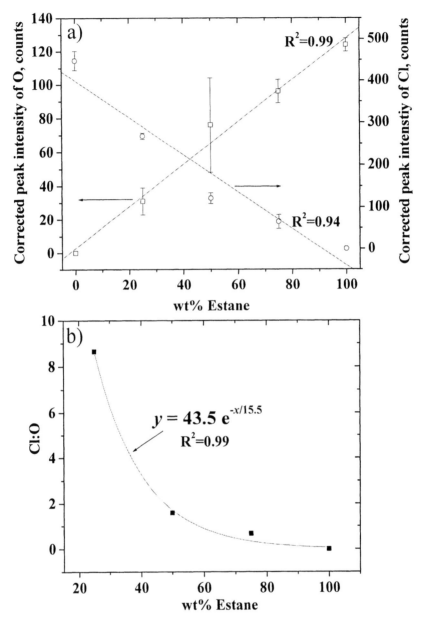

Figure 5 a) Linear variation of peak intensity of O and Cl and b) variation of the ratio of the average peak intensities of Cl and O as a function of composition in the solvent cast blend films of PVC/Estane®. The data in b) is fitted to an exponential decay, as indicated.

data points to an exponential decay does not bear any particular physical significance. Recall that the ratio of peak intensity of Cl:O in the PVC/Estane® bicomponent fiber varied from 4.6 to 7.7 along its length. These correspond to a variation in wt% Estane® from 35 to 27 respectively. A similar exercise could not be performed on PVC/PVDF as phase separation occurred in the solvent cast films from its blends due to their immiscibility.

Conclusions

Two polymer solutions were electrospun simultaneously in a side-by-side fashion to produce bicomponent fibers that had diameters in the range of 100 nm to a few microns. A miscible, PVC/Estane®, and an immiscible, PVC/PVDF were electrospun in this fashion. For each of the pairs investigated, PVC/Estane® and PVC/PVDF, EDS was utilized to identify the respective components by detecting the signal corresponding to chlorine, oxygen and fluorine from PVC, Estane® and PVDF respectively. It was found that the ratio of the peak intensities of Cl and O in PVC/Estane® and Cl and F in PVC/PVDF varied along the length of the fibers, viz. 4.6 to 7.7 for Cl:O in PVC/Estane® and 1.8 to 2.5 for Cl:F in PVC/PVDF. The ratio of the peak intensities corresponding to Cl and O in the miscible PVC/Estane® was calibrated to the actual wt% of Estane® (34 to 27 wt% respectively). Utilizing the methodology described in this study, the feasibility of electrospinning a bicomponent fiber has been demonstrated.

Acknowledgements

This material is based upon work supported by, or in part by, the U.S. Army Research Laboratory and the U.S. Army Research Office under grant number DAAD19-02-1-0275 Macromolecular Architecture for Performance (MAP) MURI. The authors would like to thank Prof. Chip Frazier, Wood Science Department, Virginia Tech, for allowing the use of AR-1000 Rheometer for viscosity measurements.

References

1. Doshi, J. N. The electrospinning process and applications of electrospun fibers. 1994.
2. Fong, H.; Chun, I.; Reneker, D. H., Beaded nanofibers formed during electrospinning. *Polymer* **1999**, 40, (16), 4585-4592.

3. Kim, J.-S.; Reneker, D. H., Polybenzimidazole nanofiber produced by electrospinning. *Polymer Engineering and Science* **1999**, 39, (5), 849-854.

4. Deitzel, J. M.; Kleinmeyer, J. D.; Hirvonen, J. K.; Beck Tan, N. C., Controlled deposition of electrospun poly(ethylene oxide) fibers. *Polymer* **2001**, 42, (19), 8163-8170.

5. Srinivasan, G.; Reneker, D. H., Structure and morphology of small diameter electrospun aramid fibers. *Polymer International* **1995**, 36, (2), 195-201.

6. Taylor, G. I., *Proceeedings of the Royal Society, London* **1964**, 280, 383-397.

7. Yarin, A. L.; Koombhongse, S.; Reneker, D. H., Taylor cone and jetting from liquid droplets in electrospinning of nanofibers. *Journal of Applied Physics* **2001**, 90, (9), 4836-4846.

8. Reneker, D. H.; Yarin, A. L.; Fong, H.; Koombhongse, S., Bending instability of electrically charged liquid jets of polymer solutions in electrospinning. *Journal of Applied Physics* **2000**, 87, (9, Pt. 1), 4531-4547.

9. Hohman, M. M.; Shin, M.; Rutledge, G.; Brenner, M. P., Electrospinning and electrically forced jets. I. Stability theory. *Physics of Fluids* **2001**, 13, (8), 2201-2220.

10. Wang, X.; Lee, S.-H.; Drew, C.; Senecal, K. J.; Kumar, J.; Samuelson, L. A., Electrospun nanofibrous membranes for optical sensing. *Polymeric Materials Science and Engineering* **2001**, 85, 617-618.

11. Zhang, Y.; Dong, H.; Norris, I. D.; MacDiarmid, A. G.; Jones, W. E., Jr., High surface area chemosensor material by electrospinning of fluorescent conjugated polymer. *Abstracts of Papers, 222nd ACS National Meeting, Chicago, IL, United States, August 26-30, 2001* **2001**, PMSE-369.

12. Fong, H.; Liu, W.; Wang, C.-S.; Vaia, R. A., *Polymer* **2002**, 43, 775-780.

13. Deitzel, J. M.; Kosik, W.; McKnight, S. H.; Tan, N. C. B.; DeSimone, J. M.; Crette, S., Electrospinning of polymer nanofibers with specific surface chemistry. *Polymer* **2002**, 43, (3), 1025-1029.

14. Boland, E. D.; Bowlin, G. L.; Simpson, D. G.; Wnek, G. E., Electrospinning of tissue engineering scaffolds. *Abstracts of Papers, 222nd ACS National Meeting, Chicago, IL, United States, August 26-30, 2001* **2001**, PMSE-031.

15. Matthews, J. A.; Wnek, G. E.; Simpson, D. G.; Bowlin, G. L., Electrospinning of Collagen Nanofibers. *Biomacromolecules* **2002**, 3, (2), 232-238.

16. Kenawy, E.-R.; Bowlin, G. L.; Mansfield, K.; Layman, J.; Sanders, E.; Simpson, D. G.; Wnek, G. E., Release of tetracycline hydrochloride from electrospun polymers. *Polymer Preprints (American Chemical Society, Division of Polymer Chemistry)* **2002**, 43, (1), 457-458.

17. Gibson, P.; Schreuder-Gibson, H.; Pentheny, C., Electrospinning technology: direct application of tailorable ultrathin membranes. *Journal of Coated Fabrics* **1998**, 28, (July), 63-72.

18. Gibson, P.; Schreuder-Gibson, H.; Rivin, D., Transport properties of porous membranes based on electrospun nanofibers. *Colloids and Surfaces, A: Physicochemical and Engineering Aspects* **2001**, 187-188, 469-481.

19. Doshi, J.; Reneker, D. H., Electrospinning process and applications of electrospun fibers. *Journal of Electrostatics* **1995**, 35, (2&3), 151-60.

20. Fong, H. The study of electrospinning and the physical properties of electrospun nanofibers. 1999.

21. Demir, M. M.; Yilgor, I.; Yilgor, E.; Erman, B., Electrospinning of polyurethane fibers. *Polymer* **2002**, 43, (11), 3303-3309.

22. Pinto, N. J.; Johnson, A. T., Jr.; MacDiarmid, A. G.; Mueller, C. H.; Theofylaktos, N.; Robinson, D. C.; Miranda, F. A., Electrospun polyaniline/polyethylene oxide nanofiber field-effect transistor. *Applied Physics Letters* **2003**, 83, (20), 4244-4246.

23. Kahol, P. K.; Pinto, N. J., An EPR investigation of electrospun polyaniline-polyethylene oxide blends. *Synthetic Metals* **2004**, 140, (2-3), 269-272.

24. Lee, S.-W.; Belcher, A. M., Virus-Based Fabrication of Micro- and Nanofibers Using Electrospinning. *Nano Letters* **2004**, 4, (3), 387-390.

25. Jin, H.-J.; Fridrikh, S. V.; Rutledge, G. C.; Kaplan, D. L., Electrospinning Bombyx mori Silk with Poly(ethylene oxide). *Biomacromolecules* **2002**, 3, (6), 1233-1239.

26. Jiang, H.; Fang, D.; Hsiao, B. S.; Chu, B.; Chen, W., Optimization and Characterization of Dextran Membranes Prepared by Electrospinning. *Biomacromolecules* **2004**, 5, (2), 326-333.

27. Kim, K.; Yu, M.; Zong, X.; Chiu, J.; Fang, D.; Seo, Y.-S.; Hsiao, B. S.; Chu, B.; Hadjiargyrou, M., Control of degradation rate and hydrophilicity in electrospun non-woven poly(d,l-lactide) nanofiber scaffolds for biomedical applications. *Biomaterials* **2003**, 24, (27), 4977-4985.

28. Loscertales, I., G.; Barrero, A.; Márquez, M.; Spretz, R.; Velarde-Ortiz, R.; Larsen, G., Electrically Forced Coaxial Nanojets for One-Step Hollow Nanofiber Design. *Journal of the American Chemical Society* **2004**, 126, (17), 5376-5377.

29. Li, D.; Xia, Y., Direct Fabrication of Composite and Ceramic Hollow Nanofibers by Electrospinning. *Nano Letters* **2004**, ACS ASAP.

30. Gupta, P.; Wilkes, G. L., Some investigations on the fiber formation by utilizing a side-by-side bicomponent electrospinning approach. *Polymer* **2003**, 44, (20), 6353-6359.

31. Cano, J. M.; Marin, M. L.; Sanchez, A.; Hernandis, V., Determination of adipate plasticizers in poly(vinyl chloride) by microwave-assisted extraction. *Journal of Chromatography, A* **2002**, 963, (1-2), 401-409.

32. In'acio, P.; Marat-Mendes, J. N.; Dias, C. J., Development of a Biosensor Based on a Piezoelectric Film. *Ferroelectrics* **2003**, 293, 351-356.

33. McKee, M. G.; Wilkes, G. L.; Colby, R. H.; Long, T. E., Correlations of Solution Rheology with Electrospun Fiber Formation of Linear and Branched Polyesters. *Macromolecules* **2004**, 37, (5), 1760-1767.

34. Buer, A.; Ugbolue, S. C.; Warner, S. B., Electrospinning and properties of some nanofibers. *Textile Research Journal* **2001**, 71, (4), 323-328.

Chapter 7

Development of Multiple-Jet Electrospinning Technology

Dufei Fang[1], Charles Chang[2], Benjamin S. Hsiao[2,*], and Benjamin Chu[2,*]

[1]Stonybrook Technology and Applied Research Inc., P.O. Box 1336, Stony Brook, NY 11790
[2]Department of Chemistry, State University of New York at Stony Brook, Stony Brook, NY 11794–3400

We have successfully developed a unique multiple-jet electrospinning technology at STAR, Inc. and Stony Brook University. The development of this technology was primarily accomplished by the incorporation of secondary electrodes to isolate the electric field distribution of the primary electrode spinnerets, the design of an individual fluid distribution system and the optimization of system design using the finite element analysis method, in combination with experimental tests. The key technological advance permits the small-scale mass fabrication of membranes with composite nanofiber/ nanoparticle hybrid morphology, tailor-designed composition variations, and 3D pattern formation from polymer solutions.

Introduction

When an external electrostatic field is applied to a conducting fluid (e.g., a charged semi-dilute polymer solution or a charged polymer melt), a suspended conical droplet is formed, whereby the surface tension of the droplet is in equilibrium with the electric field. Electrospinning occurs when the electrostatic field is strong enough to overcome the surface tension of the liquid droplet at the spinneret tip. The liquid droplet then becomes unstable and a tiny jet is ejected from the surface of the droplet. As it reaches a grounded target, the jet stream can be collected as an interconnected web of fine sub-micron size fibers. The resulting films from these nanoscale fibers (nanofibers) have very large surface area-to-volume ratios and very small pore sizes.

The electrospinning technique was first developed by Zeleny (*1*) and patented by Formhals (*2*). Up to now, there are about 60 patents on electrospinning technology. Much research has been reported on how the jet is formed as a function of electrostatic field strength, fluid viscosity, and molecular weight of polymers in solution. In particular, the work of Taylor and others on electrically driven jets has laid the groundwork for electrospinning (*3*). Although potential applications of this technology have been widely mentioned, which include biological membranes (substrates for immobilized enzymes and catalyst systems), wound dressing materials, artificial blood vessels, aerosol filters, clothing membranes for protection against environmental elements and battlefield threats, just to name a few (*4-24*), to our knowledge, no practical industrial process for electrospinning of polymer systems for fabric applications has ever been implemented. The existing commercial electrospinning process by Donaldson, Inc. is limited to the manufacture of filter membranes, not of clothing. The major technical barrier for manufacturing electrospun fabrics for clothing and other applications is the speed of fabrication. In other words, as the fiber size becomes very small, the yield of the electrospinning process becomes very low. Another major technical problem for mass production of electrospun fabrics is the assembly of spinnerets during electrospinning. A straightforward multi-jet arrangement, as in high-speed melt-spinning, cannot be used because adjacent electrical fields often interfere with one another, making the mass production scheme by this approach impractical. The main objectives of this paper are thus to discuss the underlying physics of multiple-jet electrospinning operations as well as to demonstrate a prototype multiple-jet device designed and constructed by the Stony Brook scientists and engineers.

Principle of Single-Jet Electrospinning

Electrospinning is a unique fiber spinning technique, which is capable of generating nano-sized fibers. The operation of electrospinning can be divided into three stages: (1) jet initiation, (2) jet elongation, and (3) solidification of jet (nanofiber formation). During the jet initiation stage, as the surface tension of a polymer solution or a polymer melt was overcome by electrical forces at the surface of a polymer droplet, a charged jet was ejected. After ejection, the jet travels for a certain distance in a straight line (the stage of stable jet), and then bends and follows a looping spiral course (the stage of unstable jet). During the jet elongation (mostly in the unstable jet stage), the electrical force stretches the jet stream from thousands to millions of times longer. During the first two stages, the solvent continues to evaporate. Finally, with enough solvent being evaporated, the next stage is a solidification of the jet where the viscosity of the polymer solution becomes so high that stretching of the jet stream is basically non-existent. With further evaporation of the solvent, the resulting nano-fibers can be collected on an electrically grounded collector, such as a metal drum, a screen, or a coagulating bath. The resultant nano-sized fibers are of substantial scientific and commercial interest because their unique morphologies and properties are very different from conventional fibers having one to two orders of larger diameter than those of the electrospun nanofibers. Electrospinning can be affected by many parameters such as electric field, solution viscosity, conductivity, surface tension, polymer chain relaxation time, and the electrical charges carried by the jet.

The mechanical forces acting on the conducting fluid, which must be overcome by the interactions between an electrostatic field and the conducting fluid to create the jet, can be understood by examining a fluid droplet to be formed at the tip of a capillary tube (Figure 1). In this droplet, a higher pressure is developed due to molecular interactions. This excess pressure Δp inside the droplet, which acts upon the capillary cross-section area d, is counterbalanced by the surface tension Y acting on the circumference πr^2, i.e.

$$\Delta p \cdot \pi r^2 = Y \square 2\pi r, \text{ or}$$

$$\Delta p = \frac{2Y}{r} \tag{1}$$

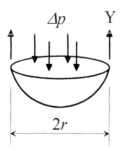

Figure 1. Schematic of a fluid droplet created from a capillary.

Equation 1 indicates that both the droplet excess pressure Δp and the surface energy per unit drop volume $(4\,\pi r^2 / [(4\pi/3)\,r^3]) = 3Y/r)$ become large when r is small. When the liquid droplet is suspended from a capillary tip (pendant droplet, as shown in Figure 2), the surface tension of the droplet can be derived from the droplet shape and the balance of all the forces acting upon the droplet, including the gravity. The relationship can be expressed as follows.

$$Y = g\Delta\rho r_0^2 / \beta \tag{2}$$

where $\Delta\rho$ is the density difference between the fluids at the interface ($\Delta\rho = \rho$ for the droplet having a liquid/air interface), g is the gravitational constant, r is the radius of drop curvature at the apex and β is the shape factor, which can be defined as:

$$dx / ds = \cos\phi$$
$$dzds = \sin\phi \tag{3}$$
$$d\phi / ds = 2 + \beta z - \sin\phi / x$$

The value of β can be accurately determined by numerical calculation.

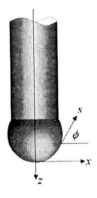

Figure 2. Schematic of a liquid drop suspended from a capillary.
(See page 5 of color inserts.)

A droplet from a single spinneret in an electrostatic field E is shown schematically in Figure 3. If a liquid has conductivity other than zero, the electric field will cause an initial current flow or charge rearrangement in the liquid. The positive charge will be accumulated at the surface until the net electric field in the liquid becomes zero. This condition is necessary for the current flow to be zero in the liquid. The duration τ of this flow can be typically expressed as $\tau = \varepsilon / \sigma$, where ε is the permittivity and σ is the conductivity of the liquid. With a surface charge density (per unit area) ρ_s, the (surface) force F_s exerted on the surface by the electrostatic field E on the droplet per unit area is:

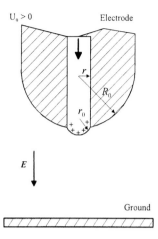

Figure 3. Schematic of a droplet from a single spinneret (electrode) in an applied electric field.

$$F_s = \rho_s(\sigma)E \qquad (4)$$

The conductivity of the liquid can be adjusted, e.g., by adding an ionic salt. Thus, the surface charge density per unit area can be tuned accordingly. With a sufficiently strong electrostatic field at the tip, the surface tension Y can be overcome, i.e.,

$$F_s = \rho_s(\sigma)E \geq Y - \rho_0 Vg \qquad (5)$$

with ρ_0, V, and g being the density, the volume of the droplet and the gravitational acceleration, respectively. If this condition is met, the droplet shape will change at the tip and become the "Taylor" cone and a small jet of liquid will be emitted from the droplet. If the electrostatic field remains unchanged, the liquid moving away from the surface of the droplet will have net charges. This net excess charge is directly related to the liquid conductivity. Furthermore, the charged jet can be considered as a current flow, $J(\sigma, E)$, which will, in turn, affect the electric field distribution on the tip of the droplet, i.e.,

$$E = E_0 + E'(J) \qquad (6)$$

with E_0 being the applied field threshold in the absence of fluid flow. For polymer solutions above the overlap concentration, the evenly distributed charges in the jet repel each other while in flight to the target (ground). Thus, the polymer chains are continuously being "stretched" in flight until the stretch force is balanced by the chain restoring force or the chains are landed on the target, whichever comes first.

In the electrospinning process, a key requirement is to maintain the droplet shape. This requirement involves the control of many parameters including liquid flow rate, electric and mechanical properties of the liquid, and the electrostatic field strength at the tip. In order to achieve high field strength, the curvature of the electrode at the tip has to be sharp (small radius R_0). However, since a stable pendant droplet is controlled by the shape factor β, the curvature r_0, and thus R_0, could not be too small. Figure 4 shows, as an example, estimates of equal potential lines of a single electrode configuration with a set of specific geometric parameters and the force line for a charge particle in the trajectory that is normal to the equal potential lines. This computer simulation (finite elements analysis) was carried out by *ESTAT* and *VESTAT* to simulate the electric field distribution.

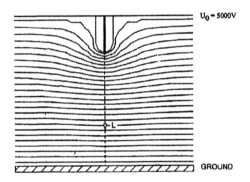

Figure 4. Potential trajectory of a charged fluid jet from a single-jet spinneret.

Figure 5 shows the estimated electric field strength along the jet direction, from the tip of the electrode to the ground (plate). The conductivity of the fluid can be increased by adding an ionic salt to create excess charge, which will facilitate the electrospinning process. Examples of suitable salts include NaCl, KH_2PO_4, K_2HPO_4, KIO_3, KCl, $MgSO_4$, $MgCl_2$, $NaHCO_3$, $CaCl_2$ or mixtures of those salts (*24*).

Principle of Multiple-Jet Electrospinning

For a configuration of electrospinning using multiple jets, two main concerns have to be addressed: (1) the liquid should be delivered to   each

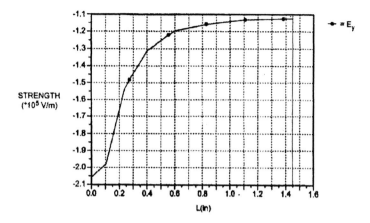

Figure 5. Electric field strength as a function of distance from the tip of a single jet.

separate spinneret either at constant pressure or at constant flow rate, and (2) the electrostatic field strength at the tip of each electrode (spinneret) should be strong enough to overcome the liquid surface tension at that tip. The first concern can be resolved by careful mechanical design of the solution distribution systems to the spinnerets. The second concern can be addressed as follows. With electrodes being placed close to one another, the electrostatic field distribution is changed and the field strength at the tip is normally weakened because of the interference from nearby electrodes, i.e.,

$$E_i = E_i^0 + \sum_{j \neq i} E_{ij} + \sum_j E'_{ij}(J_j) \qquad (7)$$

where E_i^0 is the unperturbed electric field strength due to the single electrode i. E_{ij} is the electric field at location I, contributed by electrode j, and $E'_{ij}(J_j)$ is the interference electric field caused by the current J of jet j. Figure 6 shows the equal potential line of a double jet configuration with the electrodes having the geometrical parameters being the same as that of a single jet.

Based on Equation (5) for a single jet, the criterion for the multiple-jet operation is that each jet (i) has to meet the following condition:

98

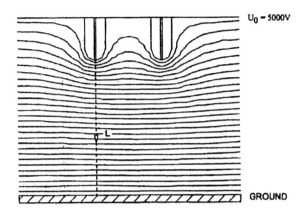

Figure 6. Potential trajectory of charged fluid jets from multiple-jet spinnerets.

$$\rho'_s(\sigma_i)E_i \geq Y_i - \rho'_0 V_i g \tag{8}$$

In fact, both conditions for Equations (7) and (8) should be met for multiple-jet operation. The multiple-jet apparatus of the present invention was based on these two criteria. For example, Figure 7 shows the estimated electric field strength along the direction from the tip to the ground. In comparison with Figure 5, the field strength is less in absolute value. A separate calculation shows that in order to achieve the same field strength as the original unperturbed single jet, the electric potential has to increase from 5.0 kV to 5.6 kV. This estimate demonstrates that the electric field strength for multiple jets can be calculated by using Equation (7). Furthermore, a shielding system or a specially shaped electrode to produce a different electric potential may be used to partially screen out the interference from nearby electrodes, making the scale-up operation practical. Numerical simulations, including jet effects based on Equation (7) can be used to guide and to obtain an optimal design for specific operations.

With multiple jets, as the electrodes are placed close to one another, the electrostatic field distribution is changed and the field strength of the spinneret i at the tip is altered by the presence of nearby electrodes. The net field strength at the tip i can be represented by three combinations: (1) the unperturbed electric field strength due to the single electrode i, (2) the sum of the electric field strength at location i due to all other electrodes, and (3) the electric field strength at location i generated by all jets (including i). The net field strength at tip i(E_i) can then be used to set the criteria for electrospinning, i.e., the product of surface charge density of the conducting fluid at tip $i(S_i)$ times E_i together with the gravity effect should overcome the surface tension of that field at tip i. These rules represent the fundamental criteria for efficient multiple-jet operation and permit an optimal design for specific operations that involve multiple parameter adjustments.

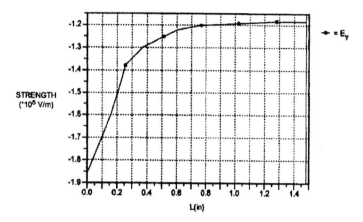

Figure 7. Electric field strength as a function of distance from the tip of a spinneret in a multiple-jet spinneret system.

Finite Element Analysis to Optimize Multiple-Jet Electrospinning Operations

Based on the above principles, we have carried out computer simulations of sample mass transport, electric field distribution and fluid flow inside a microfabricated channel network to optimize the multiple-jet electrospinning operations. Multiple-jet electrospinning involves multiple parameters with complex interactions. These interactions include the effects between adjacent electrodes, the fluid distribution of polymer solutions, and the settings of electrospinning parameters (e.g., electric field strength), where the optimization process relies on experimental means, together with the aid of computer simulation.

One of the most important criterions in multiple-jet electrospinning is the electric field distribution. As previously mentioned, successful electrospinning contains simultaneous optimization of several parameters, including solution viscosity, resistivity, and surface tension. However, all these parameters are properties of polymer solutions and cannot be easily altered. On the other hand, the applied electric field can be manipulated via various equipment settings. Therefore, the study of electric field distribution by external parameter settings is a logical way to initiate the design of a prototype multiple-jet electrospinning system.

Computer simulation of the electric field distribution in the multiple-jet system using finite element analysis (FEA) has been carried out with the aid of several commercial programs: *Turbo CAD*, *X-late*, *Mesh*, *ESTAT* and *VESTAT*. *Turbo-CAD* is a drawing software, which was used to sketch the optimal setting for the multiple-jet system. After being drawn by the *Turbo-CAT*, the system diagram was converted into a series of equations using *X-late*. Subsequently, *Mesh* was used to recreate the system drawing by using the composition of thousands of tiny triangles, which was then used as the calculating points (blocks) (the more triangles or blocks we used as calculating points, the more accurate or precise the calculations were). The last step was to use *ESTAT* and *VESTAT* to simulate the electric field distribution, according to the setting, in order for us to study the phenomenon of the multiple-jet system and to design the optimal multiple-jet system setting.

During electrospinning of a multiple-jet system, the first task is to understand the interactions between the adjacent electrodes. It is known that the electric field at the tip of the electrode can generate an electric force, pulling down the polymer solution into a jet. In order for a polymer fiber to form, the vertical electric force must be stable and aligned downward. Thus, the fiber will have sufficient time to elongate and dry up. However, if the adjacent electrodes influence each other, the vertical electric force will be disturbed and the polymer solutions will collide with each other before the polymer fibers can be formed. Based on the computer simulation (Figure 8), we discovered that the addition of ring-shape secondary electrodes with appropriate voltage under the primary electrodes can minimize the influences between the adjacent electrodes (*25*).

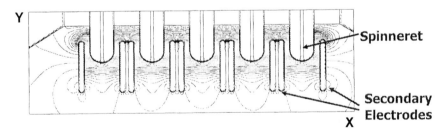

Figure 8. FEA calculation of the electric field distribution of a 5-electrode system using secondary electrodes to minimize the interactions from adjacent electrodes (See page 5 of color inserts.)

The second task is to deal with the non-equivalent distributions in multiple-jet electrospinning. In other words, non-spontaneous jet formations in the multiple-electrode assembly can take place, where some jets can come out earlier than others, thus affecting the electric field distribution on the tip of other electrodes. This problem can also be overcome by the use of secondary electrodes to shield the primary electrodes (Figure 9). However, we noticed that the presence of secondary electrodes can weaken the field strength at the electrode tip. To deal with this problem, the geometrical shape, the location and the electrical potential of the secondary electrodes, have also been optimized by the FEA simulations. The following two criteria were met simultaneously in the optimal design of multiple-jet electrospinning apparatus: (1) each electrode in the multi-jet system has the same electric field distributions, and (2) the electric field strength on the electrode tip in the multi-jet system is the same as that in the single-jet system. Figure 10 illustrates the simulation of an electric field distribution for a 5-electrode system with two jets already being formed. Results suggest that the interference of the neighboring electrodes (with and without the jet formation) of the optimized system can be reduced to about 1% by using the secondary electrodes.

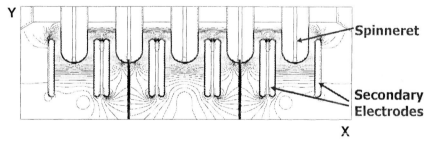

Figure 9. FEA calculation of the electric field distribution of a 5-electrode system with 2 ejecting jets using secondary electrodes to minimize the interactions from adjacent electrodes. (See page 5 of color inserts.)

Construction of a Prototype Multiple-Jet Electrospinning Apparatus

Based on the above simulations, a prototype multiple-jet electrospinning apparatus was constructed using the optimized parameters in multi-jet designs. A photograph of the prototype equipment is shown in Figure 10. Other unique features of this prototype apparatus can be summarized as follows.

Figure 10. Prototype multiple-jet electrospinning apparatus.
(See page 6 of color inserts.)

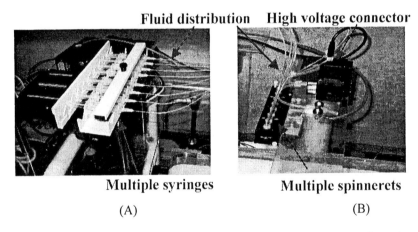

(A)   (B)

Figure 11. (A) Fluid distribution, (B) linear array electrode assembly in the
prototype multiple-jet electrospinning apparatus.
(See page 6 of color inserts.)

Figure 11 shows photographs of fluid distribution (A) and linear array 8-electrode assembly (B) of the prototype equipment constructed. The polymer solution is distributed by multiple syringes using a programmable motor to maintain the minimum pressure drop. Each syringe can be controlled independently, allowing simultaneous processing of multiple polymer solutions. In addition, a simple but robust constant pressure multiple-electrode assembly was also constructed that could allow a more sustained operation. The array spinneret system is mounted on two electrically isolated posts that are seated on a pair of precision rails (Figure 10). This allows the array system to move along the "belt" direction back and forth. The precision rails can also be mounted on a "rocking" system so that the array can move in the direction perpendicular to the "belt" direction. The heating elements are implemented underneath the belt to control the solvent evaporation rate.

The performance of this prototype multiple-jet or single-jet electrospinning apparatus has been thoroughly tested in our laboratory. Details of the test results will not be discussed here, but its excellent performance can be found in several of our recently issued patents and publications (*26-34*).

Concluding Remarks

In summary, the multiple-jet electrospinning technology (*esJets*^{TM}) that has been developed by us appears to offer great potential for future commercialization, particularly in industrial (filtration and protective fabrics) and biomedical applications. The major advantages of this technology can be summarized as follows.

1. Variations in materials parameters have been developed to control the following properties:
    (a) Controlled release rate of medications or/and growth factors (e.g., drugs, DNA, proteins).
    (b) Control of degradation rate.
    (c) Variation in degree of hydrophobicity.
    (d) Environmental responsive dimensional change.
2. Innovation in electrospinning technology:
    (a) Finite element analysis on electric field distribution of multiple spinnerets.
    (b) New spinneret designs for multiple-jet operation.
    (c) Electric field control on jet streams.
    (d) Variation in porosity, pore size, and pore size distribution of electrospun membranes.
    (e) Construction of unique hierarchical structures and pattern formation.
    (f) Ability for long-term storage in liquid nitrogen.
    (g) Scale-up production for commercialization of products.

Acknowledgement

This study was supported by a National Institutes of Health-SBIR grant (GM63283-02) administered by Stonybrook Technology and Applied Research, Inc., the SUNY-SPIR program, and the Center for Biotechnology at Stony Brook.

104

References

1. Zeleny, J. *Phys. Rev.* **1914**, *3*, 69-91.
2. Formhals, A. US Patent 1,975,504, 1934.
3. Taylor, G.I. *Proc. Roy. Soc. Lond. A.* **1969**, *31*, 453-475.
4. Deitzel, J. M.; Kosik, W.; McKnight, S. H.; Beck Tan, N. C.; DeSimone, J. M.; Crette, S. *Polymer*, **2001**, *43(3)*, 1025-1029.
5. Koombhongse, S.; Liu, W.; Reneker, D. H., *J. Polym. Sci., Part B: Polym. Phys.,* **2001**, *39(21)*, 2598-2606.
6. Shin, Y. M.; Hohman, M. M.; Brenner, M. P.; Rutledge, G. C. *Polymer,* **2001**, *42(25)*, 09955-09967.
7. Bognitzki, M.; Frese, T.; Steinhart, M.; Greiner, A.; Wendorff, J. H.; Schaper, A.; Hellwig, M. *Polym. Eng. Sci.,* **2001**, *41(6)*, 982-989.
8. Stitzel, J. D.; Bowlin, G. L.; Mansfield, K.; Wnek, G. E.; Simpson, D. G. *International SAMPE Technical Conference,* **2000**, *32*, 205-211
9. Shin, Y. M.; Hohman, M. M.; Brenner, M. P.; Rutledge, G. C. *Appl. Phys. Lett.,* **2001**, *78(8),* 1149-1151.
10. Bognitzki, M.; Hou, H.; Ishaque, M.; Frese, T.; Hellwig, M.; Schwarte, C.; Schaper, A.; Wendorff, J. H.; Greiner, A. *Adv. Mater.* (Weinheim, Ger.), **2000**, *12(9)*, 637-640.
11. Spivak, A. F.; Dzenis, Y. A. *J. Appl. Mech.,* **1999**, *66(4)*, 1026-1028.
12. Fong, H.; Chun, I.; Reneker, D. H. *Polymer*, **1999**, *40(16)*, 4585-4592.
13. Reneker, D. H.; Chun, I. *Nanotechnology*, **1996**, *7(3)*, 216-223.
14. Gibson, P., Schreuder-Gibson, H., and Pentheny, C. *J. Coated Fabrics,* **1998**, *28*, 63-72.
15. Fang, X.; Reneker, D.H. *J. Macromol. Sci.-Phys.,* **1997**, *B36(2)*, 169-173.
16. Doshi, J.; Reneker, D.H. *J. Electrostatics,* **1995**, *35*, 151-160.
17. Baumgarten, P.K. *J. Colloid Interface Sci.,* **1971**, *36*, 71-79.
18. Larrondo, L.; St. John Manley, R. *J. Polym. Sci.: Polym. Phys. Ed.,* **1981**, *19*, 909-920
19. Larrondo, L.; St. John Manley, R. *J. Polym. Sci.: Polym. Phys. Ed.,* **1981**, *19*, 921-932
20. Larrondo, L.; St. John Manley, R. *J. Polym. Sci.: Polym. Phys. Ed.,* **1981**, *19*, 933-940.
21. Jaeger, R.; Bergshoef, M.M.; Martin, C.; Schonherr, H.; Vancso, G.J. *Macromol. Symp.,* **1998**, *127*, 141-150.
22. Zachariades, A.E.; Porter, R.S.; Doshi, J.; Srinivasan, G.; Reneker, D.H. *Polym. News,* **1995**, *20*, 206-207.
23. Gibson, P. W.; Schreuder-Gibson, H. L.; Rivin, D. *AIChE Journal,* **1999**, *45(1),* 190.
24. Zong, X.; Kim, K.; Fang, D.; Ran, S.; Hsiao, B. S.; Chu, B. *Polymer*, **2002**, *43(16)*, 4403.
25. Chu, B.; Hsiao B. S.; D. Fang, U.S. Pat. 6713011 (2004).
26. Chu, B.; Hsiao, B. S.; Fang, D.; Brathwaite, C. U.S. Pat. 6685956 (2004).
27. Chu, B.; Hsiao, B. S.; Fang, D.; Brathwaite, C. U.S. Pat. 6689374 (2004).

28. Zong, X.; Kim, K.; Ran, S.; Fang, D.; Hsiao, B. S.; Chu, B. *Biomacromol.*, **2003**, *4(2)*, 416-423.

29. Luu, Y. K.; Kim, K.; Hsiao, B. S.; Chu, B.; Hadjiargyrou, M. *J. Control Release*, **2003**, *89*, 341-353.

30. Zong, X.; Ran, S.; Hsiao, B. S.; Chu, B. *Polymer*, **2003**, *44(17)*, 4959-4967.

31. Kim, K.; Yu, M.; Zong, X.; Chiu, J.; Fang, D.; Seo, Y. S.; Hsiao, B. S.; Chu, B.; Hadjiargyrou, M. *Biomaterials*, **2003**, *24*, 4977-4985.

32. Jiang, H.; Fang, D.; Hsiao, B. S.; Chu, B.; Chen, W. *Biomacromolecules*, **2004**, *5(2)*, 326-333.

33. Jiang, H.; Fang, D.; Hsiao, B. S.; Chu, B.; Chen, W. *J. Biomater. Sci. Polymer Edn*, **2004**, *15(3)*, 279–296.

33. Um, I. C.; Fang, D.; Hsiao, B. S.; Okamoto, A.; Chu, B. *Biomacromolecules*, **2004**, *5(4)*, 1428-1436.

34. Kim, K.; Luu, Y. K.; Chang, C.; Fang, D.; Hsiao, B. S.; Chu, B.; Hadjiargyrou, M. *J. Controlled Release*, **2004**, *98(1)*, 47-56.

Chapter 8

Electrostatic Effects on Electrospun Fiber Deposition and Alignment

Navin N. Bunyan[1], Julie Chen[1,*], Inan Chen[2], and Samira Farboodmanesh[1]

[1]Department of Mechanical Engineering, University of Massachusetts at Lowell, Lowell, MA 01854
[2]Consultant, 1220 Majestic Way, Webster, NY 14580

Electrospinning is a nano-fiber manufacturing process that uses an electrical potential to initiate the spinning from a charged polymer solution. The primary objective of this research is to understand the effects of electrostatics on the convergence and alignment of the fibers deposited onto various targets. Desired orientation of the fibers can be achieved by potentially addressing three aspects: jet path control, target design, and solution properties. The first two will be addressed in this paper. It is shown that the spread of the fibers on the target can be decreased by using a disc electrode at the source, which creates a nearly parallel field between the source and target. The results are also shown to agree with an ion-jet charge transport model. This electrode placed at the source can also be used to direct the fibers to different areas on the target by angling it with respect to the plane of the target. Effects of target design on fiber alignment are also demonstrated.

Introduction

Electrospinning is a method of forming polymer fibers with diameters in the micro- to nanometer range by using electric fields. The fabrication and properties of electrospun polymer fibers are gaining increased attention due to the ease of forming fibers with a high surface area/volume ratio. Some potential applications for these fibrous structures include substrates for tissue growth, catalysts and enzymes to make reactions faster, selectively permeable materials for protective clothing, highly effective thermal insulation or thermal conduction, filters for fine particles, wound dressings and artificial blood vessels. The growing fields of nanotechnology and biotechnology have given an added impetus to the creation of high-rate processes for fabricating nanofibers and nanofiber assemblies. For example, electrospun fibers of sulfonated polystyrene and enzymatically synthesized polyaniline have shown promise as conductive and photo-responsive materials for electronic devices (*1*). Nanowires and nanotubes carry charge efficiently; hence they are the ideal building blocks for nanoscale electronics and optoelectronics (*2, 3*). The key to many applications is controlled orientation and/or placement of the nanofibers to obtain the desired interconnection or anisotropy. In this paper, "orientation" refers to the alignment of the fiber rather than the molecular orientation within the fiber. Many studies have been successful in forming non-woven fiber mat from electrospinning; however, the issue of controlling the fiber path from the source to target and having control over the deposition and collection of the fibers is still a major challenge in the fiber assembly process.

Although electrospinning was patented by Formhals (*4*) in the early 1930's, it was only in the 1990's when Reneker and coworkers (*5-8*) carried out detailed experiments on the process that the research in this field received renewed interest. Many of these earlier studies in the past decade focused on demonstrating that submicron fibers could be electrospun with different polymers and also included some experimental studies on the effect of process parameters on structure and morphology of the fibers (*5-11*). The characteristics of the fibers formed were shown to be dependent upon a number of parameters such as electrical field strength, flow rate, and initial solution viscosity. Other factors that are likely to have an effect, but have not been as widely studied include molecular weight, molecular weight distribution, crystallinity, solubility, vapor pressure of the solvent, conductivity, surface tension, and charge transport. Modeling efforts are still somewhat limited. Rutledge et al. (*12-13*) proposed a stability theory for electrified fluid jets to predict the effect of process parameters on the whipping instability initiation. For their analysis they considered the surface charge density, radius of the jet, fluid conductivity and viscosity.

The research effort described in this paper concentrates on controlling the fiber spread and alignment on the target and directing the jet to specific positions on the target. The knowledge obtained in this work can be utilized for developing a design for achieving controlled fiber deposition. Since electrostatic forces dominate the electrospinning process to a large extent, it is possible to control the jet path, fiber deposition, and fiber collection by manipulating the field acting on the system. Different target geometries can be designed for fiber collection to suit specific applications. This research will open the door to obtaining controlled linear, planar and 3-dimensional fiber assemblies for various applications.

Electrospinning Setup and Process

Figure 1 shows a schematic of a typical horizontal setup. Vertical setups have also been used by the authors and by other researchers. The basic setup consists of a pipette or syringe, hereafter referred to as the source, containing a polymer solution, and an electrically grounded aluminum foil pan, hereafter referred to as the target. The target could be in various forms, including a wire mesh, plate, or rotating disc, and is mounted on an adjustable electrically insulated stand. The distance between the source and the target is generally in the range of 10-30 cm, depending on the optimal process conditions. An electrode connected to a high voltage power supply is immersed into the solution. The power supply used was from Gamma High Voltage Research Inc. (Ormond Beach, Florida), with the following specifications: Gamma high voltage model no: ES 30-0.1P, Output voltage = 0-30KV, Output Current = 100µA and Polarity = Positive. To have positive control over the flow of polymer, a flow pump connected to a syringe was used for feeding the solution.

A FLUKE 189 multimeter was connected between the ground and target to measure the current that is developed during the deposition of the fibers. This multimeter was capable of measuring current in the microampere range. Current developed at the target plays a very significant role in characterizing the spinning process. To minimize the loss of fibers due to adherence to metal parts of the setup, a grounded aluminum foil target with a large surface area was used. To determine the mass of deposited fiber, the mass of the foil before and after the spinning process was measured using a Sartorius MP1212 scale, which has a resolution of 0.001g. By knowing the current and the mass of the fibers deposited on the target, the charge per unit mass of the fibers was determined.

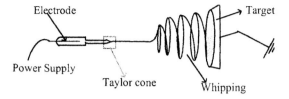

*Figure 1. A simple horizontal electrospinning setup
(See page 7 of color inserts.)*

When a high voltage is applied to the polymer solution, a jet of charged fibers is ejected from the tip towards the target. Due to the inherent viscosity of the polymer, the jet does not disintegrate into spherical droplets but forms long fibers. Originally, the jet was assumed to undergo splaying, which is the splitting of the fibers, leading to smaller diameters. Based on more recent observations, however, the jet appears to generate a rapidly-rotating spiral, which is indistinguishable from the splaying phenomenon to the naked eye (*12, 13*). Figure 1 shows a schematic of this whipping motion of the electrospinning jet. It is this motion that draws the fibers, leading to the formation of nanometer-scale diameter fibers. The solvent evaporates during the process and a nonwoven semi-dry fiber mat is formed on the target, in a relatively circular deposition pattern. Figure 2 shows a electrospun fiber mat deposited on an aluminum target and a scanning electron microscopy (SEM) image of the fibers. Note that there is no distinct directional orientation of the fibers.

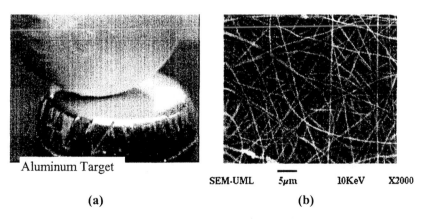

*Figure 2. (a) Electrospun polyethylene oxide(PEO) fibers on an aluminum
Target, and (b) SEM image of the same fibers. (See page 7 of color inserts.)*

Ion-Jet Charge Transport Model of Disc Electrode Effects

Part of the challenge of controlling the fiber deposition is the divergence that occurs due to the whipping motion of the jet (see Figure 1). It was hypothesized that adding a disc electrode to create a more parallel electric field (rather than a diverging one generated by a point source) would help to focus the jet. To guide the experiments and determine if the jet response could be predicted by using an ion-jet charge transport model, a model was developed using the steady state charge distribution for two parallel plate electrodes, namely the voltage source and the grounded target. Equation 1 shows the conservation law for current continuity under steady state conditions,

$$\frac{\partial q(r,z)}{\partial t} + \nabla J(r,z) = 0 \tag{1}$$

where q represents the charge density (in C/m^3) and J the current density (in A/m^2). Figure 3 shows the model representation of the electrospinning setup. The electrode at the source is called the disc electrode with radius R_d. The disc electrode radius can be varied from $R_d = R_s$ (source radius), where there is only the syringe tip and no disc electrode, to a disc electrode as large as the target.

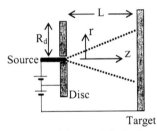

Figure 3. Schematic of the modified electrospinning setup

The voltage V_d applied to the disc electrode can be the same as or lower than that applied to the source electrode V_s. For modeling purposes, the source radius was fixed at 0.01L, where L is the distance between the source and the target. The radius of the disc electrode is varied from 0.01L (equal to R$_s$) to 0.2L. The same voltage is applied to both the source and disc, $V_s = V_d$.

On applying the boundary conditions the equations were solved by numerical iteration. Figure 4 shows the results in the form of current density versus the radial spread of the fibers on the target (i.e., at $z = L$). In these plots, if it is assumed that the jet carries the charge to the target, then the current density (J_z) is representative of the mass of fiber deposited on the target. The radial distance (x-axis values) represents the radius of the spread of the fibers on the target. Note that all the dimensions given are with respect to L (0.2→ 0.2 L). It is seen that when the radius of the disc electrode is increased from 0.01L to 0.2L, there is a four-fold decrease in the spread of the fibers at the target. For example, in Figure 4a, the radius of fiber deposition decreases from ~0.4L for a point electrode to ~0.1L as the disc electrode radius increased. This is not surprising when one considers the electric field lines generated by a point-plate arrangement (field lines diverging from point source) to those of a plate-plate arrangement (parallel field lines), as shown by Bunyan (14).

Figures 4a and 4b also demonstrate the effect of the injection parameter, γ, which is defined in equation 2,

$$\gamma = \frac{\varepsilon \mu V_o}{L^2} >> 10^{-14} (S/cm) \qquad (2)$$

where ϵ = permittivity, μ = mobility, V_o = applied voltage (or V_s) and L = source-target distance. This injection parameter corresponds to the strength of the ion source, and represents the effects of process and material parameters, such as the flow rate, applied voltage, target distance, polymer and solvent species, and concentration. In comparing Figure 4a ($\gamma = 100$) and Figure 4b ($\gamma = 10$), an order of magnitude increase of γ leads to a doubling of the spread of the fibers on the target. A third outcome of the modeling effort, the relatively horizontal current density curves with rapid dropoffs, indicates that for a given disc electrode, the amount of fiber mass collected within the fiber deposition region would not be a function of radial distance from the center of the target.

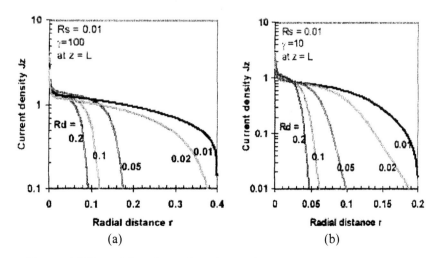

Figure 4. Effect of disc electrode radius and injection parameter on current density as a function of radial distance on the target.
(See page 7 of color inserts.)

Experimental Studies using Disc Electrodes

Validation of Ion-Jet Charge Transport Model with Disc Electrode

Because the purpose of these experiments was to validate the model and understand the effects of the disc electrode, polyethylene oxide (PEO) in ethyl alcohol and distilled water was chosen as the polymer solution. PEO is easy to electrospin and is the material for which there is the most published data available on the influence of process parameters.

The existing electrospinning setup was modified to correspond to the modeling assumptions (see Figure 5). An aluminum disc was placed about 1cm behind the tip of the syringe, so as not to physically disrupt the jet. The disc was held parallel to the target surface. Electric potential from the power supply was applied to both the disc and the syringe needle. The applied voltage was 17.5KV and the target was placed at a distance of 16cm from the tip of the needle. The spread of the fibers on the target was measured for tests with and without the disc electrode.

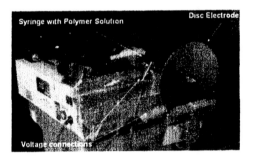

Figure 5. Electrospinning setup with the disc electrode (target not shown)
(See page 8 of color inserts.)

Figure 6 is a plot of the diameter of the circular fiber deposition pattern on the target as a function of the disc diameter for two flow pump rates. The two horizontal lines on the top of the graph are plotted simply to serve as a reference for the fiber spread when no disc electrode is used. Note that the addition of even a 5 cm diameter disc leads to the spread decreasing from roughly 20 cm to roughly 10 cm. As the diameter of the disc increases to 10 cm, there is continual reduction in the fiber spread. Figure 7 shows a schematic of the effect of the disc electrode on the electric field. During the experiment, the jet bending instability or whipping motion was observed to initiate at distance farther from the source as the disc diameter increased, leading to a decrease in fiber spread on the target. In Figure 6, it can also be seen that an increase in the polymer flow rate, which is related to the injection parameter 'γ' of the ion jet transport model, from 0.06 ml/min to 0.09 ml/min, led to an increase in the spread of fibers on the target.

Similar increases due to an increase in the electric field were also observed (Bunyan (*14*)).

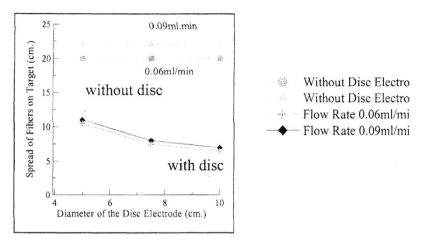

Figure 6. Plot of spread of fibers on target versus the disc diameter (applied voltage = 17.5KV, source to target distance = 16cm) (See page 8 of color inserts.)

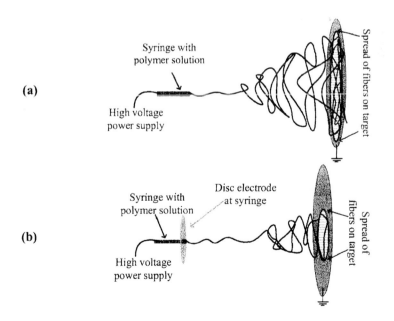

Figure 7. Schematic of effect of disc electrode on the instability initiation and fiber spread, (a) without the disc electrode and (b) with the disc electrode. (See page 9 of color inserts.)

To compare these results more quantitatively with the ion-jet charge transport model predictions, the fiber spread results from the 5 cm disc electrode were used to identify a value for the injection parameter, γ. This same value of γ was then used in the model to predict the fiber spread values for the 7.5 cm and 10 cm disc electrodes. The results shown in Figure 8 suggest that the ion-jet charge transport model captures some of the essential physics of the relationship between the charge transport and the fiber spread on the target.

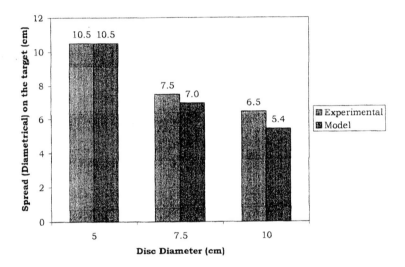

Figure 8. Comparison of fiber spread results – experimental versus model – with 5 cm data used to determine injection parameter value (See page 9 of color inserts.)

Of course, using the disc electrode to focus the fiber deposition is not beneficial if it also leads to poor fiber formation. This was a concern since the observation of a delayed whipping initiation and reduced diameter of the whipping motion could lead one to assume less time for drawing of the fiber diameter. To address this concern, scanning electron microscopy images of the fibers formed under the same conditions, with the exception of the presence or absence of a 10 cm disc electrode, were obtained (Figure 9). The SEM images demonstrate that in fact, the fibers formed using the disc electrode were even smaller in diameter. The reason for this is not known at this time, but may be tied to the overall reduction in jet diameter as it initiates from the syringe tip, well before any whipping instability occurs.

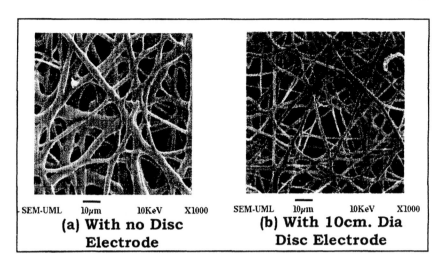

SEM-UML   10μm   10KeV   X1000   SEM-UML   10μm   10KeV   X1000

(a) With no Disc Electrode   **(b) With 10cm. Dia Disc Electrode**

Figure 9. SEM images of electrospun PEO fibers (a) without and (b) with a 10 cm disc electrode.

Experiments on Direction Control with Disc Electrode

Along with control of the fiber spread, it is desirable to have control of the direction of the fiber path such that fibers could be directed to specific areas on the target for collection. This could be achieved by controlling the electric fields near the source by angling the disc electrode (Figure 10). In this experiment, the disc was rotated about the horizontal axis (the axis parallel to the target plane) by $10°$ increments and the shift in the centers of the fiber deposition was measured. The experiments were repeated for different disc diameters and source-to-target distances to understand the effect on the jet path.

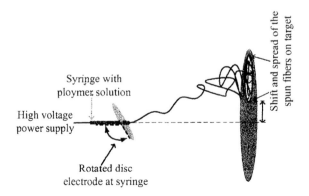

Figure 10. Directing the electrospinning jet onto the target (See page 10 of color inserts.)

There was not much variation in the spread of the fibers as this was controlled by the diameter of the disc electrode. The experiments were carried out for 5, 10, 15, and 20 cm source to target distance and 5, 10, and 15 cm disc diameters. The applied voltage was 15KV and polymer flow rate was 0.1ml/min. The solid lines in Figure 11 represent the data obtained from the experiments for the 10 cm disc. For example, for a source to target distance of 15cm, a 30° rotation shifted the centerpoint of the deposition by ~6 cm from the original position. The rotation of the disc shifts the electric field, which leads to a shift in the motion of the fibers to the target. As expected, as the source to target distance increases or the angle increases, the shift increases. Increasing disc diameter also led to increases in the shift, but not as pronounced as for the angle change *(14)*. The dotted lines in Figure 11 represent the predicted shift according to the intersection of the axis of the disc with the target (equation 3).

$$\textbf{\textit{Shift in fiber deposition}} = \textbf{\textit{L}} * \tan(\theta) \tag{3}$$

where L = source to target distance and θ is the rotation angle.

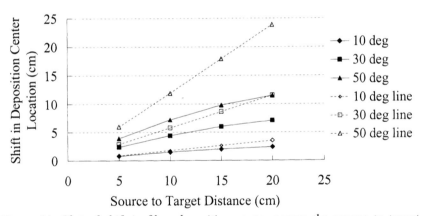

Figure 11. Plot of shift in fiber deposition center versus the source to target distance. Solid lines presents the experimental results and dashed line are the predicted results. (Applied voltage = 15KV, Flow rate = 0.1 ml/min)

For smaller angles, the predicted shift follows the experimental results fairly well. As the angle gets larger, however, the experimental results fall well below the predicted results. This inconsistency is basically caused by the dimensions of the target size, target distance, and disc diameter being in the same range. As a result, the electric field for the larger angle, which draws the fiber jet to the section of the target that is closer, leads to less shift than a straight line result.

These results suggest some limitations in steering the jet, but also confirms that changes in the electric field lines can be used to affect the jet path.

Alignment Experiments with Wire Mesh and Rotating Targets

Replacement of the aluminum foil pan target with a metal wire mesh resulted in a visible change in the deposition of the fibers. The fibers were attracted to the grounded wire, resulting in much denser accumulations in these regions (Figure 12). Similar experiments conducted by Bunyan *(14)* with oppositely charged squares on a solid plate demonstrated the same effect of the field on the deposition location, with the fibers drawn to regions of opposite charge and diverging from regions of like charge. SEM images of the wire mesh samples showed, however, that the electrospun fibers on the wires did not exhibit any particular alignment. The ratio of the wire diameter to the fiber diameter was too large to promote alignment of the deposited fibers along the wire axis.

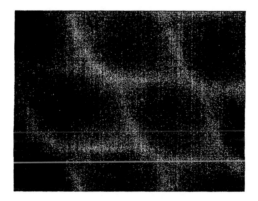

Figure 12. Optical microscopy image of fibers electrospun onto a grounded 3.2 mm side-length square metal wire mesh

As an alternative approach, following the example of Theron et al.*(15)*, who spun onto the sharp edge of a rotating disc, two cone targets separated by a gap of 8 cm between their tips were used as the target. The axis of the cones was placed 16 cm from the source. Almost immediately after starting the electrospinning, fibers bridge the gap between the two tips. By attaching one cone target to a motor, the fibers can be twisted into a more consolidated yarn-like form. SEM images were obtained to determine the effect on fiber alignment.

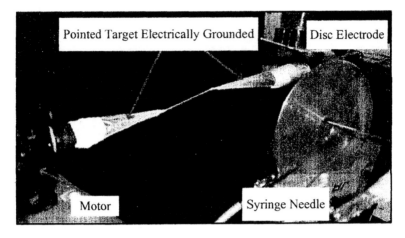

Figure 13. Cone target setup with one cone attached to a motor
(See page 10 of color inserts.)

Analysis of the SEM images to determine the fiber orientation analysis was done by M-Base Engineering + Software GmbH (Aachen, Germany) using their FiberScan software. In Figure 14, the SEM image is rotated with the fibers aligned at 55°, and more than 45% of the fibers showed directional orientation within ±20° angle of this central angle.

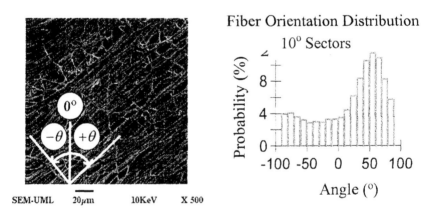

Figure 14. Analysis of fibers collected between two cones for directional alignment (~100rpm).

A sharp projection creates a strong electrostatic field around it. In the cone target configuration, the fibers were attracted to the sharp points due to charge concentration and were deposited in-between the cones. Better results in terms of fiber orientation were achieved with one of the cones rotated by a motor (e.g., at 100 rpm) while the other cone was fixed. The rotational action created by the motor also contributed to further consolidation of the fibers into a tighter yarn-like form. Insufficient solvent evaporation did lead to merging or adhesion of fibers in the yarn. Similar results for fiber alignment were obtained when using a sharp-edged rotating disc as the target (14).

Conclusions

The results suggest that the path of the electrospinning jet can be controlled by an effective electric field. One aspect of the electrospinning process was modeled using ion-jet charge transport model. The model predicted that spread of the fibers on the target could be decreased by placing a disc electrode at the source. As the radius of the disc increases the spread decreases, whereas an increase in the injection parameter leads to an increase in the spread of the fibers on the target. Experimental results correlated well with the model predictions, suggesting that the charge transport model is a reasonable approach to representing some of the governing physics. Increasing the flow rate, which corresponds to an increase in the injection parameter, led to an increase in the spread of the fibers. In addition, it was also demonstrated that the fibers could be directed to different locations on the target by angling the disc electrode and thus affecting the electric field. Thus, the use of the disc electrode at the syringe tip allows for controlling the path and the spread of the fibers through which it would be possible for effective fiber deposition, akin to writing with the jet. The second part of the study considered the effect of the local electric field at the target. A 3.2 mm wire mesh target demonstrated the ability to control deposition density at the macroscale, but did not result in significant fiber alignment. In contrast, experiments using rotating cone-tips and sharp-edged discs as the targets resulted in highly aligned fibers. The combination of source and target modifications discussed in this paper provides some potential tools for controlling the focusing, positioning, and alignment of electrospun fibers, as well as providing some insight into additional modeling and experimental efforts in electric field and charge transport effects.

Acknowledgements

This work was supported by the National Science Foundation, Division of Design, Manufacture and Industrial Innovation - Nanomanufacturing Program under the grant DMI-0200498.

References

1. Schreuder-Gibson, H., Senecal, K., Sennett, M., Huang, Z., Wen, J., Li, W., Wang, D., Yang, S., Tu, Y., Ren, Z., and Sung, C., "Characteristics of Electrospun Carbon Nanotube-Polymer Composites" Proceedings of the 197th Meeting of Electrochemical Society, Toronto, Canada, May 14-18, 2000.

2. Duan, X., Huang, J., and Lieber, C. M., "Indium phosphide nanowires as building blocks for nanoscale electronic and optoelectronic devices", Nature, 409, 66-69, 2001.

3. Cobden, D. H., "Molecular electronics: Nanowires begin to shine", Nature 409, 32-33, 2001.

4. Formhals, A., Titled "Method and Apparatus for Production of Fibers" U.S. Patent No. 1,975,504, filed 1934.

5. Doshi, J., Ph.D thesis, University of Akron, 1994.

6. Reneker, D.H. and Chun, I., "Nanometer Diameter Fibers of Polymer, Produced by Electrospinning", Nanotechnology 7, 2166, 1996.

7. Kim, J.S., Reneker, D. H., "Polybenzimidazole nanofiber produced by electrospinning" Polymer Engg. and Science, Vol. 39, No.5, 1999.

8. Koombhongse, S., Liu, W., Reneker, D.H., "Flat Ribbons and other shapes by electrospinning", Journal of Polymer Science, Polym Phys Ed, 39, 2001.

9. Deitzel, J.M., Kleinmeyer, J.D., Harris, D., Beck Tan, N.C., "The effect of processing variables on morphology of electrospun nanofibers and textiles," Polymer, 42, 261-272, 2001.

10. Warner, S.B., Buer, A., Ugbolue, S.C., Rutledge, G.C., Shin, M.Y., "A fundamental investigation of the formation and properties of electrospun fibers", National Textile Center Annual Report 1998, (www.ntcresearch.org)

11. Gibson, P.W., Gibson, H.L., Rivin, D., " Electrospun fiber mats: transport properties", AIChE Journal, 45, 1, 190-195, 1999.

12. Shin, Y.M., Hohman, M.M., Brenner, M.P., Rutledge, G.C., " Electrospinning and electrically forced jets. I Stability theory" Phys Fluids, 13: 2201-20, 2001.

13. Shin, Y.M., Hohman, M.M., Brenner, M.P., Rutledge, G.C., " Electrospinning and electrically forced jets. II Applications" Phys Fluids, 13: 2221-36, 2001.

14. Bunyan, N., "Control of Deposition and Orientation of Electrospun Fibers", MS Thesis, Dept of Mechanical Engineering, University of Massachusetts Lowell, 2003.

15. Theron. A., Zussman. E. and Yarin, A.L., "Electrostatic field-assisted alignment of electrospun nanofibers", Nanotechnology 12, Sept 2001.

Chapter 9

Applications of Electrospun Nanofibers in Current and Future Materials

Heidi L. Schreuder-Gibson and Phil Gibson

U.S. Army Research, Development and Engineering Command, Natick Soldier Center, Kansas Street, Natick, MA 01760

Abstract

The search for practical applications of electrospun microfibers and nanofibers is primarily concentrated on filters, membranes, and biomedical devices. This chapter reviews current progress towards developing new applications of electrospinning. The production of submicron as well as nanometer sized fibers by electrospinning was reported early in the 20th century (1-5). Over the intervening decades, research has continued around the world, resulting in over one hundred patents and publications. Research today continues to focus on new methods of producing nanofibers, including an increased understanding of electrospinning and its applications. These studies have resulted in process improvements that have increased the speed of electrospinning, increased the strength of the resulting nanofiber web, and may lead to better control of electrospun membrane features such as fiber size, fiber orientation and porosity for future applications.

Introduction

Most commonly reported nanofibers are solid continuous fibers of material with diameters ranging from a few hundred nanometers down to tens of nanometers, exhibiting lengths of a few microns up to meters, depending upon the manufacturing process. Such fine diameters result in fibers that are at least an order of magnitude thinner than commercial microfiber fabrics, as shown in Figure 1. Fabrics of nanofibers may have the appearance and the properties of porous membranes in the case of many organic nanofibers (6), or may exhibit the consistency of high loft batting materials in the case of rigid organic fibers and inorganic fibers (7). Nanofibers have been reported to be hollow (8,9,10), but are usually produced as dense solid fibers or multi-component composite fibers consisting of more than one material within the fiber (11,12). Organic nanofibers have been produced by various methods, but one easy fiber forming process is electrospinning (13). The application of an electric force to the surface of a polymer solution produces a continuous stream of extremely fine fiber without the use of mechanical extrusion or temperature changes. Methods of applying the field, influencing the spinning by external forces and collecting the fibers continue to be developed, and some recent processing discoveries are discussed in other chapters of this book.

Advantages of Electrospun Fibers

One of the most cited advantages of electrospun fibers that motivate research and the search for applications is the high surface to volume ratio of nanofibers. Based on the surface area of the cylinder, a fabric of 10 nm fibers will have a surface area of roughly 350 m^2/g, whereas 10 μm fibers will be 0.35 m^2/g. Although high surface areas in excess of 2,000 m^2/g can be realized with meso and nanoporous materials such as adsorbent granules and powders, fibers are more easily handled and manipulated than powders. Another advantage of these high surface area organic fibers is the filtration efficiency that has been observed with nanofibers. Electrospun nanofibers can be produced from a wide range of polymers allowing a range of fiber types for filtration and other applications.

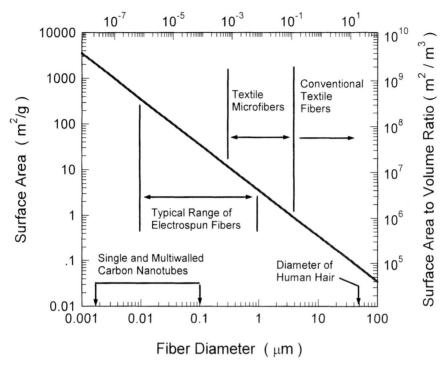

Figure 1. Inverse relationship between fiber diameter and surface area of fabrics calculated for fibers of density 1.14 g/cc. (Reproduced with permission from reference 18. Copyright 2002 Journal of Advanced Materials.)

One of the most significant advantages of electrospun fibers is the filtration efficiency that can be realized from the porous membrane-like fabrics and the high surface area available for airborne particle capture. In the case of filtration, electrospun fibers of nylon and polycarbonate have been applied in thin layers to the surface of typical filter fabrics of natural and synthetic nonwovens (*14-16*). High filtration efficiencies result from low levels of electrospun fibers (*17,18*). However, electrospun fibers of glassy and semi crystalline polymers are fragile and require supporting substrates for reinforcement during use; thus rigidly held filters have been the first successful commercial application for electrospun fibers. Shown in Figure 2, thin layers of electrospun nylon fibers coated onto the surface of a carbon-loaded foam exhibit good airflow properties while still providing high aerosol filtration

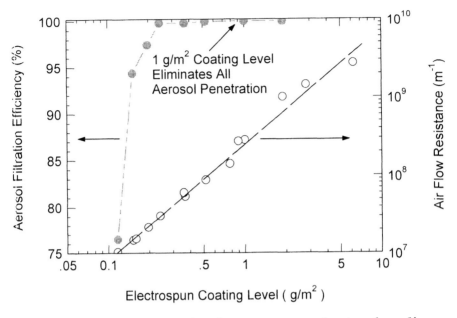

Figure 2. Aerosol filtration and air flow resistance as a function of nanofiber coating level on carbon foam substrate. Aerosol consists of potassium iodide particles with diameters of 0.5-20μm. (Reproduced with permission from reference 18. Copyright 2002 Journal of Advanced Materials.)

Another advantage of electrospinning is the ability to produce submicron fibers from a wide range of organic polymers. This enables the production of microporous membranes comprised of nanofibers from elastomers, glasses, and semi crystalline materials, and, recently the preparation of ceramic nanofibers (*19, 20*). The advantage to producing a microporous membrane by electrospinning is the simplification of manufacturing steps; instead of multiple processing steps involving film formation and manipulations such as expanding, foaming or partially dissolving components in the films to produce micro pores, electrospinning allows a direct spray-on application of the final microporous structure onto a wide range of substrates, including release papers, screens, fabrics, and even living tissue. Recent patent applications have proposed the formation of electrospun membranes onto expandable wire frames for biomedical applications, as well as tubes and stents (*21-23*). Chemical surface modification, crosslinking and other chemical treatments of electrospun microporous membranes can be easily achieved in order to produce functionalized surfaces and tailored properties within the membranes (*15, 16, 24*). Surfaces have been functionalized with enzymes (*25*), and recently the cospinning of electrospun membranes with cells for tissue growth has been reported (*26, 27*). Other advantages to electrospinning as a fiber manufacturing method include the ability to cover complex three-dimensional surfaces with a uniform fibrous layer. The mild processing temperatures involved in solution electrospinning enables the blending of temperature sensitive components such as drugs, biological materials aqueous domains into the fibers (*28, 29*).

Disadvantages of Electrospinning

Slow fiber production speed has been one of the most difficult technical problems to overcome. Whereas typical fabric coating line speeds of up to 60 meters/min produce adhesive patterns of 2 g/m^2 weight onto substrates, electrospinning, in contrast, produces only 100mg fiber/min for a 100-hole spinneret system. This translates into a line speed of approximately 120 g/min for adhesive patterning, and 0.1 g/min for electrospinning onto a 1-meter wide web. Recent improvements in electrospinning with the use of forced air at the spinnerets have been reported, resulting in electrospinning speeds of up to 1 g/min from 100 spinnerets (*26*). Another approach to increase production rates of electrospun fibers has been recently patented, utilizing a method of charge injection into a polymer solution or melt in order to produce submicron and nanometer-sized fibers at rates of 60g fiber/min from a single orifice (*30*), resulting in a rate of nearly 1kg/min at production widths of 1 meter. The phenomenon of electrospinning is not limited to the use of spinnerets or nozzles to generate the nanofibers. Using electric and magnetic fields normal to the free surface of a polymer solution layered on a magnetic fluid, Yarin and Zussman (*31*) report that fiber production can be increased 12-fold over electrospinning from multiple nozzles.

While production speeds have been reportedly improving, other disadvantages to the electrospinning process that have been deterrents to manufacturing on a large scale include safety and environmental issues of the solvents used in the spin dopes and the voltages involved in the electrospinning process. Solution spinning has been shown to be the only way to achieve fiber diameters below 1μm. However, many polymers dissolve in highly volatile and toxic solvents that pose explosion hazards in the presence of the high voltages required for electrospinning. Solvent removal and recovery increase manufacturing costs. The best approach is to develop polymer/solvent systems that are benign in terms of safety and environmental impact. In recent patents, Donaldson Company cites preferred solvents of water, alcohol, acetone and N-methyl pyrrolidone in their electrospinning production of filtration media (*15-17*).

Web strength is another disadvantage for electrospun fibers. Selection of polymers that dissolve in nonhazardous solvents can limit the fiber strength and toughness in the final electrospun web, or membrane. The use of crosslinking chemistry has improved strength of the final membrane and improved bond strength between the electrospun fibers and the substrate material (*16, 17*). In the case of electrospun Pellethane®, a thermoplastic polyurethane supplied by Dow Chemical, it has been found that solvent type can influence the strength of the final random fiber mat in the membrane-like layer, as shown in Table I, reprinted from the Journal of Advanced Materials (*18*).

Table I. Effect of Electrospinning vs. Cast Film on Tensile Properties

Sample	Specific Modulus* (MPa)	Ultimate Tensile Strength* (MPa)	Elongation (%)
Pellethane® Film	3	15.5	979
Espun Membrane -(Spun from tetrahydrofuran)	3	9.50	360
Pellethane® Film	3	24.7	730
Espun Membrane -(Spun from n-dimethyl formamide)	12	54.5	160

*Normalized by web density (30% fiber) of the electrospun membrane.

Data in Table I suggest that the more volatile solvent, tetrahydrofuran produces a slightly weaker fiber mat than the lower volatility solvent, n-dimethyl formamide, possibly due to the fusing of wetter fibers during the formation of the electrospun mat.

Prospects for Future Research and Applications

The utility of electrospun fibers to provide high surface areas in applications such as filtration can also be an advantage for biomedical devices and for catalyst supporting membranes, as evidenced in recent patent applications (*32-*

41). Some of the recent claims involve methods to electrospin highly porous fibers *(40)* to further increase surface area and to use the fiber as a carrier for a catalyst or preparing composite fibers of polymers with a mesoporous molecular sieve components *(20).*

Use of electrospun fibers in medicated stents has been proposed in recent patent applications *(36, 37)*, and in other medical applications involving cell implantation and growth, the characteristic high porosity and high surface area of the fibrous layers is a desirable feature. However, there are few methods to control the pore size of electrospun fiber mats. We speculate that one method would be to incorporate two materials into the electrospun fiber structure, followed by the removal of one material. Spinning blends of two polymers, or co-mingling two fiber streams into a single mat could be two possible approaches, but there have been difficulties in accomplishing this blending of permanent and removable materials in electrospinning. Co-spinning separate polymer solutions from different spinnerets causes field interference between the streams and prevents intermingling of the fiber streams into a final fiber mat, as shown in Figure 3. In this figure, we see polycarbonate fibers collecting into a small spot while the polyacrylonitrile fibers remain segregated from the polycarbonate. This repulsion between different polymeric fibers electrospun simultaneously is an effect that could be caused by a difference in charge density of the different fibers. A higher charge density can result from fibers of smaller diameter, so the higher charge density of the smaller polyacrylonitrile fibers results in strong repelling forces acting against larger fibers of lower charge density.

However, the use of a single spinneret that has been divided to accommodate two polymer solutions has been reported to be a successful design that combines two polymers into single fibers and also produces fiber mats containing separate fibers of each material *(9, 11, 43)*.

Other approaches to controlling pore size and total porosity in an electrospun mat involve simply collecting fibers of different diameter. Measurements done on melt blown nonwovens have shown that larger fiber diameters produce larger pore sizes within fiber mats *(43)*. This relationship, between fiber size and pore size has been derived and is useful in the indirect measurement of an effective fiber size from air permeability measurements, rather than microscopic measurements (43).

Figures 4-6 show the dramatic differences in apparent fiber size in two electrospun examples and the corresponding difference in mean pore size compared to a melt blown sample of the polyurethane, Estane®, produced by Noveon Corporation.

Figure 3. Electrospinning of polycarbonate solution in the left collection spot and polyacrylonitrile in the right. (Reproduced with permission from reference 42. Copyright 2003.)

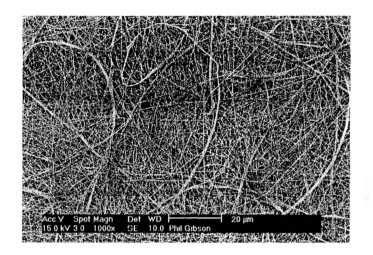

Figure 4. Electrospun Nylon® 6,6 average fiber diameter 0.13µm and mean pore size 0.1-0.2 µm.

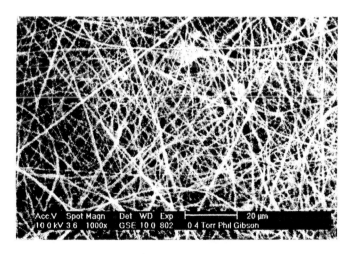

Figure 5. Electrospun Estane®, average fiber diameter 0.43µm and mean pore size 0.4µm

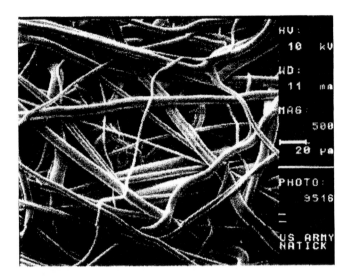

Figure 6. Melt Blown Estane®, average fiber diameter 10-20μm and mean pore size 30μm.

Combining high surface area nanofibers with larger diameter textile fibers would be expected to accomplish two objectives: achieve a controlled average pore size based upon the average fiber size and provide reinforcement of the nanofiber mat with larger fibers integrated into the structure. Research is ongoing in an effort to combine electrospinning with large nonwoven spunbonding or melt blowing processes to achieve the intermingling of electrospun and nonwoven fiber sizes, but success has been limited and more work needs to be done on fiber intermingling of these two processes.

Charged electrospun fibers can be organized into a patterned arrangement through the use of secondary electric fields such as electrostatic lenses (*44-47*) and patterned collectors (*48*). These techniques can be used to develop new fibrous architectures and are being applied in an effort to develop processing methods to control the porosity of electrospun membranes and provide mechanical reinforcement to increase toughness of the electrospun membrane. Figure 7 shows that different patterns from a grounded collector surface will influence the deposition of charged electrospun fibers and impose a fibrous pattern on the membrane.

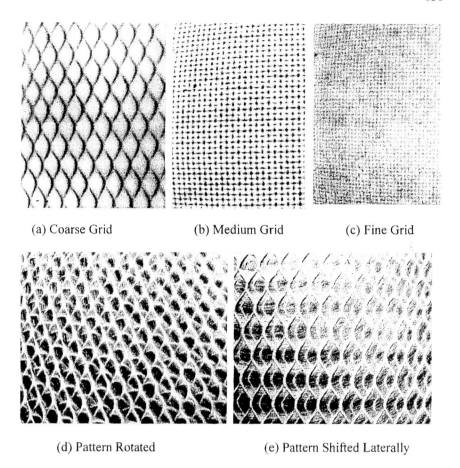

(a) Coarse Grid   (b) Medium Grid   (c) Fine Grid

(d) Pattern Rotated   (e) Pattern Shifted Laterally

Figure 7. Grid patterns used for electrospun polyurethane fiber collection. These images are electrospun membranes that display the original grid pattern. (Reproduced with permission from reference 48. Copyright 2003 e-Polymer.)

The fiber patterns in Figure 8 show that the fibers orient in some areas in the grid, and remain random in other areas. Figure 8 (a) shows a magnified image of the fiber orientation on the medium grid of Figure 7 (b). Figure 8 (b) shows orientation of the electrospun polyurethane fibers near the grid junctions, while Figure 8 (c) shows that the electrospun fibers are random in areas between the grid junctions.

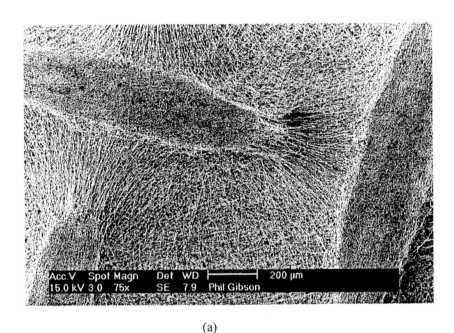

(a)

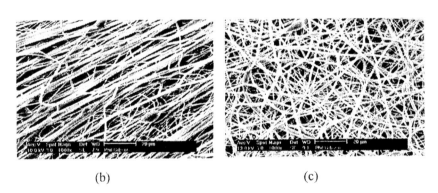

(b)   (c)

Figure 8. (a) Close up image of fiber patterns in membrane produced by medium grid; (b) Fiber orientation at junctions of grid pattern and random fiber collection and (c) farther away from grid element. (Reproduced with permission from reference 48. Copyright 2003 e-Polymer.)

Figure 9 shows that grid size appears to affect tear propagation in bursting of the elastic membrane after inflation. Coarse and medium grid patterned electrospun elastic membranes can be inflated, as shown in figure 9 (a) and (b). After inflation and bursting, the coarse grid pattern results in a small tear (circled for clarity) in Figure 9 (a), while the medium grid pattern results in a large catastrophic tear across numerous grid junction points in Figure 9 (b).

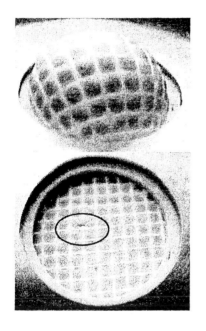
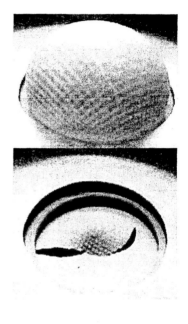

(a)   Coarse grid membrane.   (b) Medium grid membrane.

Figure 9. Examples of electrospun polyurethane membranes that have been inflated in the top images, and torn after burst testing in the bottom images. (a) Membrane with coarse grid pattern. (b) Membrane with medium grid pattern. (Reproduced with permission from reference 48. Copyright 2003 e-Polymer.)

This recent study has shown that electrospun fibers can be collected into organized patterns that exhibit fiber alignment near grid junctions. These fiber aligned regions could lead to larger pore dimensions in patterned membranes. It has also been found that pattern size appears to result in membrane tear reinforcement. Continuing research on elastic and reinforced electrospun membranes will lead to new applications in the new future such as inflatable membranes and controlled porosity structures for medical devices.

Summary

Electrospun materials are highly porous and have high surface areas that are attractive for many applications, including filtration, biomedical materials and devices, and microporous membranes. Recent improvements of electrospinning production rates have been reported in the patent literature. Progress has also been made with respect to web strength and controlled properties such as porosity. Elastic membranes produced by electrospinning may have utility as wearable materials in applications such as protective garments in the future: they exhibit high strength and are tough and flexible. Properties of elastic fibrous membranes can be improved with crosslinking and fiber patterning to arrest tear crack propagation. Combined with woven as well as knit stretchable fabrics, elastic electrospun membranes are easily post treated for increased adhesion to substrate layers. The electrostatic fields used in electrospinning allow control of the variation of fiber deposition and orientation across the area of a web. This is an advantage over other nonwoven processes such as meltblowing, spinlacing, spunbonding, hydroentanglement, etc., which produce a fairly uniform fiber distribution across the area of the manufactured web. Controlled variations of the electric field in electrospinning or combined melt blowing/electrospinning can produce patterns across the web width that result in distict fiber pattern arrays that are useful for strength enhancements and could potentially produce porosity variations for future applications of electrospun membranes.

References

1. Formhals, A. U.S. Patent 1,975,504, 1934.
2. Formhals, A. U.S. Patent 2,160,962, 1939.
3. Formhals, A. U.S. Patent 2, 187, 306, 1940.
4. Formhals, A. U.S. Patent 2,160,962, 1943.
5. Formhals, A. U.S. Patent 2,160,962, 1944.
6. Gibson, P.; Schreuder-Gibson, H.; Rivin, D., *Colloids and Surfaces, A.*, **2001**, *187-188*, 469-481.
7. Webb, G.W. *J. Alloys and Compounds*, **1992**, *183*, 109-115.

8. Loscertales, I.G., Barrero, A.; Marquez, M.; Spretz, R.; Velard-Ortiz, R.; Larsen, G. *J. Am. Chem. Soc.,* **2004**, *126*(17), 5376-5377.
9. Sun, Z. , Zussman, E., Yarin, A.L., Wendorff, J. H., and Greiner, A., Adv. Mat., **2003**, *15*, 1929-1932.
10. Li, D., Xia, Y., NanoLetters, **2004**, *4*(5), 933-938.
11. Gupta, P.; Wilkes, G.L. *Polymer*, **2003**, *44*(20), 6353-6359.
12. Fong, H., Vaia, R., Sanders, J.H., Lincoln, D., Vreugdenhil, A.j., Liu, W., Butltman, J., Chen, C., Chem. Mat., **2001**, *13*(11), 4123-4129.
13. Reneker, D.; Chun, I. *Nanotechnology*, **1996**, *7*, 216-223.
14. Weghmann, A.; *Nonwovens Ind.*, **1982**, November, 24-32.
15. Gogins, M.A.; Weik, T.M; U.S. Patent 6,716,274, 2004.
16. Gillingham, G.R.; Gogins, M.A.; Weik, T.M.; U.S. Patent 6,673,136, 2004
17. Grafe, T.; Graham, K.; *Intl. Nonwovens J.*, **2003**, 12(1), 51-55.
18. Schreuder-Gibson, H.; Gibson, P.; Senecal, K.; Sennett, M.; Walker, J.; Yeomans, W.; Ziegler, D.; *J. Adv. Mat.*, **2002** *34*, 3, 44-55.
19. Lo, D.; Xia, Y.; *Nano Letters*, **2004**, ASAP Article, March 30, 2004 URL: http://pubs.acs.org/cgi-bin/jcen?nalefd/asap/html/n1049590f.html.
20. Balkus, K.J.; Ferraris, J.P.; Madhugiri, S.; U.S. Patent Application 20030168756, 2003.
21. Bornat, A. U.S. Patent 4,689,186, 1987.
22. Greenhalgh, K. U.S. Patent Appl. 20040013873, 2004.
23. Bowlin, G. L. U.S. Patent Appl. 20030088266, 2003.
24. Bamford, C.H.; Al-Lamee, K.G. *J. Chrom.* **1992**, *606*, 19-31.
25. Kenawy, E.; Bowlin, G.L.; Mansfield, K.; Layman, J.; Simpson, D.G.; Sanders, E.H.; Wnek, G.E. J. *Contr. Release*, **2002**, *81*, 57-64.
26. Chu, B.; Hsiao, B.S.; Hadjiargyrou, M.; Fang, D.; Zong, X.; Kim, K. U.S. Patent Appl. 20030054035, 2003.
27. Wnek, G.E.; Simpson, D.G., Bowlin, G.L.; Yao, L; Kenawy, E.; Layman, J.M.; Sander, E.H.; Fenn, J. U.S. Patent Appl. 2004018226, 2004.
28. Chu, B.; Hsiao, B.S.; Fang, D.; U.S. Patent Appl. 20020175449, 2002.
29. Katti, D.S., Robinson, K.W., Ko, F.K., Laurencin, C.T., *J. Biomed. Mat. Res., B. Appl. Biomat.*, **2004**, *70B*(2), 286-296.
30. Kelly, A. J., U.S. Patent 6,656,394, 2003.
31. Yarin, A.L., Zussman, E., Polymer, **2004**, *45*, 2977-2980.
32. Reneker, D.; Chase, G.; Smith, D.; U.S. Patent Application 20030232195, 2003.
33. Hou, H.; Averdung, J.; Czado, W.; Greiner, A.; Wendorff, J.H.; U.S. Patent Application 20040013819, 2004.
34. Laksin, O.; Du, G.W.; U.S. Patent Application 20030100944, 2003.

35. Pavlovic, J.L.; U.S. Patent Application 20020128680, 2002.
36. Dubson, A.; Bar, E.; U.S. Patent Application 20040030377, 2004.
37. Dubson A.; Bar, E.; U.S. Patent Application 20040054406.
38. Laksin, O.; Du. G.W.; U.S. Patent Application 20030100944, 2003.
39. Ignatious, F.; Baldoni, J.M.; U.S. Patent Application 20030017208, 2003.
40. Wendorff, J.H.; Stainhart, M.; U.S. Patent Application 20040013873.
41. Lee, W.S.; Jo, S.M.; Chun, S.W.; Choi, S.W.; U.S. Patent Application 20020100725.
42. Tsai, P.; Schreuder-Gibson, H.L.; *Proceedings of INTC2003, International Nonwovens Technical Conference*, **2003**, September, 403-413.
43. Wadsworth, L.C.; Sun, C.; Zhang, D.; Ronguo, Z.; Schreuder-Gibson, H.L.; Gibson, P.W.; Proceedings of the Joint INDA-TAPPI Conference, **2001**, September, 179-195.
44. Deitzel, J.M.; Kleinmeyer, J.; Harris, D; Tan, N.C.B; *Polymer*, **2001**, *43*, 3303-3309.
45. Theron, A., Zussman, E., Yarin, A.L., Nanotechnology, **2001**, *12*, 384-390.
46. Zussman, E., Theron, A., Yarin, A.L., Appl. Phys. Lett. **2003**, *82*, 973.
47. Bunyan, N., Chen, J. Chen, I. Farboodmanesh, S., ACS Symp. Ser. "Polymeric Nanofibers," **2005**, in press.
48. Gibson, P.; Schreuder-Gibson, H.; *e-Polymers*, **2003**, *T-002*, www.e-polymers.org.

Chapter 10

Electrostatic Assembly of Polyelectrolytes on Electrospun Fibers

Christopher Drew[1], Xianyan Wang[1], Lynne A. Samuelson[2], and Jayant Kumar[1]

[1]Center for Advanced Materials, University of Massachusetts at Lowell, Lowell, MA 01854
[2]U.S. Army RDECOM, Natick, MA 01760

Electrostatic layering of polyelectrolytes and charged titanium dioxide nanoparticles was demonstrated on the surface of polyacrylonitrile nanofibers. The surface chemistry of the electrospun fibers was modified by immersing them in a heated aqueous solution of 1.0 N sodium hydroxide. The resulting surface groups of carboxylic acid were then used to build alternating layers of polycations and polyanions to form nanometer-scale composite archetechtures. Since the polyacrylonitrile in the nanofibers were not entirely converted to acrylic acid, the fibers remained insoluble in water, unlike electrospun polyelectrolytes. This difference allows layer-by-layer assembly that pure polyelectrolyte nanofibers do not.

Introduction

Recognizing that electrospinning alone cannot produce nanometer-scale structures for a number of applications, this work has sought to expand the utility of electrospinning by combining it with the electrostatic layer-by-layer assembly technique. Polyelectrolytes and other electrostatically charged entities can be permantly adsorbed onto the surface of an electrospun fiber and complex nanoscale layers on a nanofiberous membrane can result.

Electrospinning

Electrospinning uses a high voltage applied to a viscous polymer solution to create electrically charged jets . These jets dry to form very fine polymer fibers, which are collected on a target as a non-woven membrane. The diameters of electrospun fibers can range anywhere from several micrometers to as narrow as tens of nanometers depending on the polymer and experimental conditions. As a result, electrospun nanofibrous membranes can have a surface area per unit volume of up to two orders of magnitude higher than that of continuous thin films.

Many polyelectrolytes are difficult to electrospin as they require significant amounts of water to solvate them. Because of its high surface tension, water can be problematic as a spin-dope solvent, causig the jet to collapse. A high surface tension solvent tends to cause the spinning jet to collapse, or partially collapse into droplets. This can be ameliorated somewhat by mixing water with a lower surface-tension solvent such as ethanol or N,N dimethyl-formamide (DMF). Polyacrylic acid and sulfonated polystyrene have been successfully electrospun using mixed solvents of water and organics.

Electrospun nanofibers of polyelectrolytes are, however, unsuitable for subsequent layer-by-layer LBL assembly because they dissolve when immersed in the appropriate counter-ion solution. Because of this observation, an alternative approach of functionalizing the nanofiber surface was used in this research. While this added a processing step, it has the advantage of creating an insoluble nanofiberous membrane with electrolyte groups on the surface suitable for electrostatic assembly.

Electrostatic layer-by-layer Assembly

Layers of polyelectrolytes may be deposited onto an electrostatically charged substrate by alternately dipping it in polycationic and polyanionic solutions. As the substrate is immersed in each solution, the polyions complex together to form an insoluble polymeric salt on the substrate. A diagram illustrating the layer-by-layer technique is illustrated in Figure 1. The illustration shows an anionic substrate, a cationic substrate is equally feasible, but would be dipped in the solution in the opposite sequence. Because not all of the ionic groups in any one layer are complexed with the underlying layer, the new surface retains the charge of the last electrolyte applied. Electrostaticly assembled structures are often described in terms of numbers of bilayers, where a bilayer is a cationic layer paired with an anionic layer. This method was first reported by Decher et al.[i,ii] and has garnered much interest and research attention in the last decade, including several review articles.[iii]

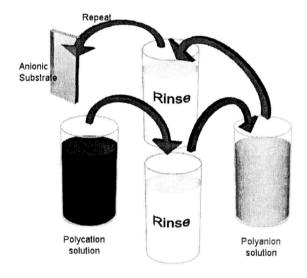

Figure 1. Diagram of electrostatic layer-by-layer assembly dipping sequence, shown for an anionic substrate.

Dubas and Schlenoff presented a study of the factors affecting the multilayer growth process[iv]. They used the strong polyelectrolytes polystyrenesulfonate (sodium salt) and poly(diallyl-dimethylammonium chloride). Low molecular weight salts were used to shield some of the charged groups on the polyelectrolytes. Film thickness was found to scale proportionately with the number of bilayers deposited and with salt concentration, *i.e.* increased salt concentration caused the individual layers to be thicker. The shielded ionic groups on the polymers do not complex with each other, resulting in longer segments of the polyelectrolyte chain free to adopt conformations with larger interaction volumes. They also indicated that localized rearrangements of the adsorbed polymer were possible due to the low molecular weight salt adsorbing and desorbing onto the substrate at the ionic sites. However, they used labeled polyelectrolytes to show that the polymers, once adsorbed do not desorb and exchange with polymers in solution.

This was different from the approach taken by Michael Rubner and coworkers in using weak polyelectrolytes and controlling the pH of the solution, rather than strong electrolytes and salts to impart the desired properties in the final film.[v,vi]

Techniques

Electrospinning

A spin-dope of 8 weight percent polyacrylonitrile in DMF was prepared and electrospun at 20 to 25 kV. The spin dope was held in a horizontal glass pipette with the high-voltage electrode inserted in the top. Electrospinning targets used were copper mesh or tin-oxide coated glass slides. In samples for TEM imaging, electrospun membranes were prepared directly on copper TEM sample grids. Figure 2 illustrates the electrospinning apparatus used for this work.

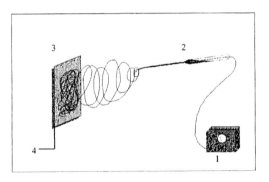

Figure 2. Electrospinning set-up (1: power source; 2:pipette and polymer spin-dope solutions; 3: collection target; 4: ground)

Surface Functionalization

The electrospun membranes of PAN nanofibers were immersed in a 1.0N sodium hydroxide (NaOH) solution at 80°C for 20 minutes to functionalize the surface with carboxylic acid groups. Upon removal, the membranes were rinsed with de-ionized water until the rinse water had a pH of 7 as measured by an electronic pH meter.

The surface functionalization reaction used was the well known conversion of cyano groups into carboxylic acid by exposure to strong alkali. The reaction scheme of cyano groups with hydroxy anions to form carboxylic acid groups is outlined in Figure 3. Electrospun membranes immersed in the NaOH solution turned brown after a few minutes. This color change has been reported as indicative of the presence of heterocyclic intermediates in the formation of carboxylic acid groups.[vii]

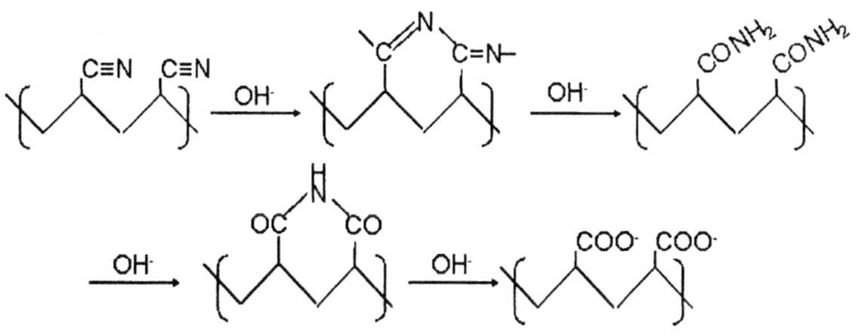

Figure 3. Reaction scheme for surface functionalization of polyacrylonitrile to polyacrylic acid (from ref. vii).

Electrospun membranes did not exhibit any morphological changes following treatment with sodium hydroxide. SEM images of electrospun membranes of PAN prior to functionalization and after being surface functionalized are shown in Figure 4. Following the surface treatment reaction, the nanofibers were completely insoluble in a number of solvents, including water, DMF, ethanol, and chloroform. Untreated PAN nanofibers dissolved immediately when immersed in DMF. This insolbility suggestes the presence of cross-linked intermediate functionality present in the fiber. It is believed the nanofibers consist of a core of unmodified poyacrylonitrile and a surface of ionic carboxylic acid (or sodium salt analogue) groups with the intermediate reaction products in the intermediate layers.

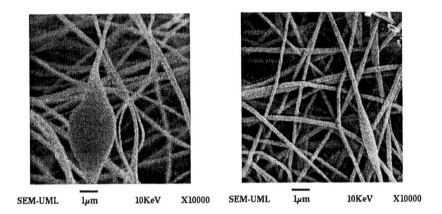

SEM-UML
1μm
10KeV
X10000
SEM-UML
1μm
10KeV
X10000

Figure 4. SEM images of polyacrylonitrile fibers, as spun (left) and of carboxylic acid-functionalized polyacrylonitrile fibers (right).

H2
H2
C
C
n
N+
Cl-
H3C
CH3
H
C
C
H2
n
SO3- Na+

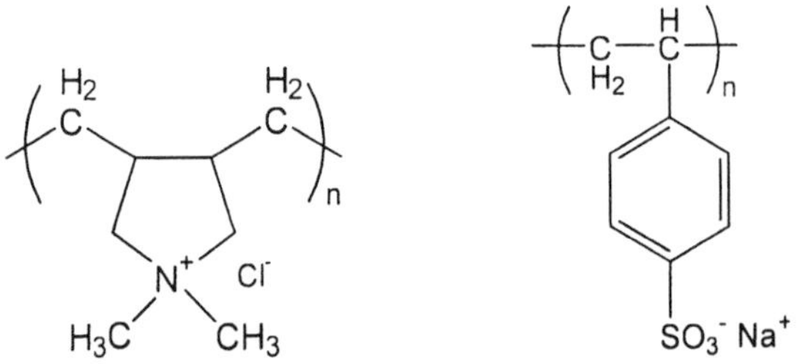

Figure 5. Chemical structure of poly(diallyldimethylammonium chloride)(PDAC)(left) and sulfonated polystyrene (SPS)(right).

Electrostatic Assembly

Charged particles of titanium dioxide, carboxylic acid functionalized hematin, and the polyelectrolytes: sulfonated polystyrene (SPS) (molecular weight 1,000,000 g/mol) and poly(diallyldimethylammonium chloride)(PDAC) (molecular weight 500,000-750,000 g/mol) were assembled on the nanofiber surface by LBL deposition. The chemical structures for PDAC and SPS are shown in Figure 5. Titanium dioxide nanoparticles were ground with a mortar and pestle together with 5N HCl (1g TiO_2 to 20ml). This solution was diluted 40 times with deionized water. The functionalized membranes were immersed in the diluted TiO_2 suspension for one minute and then rinsed with deionized water. Membranes were immersed in solutions of SPS and PDAC with concentrations of 0.1 weight percent for 10 minutes and then rinsed in de-ionized water. Carboxylic acid-functionalized hematin was prepared in a buffer solution at pH 11 and was used as an anion in place of SPS. More evidence of successful surface functionalization of PAN nanofibers was the application of electrostatic layer-by-layer assembly. This would not have been possible without ionic, or ionizable, surface groups on the nanofiber surface. Following electrostatic assembly, the surface area of the nanofibrous membrane was largely retained. Electron microscope images of 1 PDAC/SPS bilayer applied to the fiber surface can be seen in Figure 6. Figure 7 and Figure 8 show functionalized nanofibers with 5 and 10 PDAC/SPS bilayers, respectively. It was observed that the fiber surface became rougher as more bilayers were applied. The porosity and large surface-area of the original electrospun membrane remained even after 10 bilayers were applied.

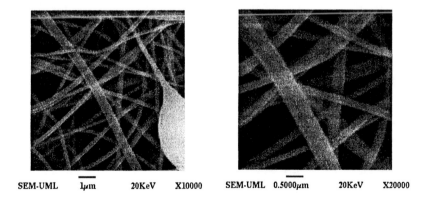

SEM-UML   1μm   20KeV   X10000   SEM-UML   0.5000μm   20KeV   X20000

Figure 6. SEM images of carboxylic acid functionalized PAN nanofibers with one PDAC/SPS bilayer applied to the surface.

144

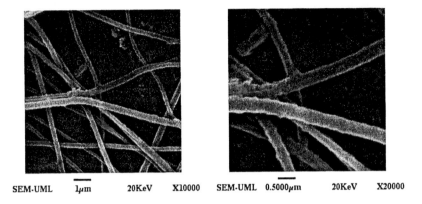

SEM-UML   1μm   20KeV   X10000   SEM-UML   0.5000μm   20KeV   X20000

Figure 7. SEM images of carboxylic acid functionalized PAN nanofibers with five PDAC/SPS bilayers applied to the surface.

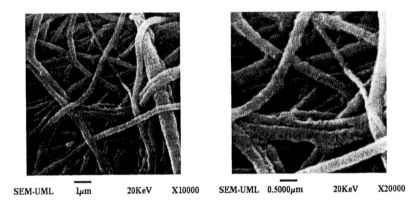

SEM-UML   1μm   20KeV   X10000   SEM-UML   0.5000μm   20KeV   X20000

Figure 8. SEM images of carboxylic acid functionalized PAN nanofibers with ten PDAC/SPS bilayers applied to the surface.

In addition to polyelectrolytes, it was possible to electrostatically assemble charged titanium dioxide particles onto the functionalized PAN surface. Cationic TiO_2 particles were found to sparsely adsorb to the surface of the anionic-functionalized PAN nanofibers. The particles were rendered cationic by grinding commercial TiO_2 particles in strong hydrochloric acid and then diluting the paste to a liquid. The pH of the liquid remained low and was likely responsible for the sparse coverage of the fibers by the TiO_2 particles. To overcome this, a new membrane was prepared and five bilayers of PDAC/SPS were assembled on the nanofiber surface prior to immersion in the titanium dioxide suspension. Since SPS remains anionic over a much wider pH range, more anionic sites should have been available to complex with the positively charged TiO$_2$ particles. This behavior was observed and is shown in SEM images in Figure 9 and in TEM images in Figure 10 showing reasonably complete particle coverage. Coverage remained incomplete in the five bilayer samples. Energy dispersive X-ray spectroscopy (EDS) of the same sample confirmed the presence of sulfur from the SPS where gaps in the TiO$_2$ particle coverage were observed.

SEM-UML   2μm   20KeV   X5000   SEM-UML   1μm   20KeV   X10000

Figure 9. SEM images of cationic TiO$_2$ assembled onto 5 bilayers of PDAC/SPS on top of functionalized PAN nanofibers.

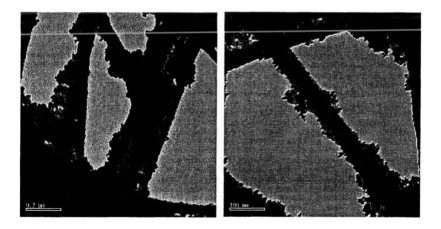

Figure 10. TEM images of cationic TiO$_2$ assembled onto 5 bilayers of PDAC/SPS on top of functionalized PAN nanofibers.

Hematin was used in LBL deposition and can be employed as a catalyst in conjugated polymer synthesis. Hematin functionalized with carboxylic acid groups can be electrostatically layered as an anion at pH 11. At this pH it is colored a brownish green and has an absorbance peak around 395 nm. As such, it was used to spectroscopically monitor successful electrostatic layer-by-layer assembly on a functionalized PAN electrospun membrane. To accomplish this, a glass slide was coated in a thin electrospun membrane of carboxylic acid-functionalized PAN and then placed in a cuvette in the UV-vis spectrometer (Perkin Elmer Spectometer Lambda 9) with a blank glass slide reference. Electrostatic assembly of the PDAC and hematin layers was done in the cuvette to minimize errors associated with changing the angle between the incident beam and the glass slide.

The absorbance spectrum is shown in Figure 11. With each additional hematin layer the 395 nm peak absorbance increased. Assembly of the PDAC cationic layer caused a slight decrease in absorbance in the same region. This was attributed to scattering of the incident beam by the PDAC layer decreasing photon access to the underlying hematin layer to a small degree. PDAC should not show any absorbance in this wavelength range. The non-zero absorbance of the first layer of PDAC on the electrospun fibers was likely due to scattering of the incident beam by the fibers themselves, since the membrane almost certainly contained fibers with diameters in the 100 to 500 nm range.

The increasing absorbance at 395 nm exhibited good linearity with sequential addition of hematin layers as shown in Figure 12. This indicated that approximately the same amount of hematin was deposited with each assembled layer. Thus, it can be inferred that the ionic character of the fiber surface was renewed with each addition of PDAC and hematin, indicating successful LBL assembly.

Conclusions

Electrostatic LBL assembly has been demonstrated on electrospun nanofiber surfaces. A variety of polymer electrolytes, charged metal oxide particles, and the biocatalyst hematin were successfully adsorbed. It is believed any ionic material reported in the literature used in electrostatic LBL assembly of films can also be applied to electrospun fibers. The key to this success was the surface-functionalization of the nanofiber with carboxylic acid groups. This simple, one-step reaction in sodium hydroxide created an insoluble nanofibrous membrane with a surface suitable for electrostatic assembly and a core rugged enough to survive the assembly process. It is postulated that similar results could be obtained by adding a crosslinking agent to a suitable polyelectrolyte spin-dope and crosslinking the nanofibers after electrospinning. However, this may involve more complicated spin-dope solutions and electrospinning processes. Thus, it may be possible to obtain nanofibers with alternate surface chemistries and gel-like fibers.

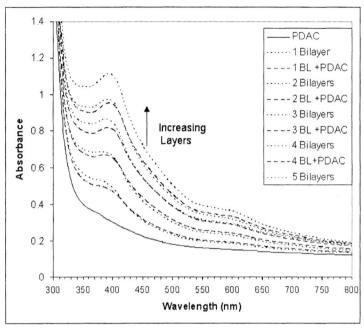

Figure 11. UV vis spectrum of electrospun functionalized PAN with PDAC and hematin adsorbed on the fiber surface.

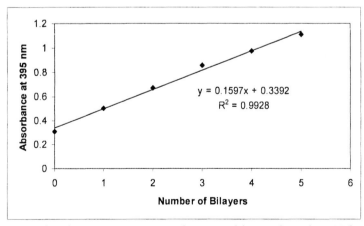

Figure 12. Absorbance at 395 nm as a function of the number of PDAC/hematin bilayers on an electrospun membrane.

Acknowledgements

The authors gratefully acknowledge financial support from the US Army. Thanks are due to Mr. Bon-Cheol Kim and Ms. Robyn Johnson for help with the hematin and the layer-by-layer work. Ms Bongwoo Kang from UMass Lowell and Mr. David Ziegler of the US Army provided invaluable assistance with SEM and TEM respectivly. Dr. Ferdinando Bruno and Dr. Ramaswamy Nagaragian were ever helpful with advice and useful discussions. This work is dedicated to Professor S. K. Tripathy.

References

[i] Lvov, Yuri; Decher, Gero; Moehwald, Helmuth. "Assembly, structural characterization, and thermal behavior of layer-by-layer deposited ultrathin films of poly(vinyl sulfate) and poly(allylamine)." *Langmuir* **1993**, *9*(2), 481-6.

[ii] Sukhorukov, G. B.; Moehwald, H.; Decher, G.; Lvov, Y. M. "Assembly of polyelectrolyte multilayer films by consecutively alternating adsorption of polynucleotides and polycations." *Thin Solid Films* **1996**, *284-285* 220-223.

[iii] Hammond, Paula T. "Recent explorations in electrostatic multilayer thin film assembly." *Current Opinion in Colloid & Interface Science* **2000**, *4*(6), 430-442.

[iv] Dubas, S.; Schlenoff, J. "Factors Controlling the Growth of Polyelectrolyte Multilayers." *Macromolecules* **1999**, *32*, 8153.

[v] Mendelsohn, J.; Barrett, C.; Chan, V.; Pal, A.; Mayes, A.; Rubner, M. "Fabrication of Microporous Thin Films from Polyelectrolyte Multilayers." *Langmuir* **2000**, *16*, 5017.

[vi] Park, S.; Barrett, C.; Rubner, M.; Mayes, A. "Anomalous Adsorption of Polyelectrolyte Layers." *Macromolecules* **2001**, *34*, 3384.

[vii] Deng, Shubo; Bai, Renbi; Chen, J. P. "Behaviors and mechanisms of copper adsorption on hydrolyzed polyacrylonitrile fibers." *Journal of Colloid and Interface Science* **2003**, *260*(2), 265-272.

Chapter 11

Preparation of Nanofibers with Controlled Phase Morphology from Electrospinning of Polybutadiene–Polycarbonate Blends

M. Wei[1], B. Kang[2], C. Sung[3], and J. Mead[1,*]

Departments of [1]Plastics Engineering and [2]Chemical Engineering,
University of Massachusetts at Lowell, One University Avenue,
Lowell, MA 01854
[3]Current address: Inje University, South Korea

This work focused on the internal morphology of nanofibers prepared by electrospinning polymer blends. Nanofibers of polybutadiene (PB)/polycarbonate (PC) blends were electrospun from polymer solutions. The composition ratio, molecular weight, and solvent type were changed to investigate how these factors affected the resulting phase morphology within the nanofiber. Co-continuous and core-sheath structures were observed by TEM. For PB/PC composition ratios above 25/75 weight percent, co-continuous morphologies were found, while core sheath morphologies were seen for composition ratios of PB/PC blends below 25/75. The molecular weight of both the polycarbonate and the polybutadiene had a significant effect on the phase morphology. Electrospinning of polymer blends offers the potential to prepare unique morphologies for use in a variety of applications.

Introduction

Electrospinning offers the ability to produce polymer fibers with diameters in the nanometer range (1,2,3). The high surface area/volume ratio yields nanofibers with enhanced properties for a variety of applications, including tissue growth, protective clothing, highly effective thermal insulation, filters for fine particles, reinforcing fibers, high surface area catalysts, wound dressings, and artificial blood vessels (4, 5). Most of the past and current research efforts (6,7) have focused on fiber formation mechanisms, electrospinning of different polymer types, blends, and control of mat properties such as fiber size, type, process, etc. Electrospinning of polymer blends offers the potential to prepare unique materials, but has received more limited investigation. The focus has been primarily on the properties of the overall mat. Blends of polyaniline and polyethylene oxide or sulfonated polystyrene and polyethylene oxide have been used to produce conductive and photo-responsive nanofibers for electronic devices (8). Nanofibers prepared from SBS triblock copolymers showed the two-phase morphology with domains elongated along the fiber axis (9). Annealing caused the phases to increase in domain size and become more ordered. Co-continuous phase morphologies were observed in electrospun polylactide/polyvinylpyrrolidone blends and specific surface topologies or fine pores were generated by selective removal of one of the components (4). In general, however, there has been little effort made towards control of the internal morphology of the nanofibers.

Morphologies formed in polymer blends are controlled by both thermodynamic and kinetic factors (10). Phase structures in polymer blends are governed by equilibrium thermodynamic factors, such as the composition, interfacial tension, etc. Under the rapid solvent evaporation of the electrospinning process it is expected that kinetic factors may give rise to nonequilibrium structures. Finally, shear and elongation effects occurring during the process may alter the shape of the resulting morphology (e.g. yielding elongated structures). As a result, the morphology formed in a polymer blend system depends on the interrelation between these different effects (11).

During the electrospinnning process the evaporation of solvent will take the polymer blend solution through the phase boundary into the two phase regime. As solvent evaporation continues, the polymer will solidify either by crystallization or reaching the glass transition temperature. This process may result in the formation of very fine morphologies because of the limited time available for the phases to coarsen and form larger domains (4). At the same time rheological processes will also occur, affecting the resulting morphology. In the case of pipe or capillary flow, the system will attempt to attain a state of minimum energy dissipation. This results in the lower viscosity fluid migrating to locations of highest shear rate, which are located at the walls. Thus, the lower viscosity fluid will attempt to encapsulate the higher viscosity material (12).

Considering the effects of both thermodynamic and kinetic factors, the electrospinnning process should allow for the preparation of unique fiber structures. The objective of this work was to develop nanofibers with controlled fiber morphology, e.g. core-sheath structure and co-continuous structures, by the electrospinning method and to understand the effect of polymer blend composition, polymer molecular weight, and solvent type on the resulting morphology. This information can be used to develop nanofibers with controlled internal morphology.

Experimental

Materials

Polycarbonate (PC) and polybutadiene (PB) were used as the materials. Tetrahydrofuran (THF), which evaporates rapidly at room temperature (boiling point: 65°C), was selected as the solvent. Four different molecular weights of Polycarbonate (PC) 21900, 23300, 27000 and 30600 g/mol (weight average molecular weight) obtained from GE Plastics were used. Polybutadiene (PB) with molecular weight 420,000 and 2,500,000 g/mol, tetrahydrofuran (THF), osmium tetroxide (OsO$_4$), chloroform (CHCl$_3$), and dimethylformamide (DMF) were purchased from Aldrich.

Blend preparation

PB/PC blends were prepared by dissolving the polymers in a flask filled with solvent using a magnetic stirrer. PC with 21,900 g/mol MW and PB with 420,000 g/mol MW were selected as the standard compositions; THF was selected as standard solvent. The following factors were varied to study their effect on the phase morphology of electrospun nanofibers: The composition ratios of PB/PC in THF studied were: 90/10, 75/25, 65/35, 50/50, 35/65, 25/75, and 10/90. At 75/25 and 25/75 weight ratios of PB/PC, the molecular weight of PC was varied from 21,900, 23,300, 27,000, to 30,600 g/mol and PB was changed from 420,000 to 2,500,000 g/mol. At the 75/25 wt% PB/PC blend, three different solvents, CHCl$_3$, THF, and THF/DMF with 90/10 ratio were used. The concentration of the solution was in the range from 6wt% to 16wt%. Figure 1 presents the solution concentration used for the electrospinning of blends, as well as the cloud points for the solutions. The cloud points were determined visually at room temperature by titrating 8% solutions of PB/PC blends in THF with THF. The cloud point was measured as the concentration at which the turbid solution became visually clear. The viscosity of the solution was measured using a HADT Brookfield digital viscometer.

Electrospinning

The electrospinning apparatus consisted of a high voltage power supply (Gamma High Voltage Research Co.), a steel syringe steel, a digitally adjusted syringe pump (Harvard Apparatus, PHD 2000, South Natick Co.), and an electrically grounded aluminum foil target. Polymer solution was fed through the needle tip by the syringe pump at a controlled flow rate. As the electrical field between needle tip and target increased, the electrically charged nanofibers were formed and collected on the target. The applied voltage was 15 kV, the flow rate of the polymer solution was 0.02 ml/min, and the distance between the needle tip and the target was 20 cm in all cases.

Transmission electron microscopy (TEM) and image analysis

The morphology of nanofibers was observed using a Philips EM 400 transmission electron microscope (TEM). For the preparation of TEM samples, the nanofibers were collected on the carbon coated copper specimen grid, followed by staining in OsO_4 vapor by suspending them over a 4.0% aqueous solution of OsO_4 for 30 minutes. Gaia Blue Software was used to calculate what is termed the nanolayer width, which indicates the domain size, of PC dispersed phases in electrospun PB/PC blend nanofibers. Each nanolayer width was the average of at least 100 independent measurements.

Results and Discussion

As discussed above, the phase morphology of electrospun polymer blends will depend on the competition between thermodynamic and kinetic factors. For the morphology formed in the electrospinning process, one must consider the combination of three processes, phase separation, solidification (or solubility of the polymer), and rheological processes. In the electrospinning process, the whipping motion of the electrospinning jet results in drawing of the fibers, increasing the surface area of the jet, leading to rapid solvent evaporation. As the solvent evaporates, the concentration will increase and the immiscibility of PB and PC phases will cause the blend system to separate into two phases. As solvent continues to evaporate, the polymers will solidify, bringing into play solvent effects, or the solubility difference of the components in the common solvent. THF is a poorer solvent for PC than PB because both PC molecules and

THF are nucleophilic, although PC has a lower solubility parameter difference than PB (*13*).

Rheological effects during flow of the mixture in the needle of the electrospinning equipment may also be present. As seen in Figure 2, PB has a much higher Brookfield viscosity than PC and with increasing solution concentration, the viscosity difference between PB and PC increases. For the capillary flow of the PB/PC solution mixture, according to the minimum energy dissipation theory of a flow system, the lower viscosity PC component would like to encapsulate the higher viscosity PB component (*11*). Therefore, in the electrospinning process where the polymer solution is fed through the needle capillary, the higher viscosity polybutadiene should be located in the center where the shear stresses are lowest and the low viscosity polycarbonate should be located close to the wall, where the shear stresses are highest. Thus, core sheath nanofibers would be expected to show PB as the core and PC as sheath. Finally, the morphology of the system may be affected by the flow field leading to elongated structures. All these factors will play a role in the morphologies discussed below.

Effect of composition ratios of PB/PC blending system

Figure 3 presents the TEM images of electrospun PB/PC nanofibers with different composition ratios of 90/10, 75/25, 65/35, 50/50, 65/35, 25/75 and 10/90 after stained by OsO_4. OsO_4 is known to preferentially stain the polybutadiene phase, so that the dark regions in the TEM images are identified as the polybutadiene phase and the light regions as the polycarbonate phase (*14*). In nanofibers with high concentrations of polybutadiene, there appears to be some spreading of the rubbery polybutadiene on the substrate, as evidenced by the dark region around the fibers. As seen in this Figure, when the weight ratio of PB/PC blends was larger than 25/75, co-continuous structures with interconnected PB and PC nanolayers or strands were formed with the nanolayers oriented along the fiber axis. When the weight ratio of PB/PC blends was smaller than 25/75, a change in the morphology occurred, presenting core sheath structures with PB located in the center of the fiber and PC located on the outside.

At low PC weight fractions, the PB is expected to play a more dominant role in the behavior. The PB has a much higher viscosity than PC at the same concentration, thus the whole blending system has high viscosity as compared to a composition where PC is the major weight fraction. Considering the processes affecting the development of the phase morphology, the high viscosity of the blending system, combined with the short time of solvent evaporation, will retard the coarsening of phase separated regions as well as migration of PB and PC due to rheological effects, resulting in a very fine phase morphology (*4*). In

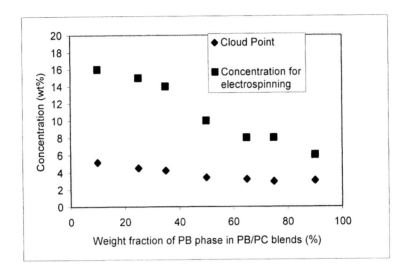

Figure 1. Elecrospinning concentrations and cloud point for blends.

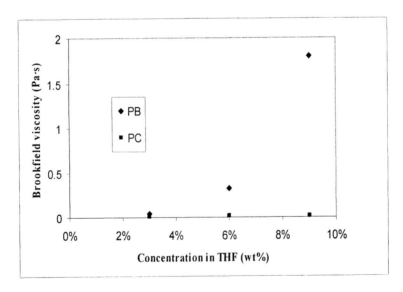

Figure 2. Brookfield viscosity of PB (420 kD) and PC (21.9 kD) at different concentrations.

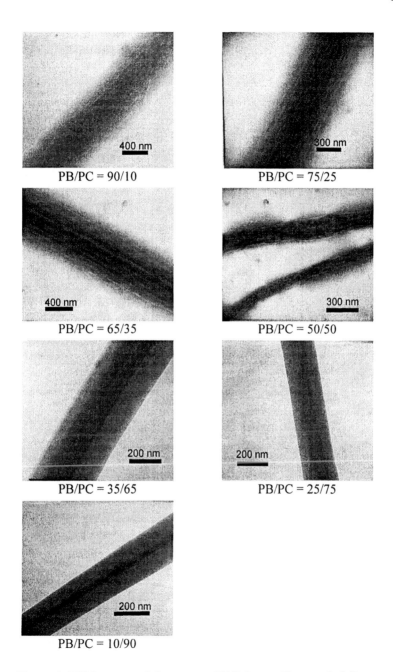

Figure 3. TEM images of electrospun PB/PC nanofibers with different composition ratios.

addition the elongated structures show the effect of shearing and elongation on the resulting morphology.

When the PC weight fraction is increased to 75 wt%, the nanofibers presented core-sheath structures with PB located at the center and PC located outside. The formation of core-sheath structures may be governed by a combination of factors. Although the viscosity of the starting solutions were kept the same, as the solvent evaporates the viscosity of the system will be lower for the blends with high PC content because of the increased fraction of the lower viscosity PC. This decreased viscosity promotes the development of a core sheath structures. At this composition ratio, PC was the major component in the blend system. As shown in Figure 1, PC had much lower solution viscosity than PB, thus the whole blend system would be expected to have lower viscosity. The decreased viscosity of the whole blending system both initially and during the rapid solvent removal provides the polymer molecules greater mobility, allowing the PC molecules to migrate, coalesce, and form large domains (i.e. stratified structures) so that the PB and PC phases are well separated. Furthermore, this decreased viscosity will provide time for the low viscosity PC phase to migrate to the outside of the fiber under the influence of rheological effects, while the PB phases are located at the center of the fiber forming a core-sheath structure.

Effect of molecular weight of components

Figure 4 and Figure 5 show the TEM images for nanofibers with different PC and PB molecular weights at 75/25 and 25/75 weight ratios of PB/PC, respectively. As shown in Figure 4, changes in molecular weight did not change the morphology for PB/PC blend ratios of 75/25 wt% with all nanofibers showing co-continuous structures. This would be expected as an increase in molecular weight of the compositions decreases the mobility of molecules and increases the viscosity of the blend system during the entire electrospinning process, impeding the ability of the polymer molecules to coalesce and form large domains in the short time of the process.

For the 25/75 weight ratio of PB/PC, as shown in Figure 5, a morphology change from core-sheath to co-continuous structures was observed when PC molecular weight was increased from 21.9 kD to 23.3 kD. Table I presents the melt flow index (MFI) and Brookfield viscosity for the different molecular weights of PC. As expected, higher molecular weights give higher Brookfield viscosity. At the 25/75 weight ratio of PB/PC blends, PC is the major component, thus increasing the PC molecular weight apparently increased the viscosity (decreased mobility) of the blending system sufficiently to prevent the two polymers from forming large stratified domains and core sheath structures. At the same time, the mobility of PC molecules would also be decreased. As a

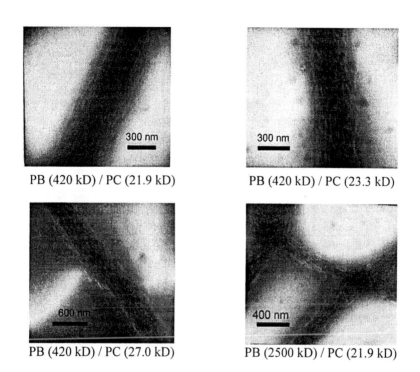

PB (420 kD) / PC (21.9 kD)

PB (420 kD) / PC (23.3 kD)

PB (420 kD) / PC (27.0 kD)

PB (2500 kD) / PC (21.9 kD)

Figure 4. TEM images of PB/PC blends with different molecular weight at weight ratios of 75/25.

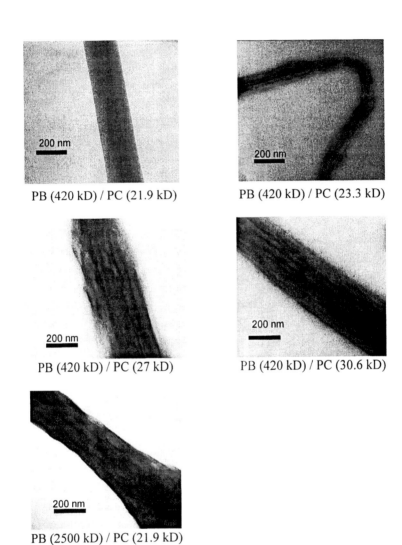

PB (420 kD) / PC (21.9 kD)   PB (420 kD) / PC (23.3 kD)

PB (420 kD) / PC (27 kD)   PB (420 kD) / PC (30.6 kD)

PB (2500 kD) / PC (21.9 kD)

Figure 5. TEM images of PB/PC blends with different molecular weight at weight ratios of 25/75.

result co-continuous structures were formed for PC molecular weights above 21.9 kD. Clearly kinetic factors are dominant as one would expect higher molecular weights to favor better phase separation under equilibrium conditions.

The effect of PC molecular weight on nanolayer width (indicative of the domain size) was also investigated. Figure 6 compares the nanolayer width of PC phase for blends with PC molecular weights of 23.3, 27, and 30.6 kD (all co-continuous structures). As seen in the Figure 6, the nanolayer width decreases with increasing molecular weight. Again, as the molecular weight increases, the mobility of the polymer chains will decrease, particularly as the solution concentration increases, leading to smaller domains (nanolayers).

For similar reasons, when the molecular weight of PB was increased from 420 to 2500 kD, the core-sheath structure was transformed to a co-continuous structure. This indicates that the morphology developed may be very sensitive to viscosity and molecular weight of compositions, and further work is required to understand these effects.

Table I. Physical properties of PC with different molecular weight and phase morphology of PB/PC blends

MW of PC (kD)	MFI (g/10 min)	Brookfield Viscosity (mPa-s) 6 wt%	Morphology of PB/PC	Nanolayer width (nm)
21.9	50	13	core-sheath	-
23.3	17.5	19	co-continuous	35
27.0	10.5	15	co-continuous	27
30.6	2.5	15	co-continuous	22

Effect of solvent type of PB/PC blending system

It is expected that the solvent evaporation and solubility will have an influence on the resulting fiber morphologies. The phase separation process will be aided by the different solubility of the components in the common solvent. Furthermore, the solvent evaporation rate in the electrospinning process would affect the phase separation process. To study these effects three different solvents were used in the electrospinning process: $CHCl_3$, THF and a 90/10 wt% THF/DMF solvent mixture. Figure 7 shows the TEM images of nanofibers spun from the three different solvents. Table III lists the evaporation rates of these solvents and the resulting nanolayer width of the PC phase. The nanofibers spun from $CHCl_3$ had the smallest nanolayer width or finest phase morphology, while nanofibers spun from THF/DMF showed nonuniform structures with a large number of PB beads along the fiber axis.

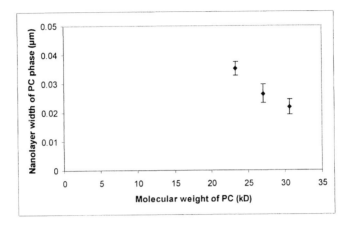

Figure 6. Effect of PC molecular weight on nanolayer width of PC phase at 25/75 wt% of PB/PC.

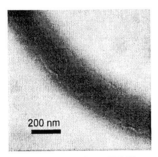

PB/PC = 75/25, CHCl₃

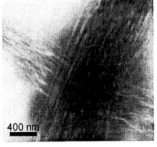

PB/PC = 75/25, THF

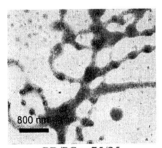

PB/PC = 75/25,
THF/DMF=90/10

Figure 7. TEM images of Nanofibers by different solvents.

Table III. The evaporation rates of solvent and the nanolayer width of nanofibers by different solvents.

Solvent	Vapor pressure (kPa)	Solubility parameter	Nanolayer width (nm)
CHCl$_3$	21.2	19.0	11
THF	17.9	18.6	21
THF/DMF (90/10 wt%)	16.2	18.6/16.0	-

Two effects could explain this behavior. First, the higher solvent evaporation rate of chloroform may not allow sufficient time for coarsening of the phases during the phase separation process, thus fine phase morphology was obtained. Second, the solubility of PC in CHCl$_3$ is better than in THF. Large solubility difference of two phases would promote the demixing of one phase from the blending system. As chloroform has a smaller solubility parameter difference with PC as compared to THF, chloroform is a better solvent for PC than THF. As a result we would expect the THF system to give larger domain sizes. In the case of the THF/DMF system, DMF is a poor solvent for PB and good solvent for PC, so the addition of DMF led to the development of large beads of PB within the fiber.

Conclusions

Nanofibers were electrospun from polybutadiene/polycarbonate blends. Co-continuous and core-sheath structures in the nanofibers were observed by TEM. The phase morphology of nanofibers from PB/PC blends was found to depend not only on the composition ratio of the blends, but also on the molecular weight of individual polymer components. When the composition ratio of PB/PC blends is above 25/75 weight percent, co-continuous morphologies were found, but when the composition ratio of PB/PC blends is below 25/75 the morphology changed to core sheath structures. The molecular weight of both the polycarbonate and the polybutadiene had a significant effect on the phase morphology. With increasing molecular weight of either the polycarbonate or the polybutadiene, the phase morphology of the nanofibers was changed to co-continuous structures. The results indicate that viscosity of the blend system and polymer mobility plays a large role in the development of nanofiber morphology under the rapid solvent evaporation that occurs during the electrospinning process.

Acknowledgment

The authors would like to acknowledge the support of the National Science Foundation (Grant DMI-0200498).

References

1. Pawlowski, K. J.; Belvin, H. L.; Raney, D. L.; Su, J. *Polymer,* **2003**, *44*, 1309.
2. Walters, D. A.; Ericson, L. M.; Casavant, M. J. *Appl. Phys. Lett,* **1999**, *74*, 3803.
3. Fong, H.; Reneker, D. H. *Structure Formation in Polymeric Fiber;* Salem, D. R., Ed.; Hanser Publishers: Cincinnati, OH, 2000; vol. 1, pp225-245.
4. Bognitzki, M.; Frese, T.; Steinhart, M. *Polym. Eng. and Sci,* **2001**, *41*, 982.
5. Norris, I. D.; Shaker, M. M.,; Ko, F. K., MacDiarmid A. G. *Synthetic Metals* **2000**, *114*, 109.
6. Doshi, J.; Reneker, D. H. *J. Elec,* **1995**, *35*, 151.
7. Baumgaarten, P. K. *J. Coll. Int. Sci.* **1971**, *36*, 71.
8. Norris, I. D. *Synthetic Metals*, **2000**, *114*, 109
9. Fong, H.; Reneker, D. H. *J. Polym. Sci B Polym. Phys,* **1999**, *37*, 3488.
10. Wolf, B. A.; Horst, R. *Rheo-Physics of Multiphase Polymer Systems;* Søndergaard, K., Ed.; Technomic Publishing Co.: Pennsylvania, PA, 1995, pp651-693.
11. Paul, D.R . *Polymer Blends;* Paul, D.R., Ed.; Academic Press: New York, NY, 1978; Vol.2, pp168-212.
12. Southern, J. H.; Ballman, R. L. *J. Appl. Polym. Sci.* **1973**, *20*, 175.
13. Des Cloizeaux, J.; Jannink, G. *Polymers in Solution;* Oxford University Press: NY, 1990, pp13.
14. Sawyer, L. C.; Grubb, D. T. *Polymer Microscopy;* Chapman and Hall Ltd: New York, NY, 1978, pp93.

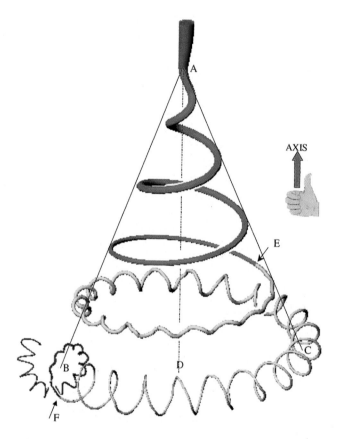

Plate 2.2. A diagram showing the end of the straight segment, followed by the onset and development of three successive generations of right handed bending instabilities. The convention used for handedness is that if the right hand is rotated in the direction of the fingers, the hand will advance along a right-handed helix in the direction of the thumb.

Color insert page 2

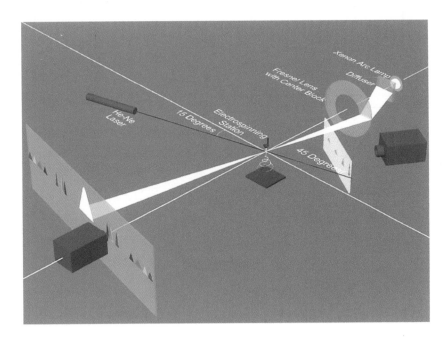

Plate 3.1. Experimental setup of the interference color technique

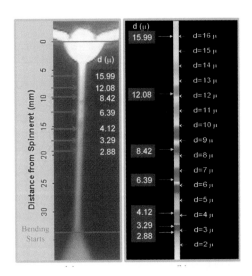

Plate 3.2. Comparison between interference colors and jet diameters measured at points along the jet, using light diffraction. The diagram at the right shows the range of colors predicted from the trichromatic theory of color vision.

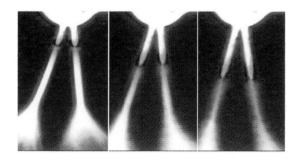

*Plate3.3a. Interference colors on a single jet during electrospinning
(Increasing voltage from left to right)*

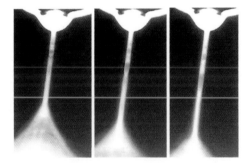

*Plate 3.3b. Interference colors on a twin jet initiated from a single droplet
(Increasing voltage from left to right)*

Color insert page 4

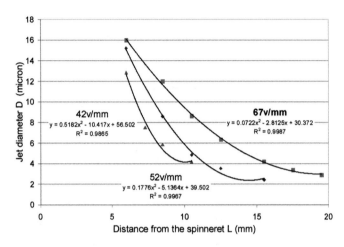

Plate 3.4. Laser diffraction measurement of electrospinning jet diameter measured at different voltages. Polynomials fitted to the data are shown.

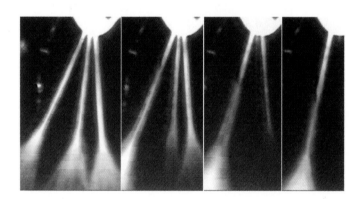

Plate 3.5. Interference colors on a triple jet initiated from a single droplet (From left to right: 34ms, 68ms, 102ms and 136ms after voltage was dropped)

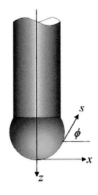

Plate 7.2. Schematic of a liquid drop suspended from a capillary.

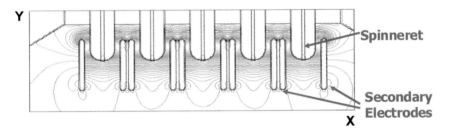

Plate 7.8. FEA calculation of the electric field distribution of a 5-electrode system using secondary electrodes to minimize the interactions from adjacent electrodes

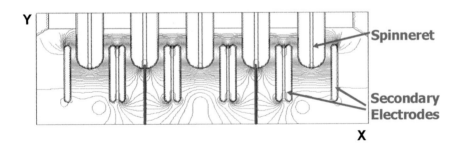

Plate 7.9. FEA calculation of the electric field distribution of a 5-electrode system with 2 ejecting jets using secondary electrodes to minimize the interactions from adjacent electrodes.

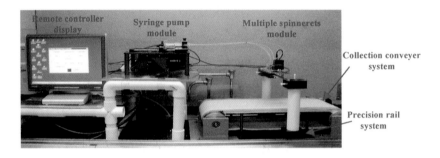

Plate 7.10. Prototype multiple-jet electrospinning apparatus.

Plate 7.11. (A) Fluid distribution, (B) linear array electrode assembly in the prototype multiple-jet electrospinning apparatus.

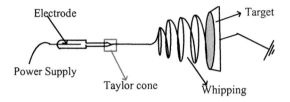

Plate 8.1. A simple horizontal electrospinning setup

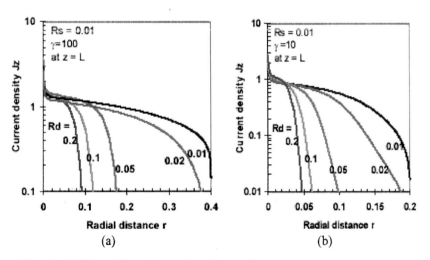

Plate 8.2. (a) Electrospun polyethylene oxide(PEO) fibers on an aluminum Target, and (b) SEM image of the same fibers.

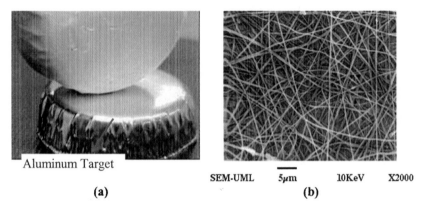

Plate 8.4. Effect of disc electrode radius and injection parameter on current density as a function of radial distance on the target.

Plate 8.5. Electrospinning setup with the disc electrode (target not shown)

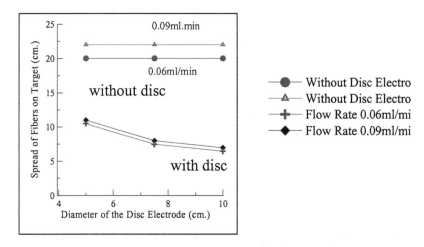

Plate 8.6. Plot of spread of fibers on target versus the disc diameter (applied voltage = 17.5KV, source to target distance = 16cm)

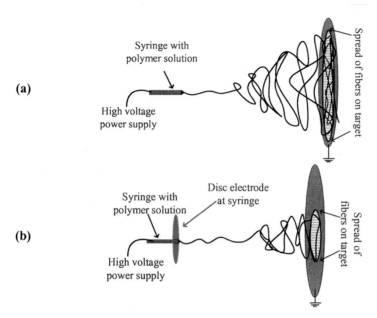

Plate 8.7. Schematic of effect of disc electrode on the instability initiation and fiber spread, (a) without the disc electrode and (b) with the disc electrode.

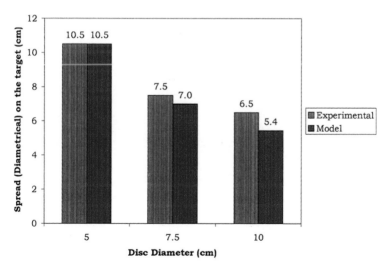

Plate 8.8. Comparison of fiber spread results – experimental versus model – with 5 cm data used to determine injection parameter value

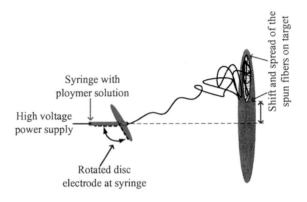

Plate 8.10. Directing the electrospinning jet onto the target

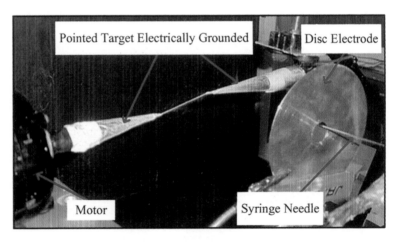

Plate 8.13. Cone target setup with one cone attached to a motor

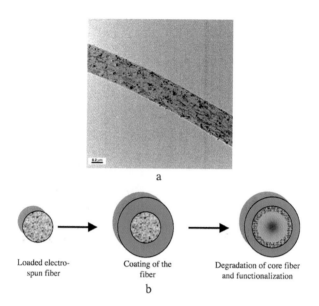

Loaded electro-
spun fiber

Coating of the
fiber

Degradation of core fiber
and functionalization

b

Plate 12.2. (a) TEM of unstained electrospun PLA / silver acetate fibers and (b) schematic process for the preparation of polymer tubes with metal nanoparticles via functionalized template fibers.

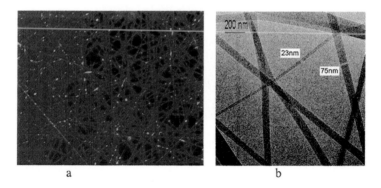

a   b

Plate 12.8. Fluorescence microscopy (a) and TEM (a) of electrospun PVA-Anth fibers.

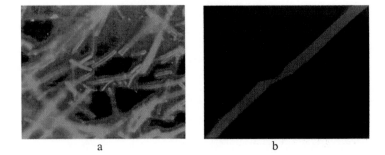

a   b

Plate 12.9. Fluorescence optical microscopy of PPX-PVA/PAA tubes functionalized by anthracene (a, magn. 480x) and necking-type defect (b)

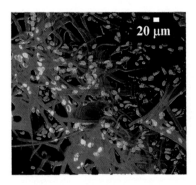

Plate 15.1. LSCM image of gelatin scaffold seeded with MG63 osteoblast cells.

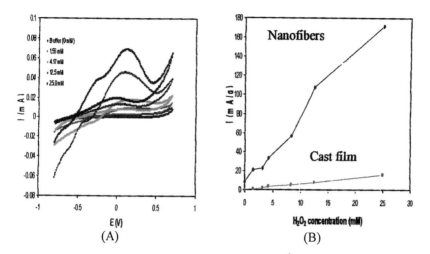

(A)   (B)

*Plate 18.7. (A) CVs of HCSA doped-PANI/PS nanofiber sensors at different
H_2O_2 concentrations. (B) The current response of HCSA-doped PANI/PS
nanofiber and spin-cast sensors at various H_2O_2 concentrations. Note that
the current response has been scaled by the weight of the polymer material
deposited on the electrodes.*

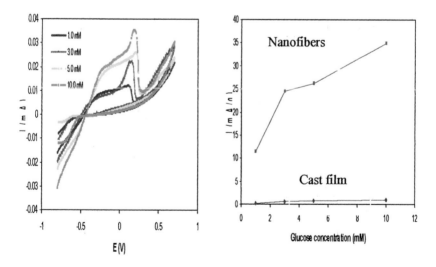

*Plate 18.8. (A) CVs showing the H_2O_2 peaks with varying concentration
of glucose solution for HCS-doped PANI/PS nanofiber sensor. (B) The current
response of HCSA-PANI/PS nanofiber and spin-cast film sensors at various
glucose concentrations. Note that the current response has been scaled
by the weight of the polymer material deposited on the electrodes
(scan rate = 100 mV/s).*

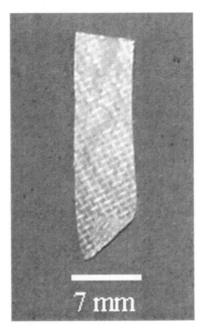

Plate 25.3. Nylon Nanofiber mat

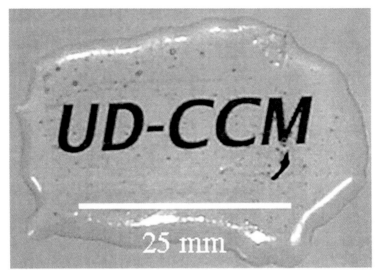

Plate 25.4. Nanofiber composite

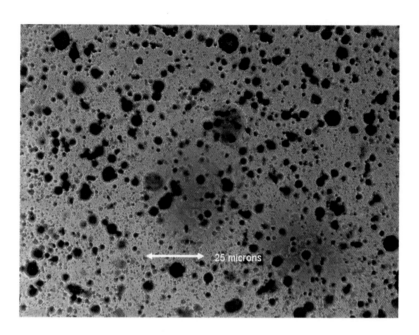

Plate 25.5. Optical micrograph of voids in the electrospun fiber composite

Plate 25.6. Nylon/Vinyl ester nanofiber composites films ~9 % fiber volume fraction

Chapter 12

Functional Polymer Nanofibers and Nanotubes via Electrospinning

J. Zeng, Z. Sun, H. Hou, R. Dersch, H. Wickel, J. H. Wendorff, and A. Greiner

Department of Chemistry and Scientific Center of Materials Science, Philipps-University of Marburg, Hans-Meerwein-Strasse, D–35032 Marburg, Germany

Functional polymer nanofibers were prepared by electrospinning of ternary polymer solutions (e. g. poly-L-lactide (PLA)) and low molecular weight metal compounds. These functionalized fibers were further modified by coating with poly(p-xylylene) (PPX) via chemical vapor deposition (CVD). Removal of PLA by decomposition or selective solvent extraction resulted in the formation of nanotubes with e. g. Pd-nanoparticles on the inside or outside of the tubes depending on the reaction conditions. Functional nanotubes available for numerous chemical modifications were obtained by the TUFT-process (Tubes by Fiber Templates). According to this process by PLA template fibers were coated with a mixture of poly(vinylalcohol) (PVA) and poly(acrylic acid) (PAA) followed by CVD of PPX as a protective coating. Heating of these composite fibers resulted in decomposition of PLA and chemical crosslinking PVA / PAA. Residual OH-groups were used as reactive sites for immobilization of functional small molecules such as anthracene derivatives. Co-electrospinning by coaxial jets of different materials resulted in fibers with concentric variation of materials concentration. For example, fibers with PLA shell and Pd core where obtained after co-electrospinning of PLA-chloroform and palladium diacetate – THF solutions followed by heating to 170 °C.

The preparation of polymer nanofibers and nanotubes is a challenge from academic as well as technical point of view. Polymer nanofibers and nanotubes can be prepared by methods involving self-organization or by template methods. A highly versatile method for the preparation of polymer nanofibers is electrospinning. Basic principles of electrospinning of polymer solutions have been described extensively in other contributions of this symposium series and related reviews (1) and therefore will be not discussed in detail here.

Our research on electrospinning of polymer nanofibers is presently focused mainly on structure formation in electrospun fibers (2), on the reduction of fiber dimensions by systematic variation of processing parameters (3), the usage of electrospun fibers as templates for polymer nanotubes in the TUFT-process (4), on core/shell fiber preparation (5), and on the chemical modification of electrospun fibers, and on the search of new applications for functionalized electrospun nanofibers. In the context of this contribution it should be not unmentioned that other groups succeeded in the preparation of nanotubes involving electrospun fibers as well (6).

Here, we want to give an overview on the work in our group on the preparation of functionalized polymer nanofibers and nanotubes via electrospinning and the chemical modification of electrospun nanofibers in order to present different conceptual approaches.

Results and discussion

Particular advantage of electrospinning is that polymer fibers can be furnished with additional functionalities simply by additives in the medium used for electrospinning, by coating of electrospun fibers, or by usage of functional polymers. In course of our investigations we have functionalized electrospun polylactide (PLA) fibers by physical vapor deposition (PVD) with different metals such as aluminum, chromium, and gold. Poly(p-xylylene) (PPX) / metal composite-tubes were prepared by coating of metal-coated PLA fibers via chemical vapor deposition with PPX according the to the so-called Gorham-procedure (7) and subsequent thermal decomposition of PLA template fibers. Inspection by transmission electron microscope (TEM) showed PPX tubes coated on the inside wall by the metals previously deposited on the template fibers (Figure 1).

Functionalized polymer fibers can also be prepared by simultaneous spinning of the polymer material with a functional material, by spinning of a functional polymer, by postprocessing chemical treatment of electrospun polymer fibers, or by co-electrospinning.

Polymer fibers can be loaded by metal compounds via electrospinning of polymer solutions, for example by electrospinning of a 1 : 1 mixture of PLA and silver acetate in dichloromethane solution. Silver is distributed all over the fibers as obvious from TEM micrograph (Figure 2a). According to the process shown schematically in Figure 2b polymer tubes with metal nanoparticles inside the tubes are readily available by electrospinnig of template fibers with metal compounds, followed by coating with the desired wall material, and subsequent degradation of the template core fibers accompanied by conversion of the metal compounds to the corresponding elemental metals. Here, metal containing template fibers were obtained by electrospinning of PLA and palladium diacetate (PdOAc) (solution used for electrospinning: 3 % PLA and 3 % PdOAc in dichloromethane). These composite fibers were coated with PPX by CVD. Subsequent heating resulted in the formation of PPX tubes with Pd nanoparticles on the inside wall of the tubes (Figure 3). Energy dispersive (EDX) analysis and electron diffractometry of these fibers clearly showed the presence of palladium in these fibers (4b).

High-resolution inspection of the "Pd-nanowires" showed the formation of Pd-nanoparticles with sizes centered at 10 nm (Figure 4) (4d). The particles are homogeneously distributed. Interestingly, the particles seem not to be agglomerated. Following the same concept PPX tubes were obtained with copper and silver nanoparticles (Figure 5).

The sizes of Pd-nanoparticles as well as their positions depend significantly on the processing parameters. For example, heating of cut PLA-PdOAc-PPX composite fibers (cut by a scalpel perpendicular to long tube axis) for S hours in vacuum resulted in the formation of PPX-Pd-composite nanotubes with Pd-nanoparticles also on the outside of the tubes (Figure 6a and Figure 6b). Otherwise uncut tubes of the same sample resulted in tubes with nanoparticles only inside the tubes (Figure 6c), which excludes the migration of Pd particles through the tube wall as possible explanation for the results with cut tubes.

Electrospinning of polymers with tailor-made properties can result in fibers with corresponding properties. An example in case is shown for fluorescent electrospun fibers. Firstly, 9-Anthracene carboxylic acid was converted to 9-carboxylic acid chloride according to scheme 1. Esterification of 9-carboxylic acid chloride and PVA (degree of hydrolysis 99 %) resulted in fluorescent PVA (2.5 mol % of OH groups of PVA were esterified) in DMF. Solution cast films showed strong blue fluorescence upon irradiation with UV-light. Fluorescence spectrum with excitation at 300 nm showed a maximum of light emission centered at 450 nm (Figure 7).

Fluorescent fibers were obtained by electrospinning of 10 wt% PVA – Anth solution in water (Fig. 8a). Inspection of the electrospun fibers by TEM showed smooth fibers with diameters in the range of 20 – 100 nm (Figure 8b). In contrast, electrospinning of PVA and other water-soluble polymers like polyethylene oxide under standard conditions results in beaded fibers (8).

Following the TUFT-process, fluorescent tubes were prepared by reaction of OH-functionalized PPX tubes with fluorescent molecules according to scheme 2. Firstly, electrospun PLA template fibers were coated by PVA / PAA (dissolved in water 75 : 25 weight ratio) using airbrushing technique. The composite fibers were coated by PPX via CVD as protective layer in order to ensure OH-functionalization only inside the tubes. The PVA / PAA blend was stabilized by thermal crosslinking via esterification as described previously for bulk samples (9). PPX tubes with crosslinked PVA / PAA on the inside were developed by extraction of PLA with chloroform. The anthracenoyl chlorid was soaked into tubes with OH- groups inside and chemically linked to the inner tube wall by esterification. The final products of this procedure were fluorescent tubes, which maintained fluorescence even after several washing procedures with tetrahydrofurane and chloroform (Fig. 9a).

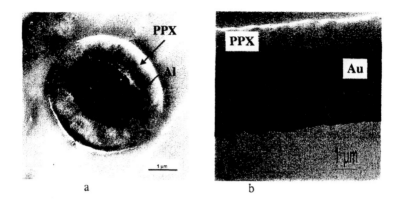

Figure 1. TEM micrograph of unstained PPX tubes with aluminum (a, reported previously in Lit. (4a)) and gold (b) coating on the inside wall prepared by coating of PLA template fibers via PVD of corresponding metals followed by CVD of PPX and subsequent thermal decomposition of PLA. (Reproduced with permission from reference 4a. Copyright 2000 Wiley–VCH–Verlag.)

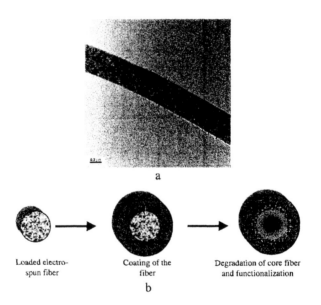

Figure 2. (a) TEM of unstained electrospun PLA / silver acetate fibers and (b) schematic process for the preparation of polymer tubes with metal nanoparticles via functionalized template fibers. (See page 11 of color inserts.)

Figure 3. TEM of PPX / Pd tubes obtained from electrospun PLA / PdOAc fibers after CVD of PPX, thermal decomposition of PLA and conversion of PdOAc to Pd. (Reproduced from reference 4b. Copyright 2002 American Chemical Society.)

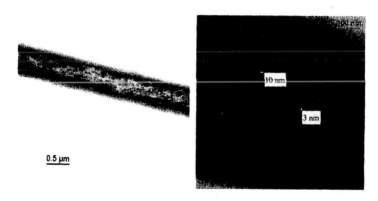

Figure 4. High resolution TEM of PPX / Pd tubes.

168

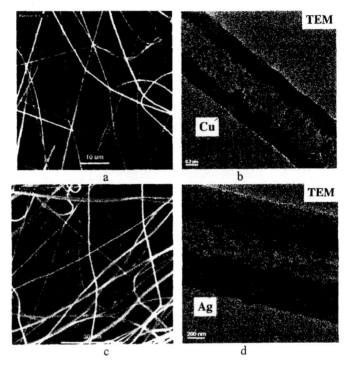

a
b

c
d

Figure 5. SEM of electrospun PLA / Cu(OAc)₂ fibers (a) and TEM of corresponding PPX / Cu tubes (b), SEM of electrospun PLA / AgOAc fibers (c) and corresponding PPX / Ag tubes (d). (Figure 5A is reproduced with permission from reference 5a. Copyright 2000 Wiley–VCH–Verlag.)

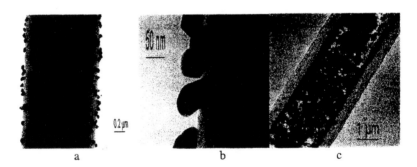

a
b
c

Figure 6. TEM of cut PPX-Pd composite tubes (a), high resolution TEM of PPX-Pd composite tubes (b), and TEM of uncut PPX-Pd tubes after heating to 385 °C for 5 hrs (a + b) and to 365 for 3 hrs (c) (4d).

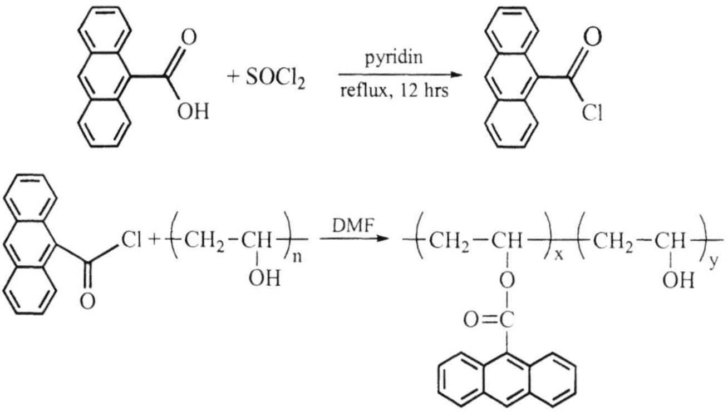

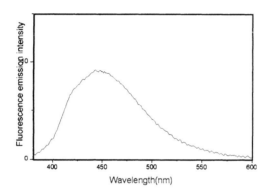

Scheme 1. Conversion of 9-anthracene carboxylic acid to 9-anthracene carboxylic acid chloride and esterification reaction with PVA

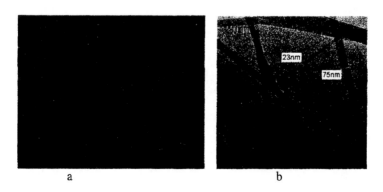

Figure 7. Fluorescence spectrum of solution cast films of PVA-Anth (excitation wavelength 300 nm).

a b

Figure 8. Fluorescence microscopy (a) and TEM (a) of electrospun PVA-Anth fibers. (See page 11 of color inserts.)

Scheme 2. Reaction of PPX-PVA/PAA composite tubes with anthracenoyl chloride.

$$a \qquad\qquad\qquad\qquad b$$

Figure 9. Fluorescence optical microscopy of PPX-PVA/PAA tubes functionalized by anthracene (a, magn. 480x) and necking-type defect (b). (See page 12 of color inserts.)

Of unknown origin are necking-type defects of the tubes which where observed with the fluorescent tubes by fluorescence microscopy (Figure 9b). A possible explanation could be post-processing mechanical damages by in proper handling but this is still speculation.

A highly versatile method for the preparation of functional nanofibers is the co-electrospinning technique, which was recently published by several groups (5, 6b,c). Here, composite fibers with coaxial variation of concentration variation are prepared by coaxial spinning of jets. For example, composite fibers of polyethylene oxide (PEO, shell) and poly(dodecylthiophene) (PDT, core) were prepared by co-electrospinning of chloroform solutions (Figure 10a). Another example in case is co-electrospinning of PLA (shell, from chloroform solution) and PdOAc (core, from THF solution). Here, PLA / Pd composite fibers were obtained after conversion of PdOAc to Pd at 170 °C for 2 hours. Using the electrospinning technique it is possible obtain fibers from materials which otherwise do not form fibers by electrospinning.

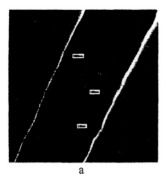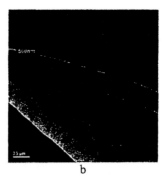

a   b

Figure 10. (a) TEM of an unstained sample of co-electrospun PEO (shell) and PDT (core) fiber and (b) TEM of annealed (170 °C / 2 h) unstained samples of co-electrospun PLA (shell) and PdOAc (core) (previously published in Lit. (8a).

Conclusions

Electrospinning is a highly versatile method for the preparation of functional polymer nanofibers and nanotubes. Functional nanofibers were prepared by electrospinning of functional polymers, by electrospinning of mixtures of polymers and functional materials or by functionalization of polymer nanofibers via post-electrospinning coating. In addition, co-electrospinning offers many possibilities for new materials combinations. All of these methods are very general and offer manifold possibilities for the preparation of new materials for various field of applications including catalysis, special filtration, membranes, biomedical applications just to name a few. However, what is still missing is the placement of functionalities at particular locations along electrospun fibers, proper control of materials density gradients, as well controllable shape of electrospun fibers, which will be challenges for future work.

Acknowledgements

The authors are indebted to Specialty Coating Systems, Indianapolis for the donation of parylene dimer and technical support, to Boehringer Ingelheim for the donation of PLA, and to Dr. A. Schaper and M. Hellwig for SEM and TEM support.

References

1. For recent reviews see: a) *Electrospinning and the formation of nanofibers;* Fong, H.; Reneker, D. H. in: *„Structure formation of polymeric fibers";* Salem, D. R.; Sussman, M. V., Eds.; Hanser **2000**; p. 225. b) Huang, Z.-M.; Zhang, Y.-Z.; Kotaki, M.; Ramakrishna, S. *Comp. Sci. Techn.* **2003**, *63,*

2223. c) *Polymer Nanofibers by Electrospinning;* Dersch, R.; Greiner, A.; Wendorff, J. H. in: *Dekker Encyclopedia of Nanoscience and Nanotechnology;* Schwartz, J. A.; Contescu, C. I.; Putyera, K., Eds.; Marcel Dekker; New York **2004**; p. 2931.

2. a) Bognitzki, M.; Frese, T.; Steinhart, M.; Greiner, A.; Wendorff, J. H., *Polymer Engineering and Science* **2001**, *41*, 982. b) Bognitzki, M.; Czado, W.; Freese, T.; Schaper, A.; Hellwig, M.; Steinhart, M.; Greiner, A.; Wendorff, J. H., *Adv. Mater.* **2001**, *13(1)*, 70. c) Dersch, R.; Liu, T.; Schaper, A. K.; Greiner, A.; Wendorff, J. H., *J. Polym. Sci.: Part A: Polym. Chem. Ed.* **2003**, *41*, 545.

3. Jun, Z.; Hou, H.; Schaper, A.; Wendorff, J. H.; Greiner, A., *e-Polymers* **2003**, No. 9.

4. a) Bognitzki, M.; Hou, H.; Ishaque, M.; Frese, ,T.; Hellwig, M.; Schwarte, C.; Schaper, A.; Wendorff, J. H.; Greiner, A., *Adv. Mater.* **2000**, *12*, 637. b) Hou, H.; Zeng, J.; Schaper, A.; Wendorff, J. H.; Greiner, A., *Macromolecules* **2002**, *35*, 2429. c) Caruso, R. A.; Schattka, J. H.; Greiner, A., *Adv. Mat.* **2001**, *13*, 1577. d) Sun, Z.; Zeng, J.; Hou, H.; Wendorff, J. H.; Greiner, A., submitted.

5. a)Sun, Z.; Zussman, E.; Yarin, A. L.; Wendorff, J. H.; Greiner, A., *Adv. Mat.* **2003**, *15(22)*, 1929

6. a) Dong, H, Jones, W. E., *Polymeric Materials: Science & Engineering* **2002**; *87*, 273; b) Li, D.; Xia, Y., *Nanoletters* **2004**, *4*, 933; c) Loscertales, I. G.; Barrero, A., Marquez, M.;M Spretz, R.; Velarde-Ortiz, R.; Larsen, G., *J. Am. Chem. Soc.* **2004**, *126*, 5376.

7. Gorham, W. F., *J. Polym. Sci. Part A-1* **1966**, *4*, 3027.

8. a) Ding, B.; Kim, H. Y.; Lee, S. C.; Lee, D. R.; Choi, K. J., *Fibers and Polymers* **2002**, *3*, 73. b) Fong, H.; Chun, I.; Reneker, D. H., *Polymer* **1999**, *40*, 4585.

9. Hassan, C. M.; Peppas, A. N., *Adv. Polym. Sci.* **2000**, *153*, 37.

Chapter 13

Production of Cross-Linked PET Fibrous Substrates via Electrospinning

Darren A. Baker and Philip J. Brown[*]

School of Materials Science and Engineering, Clemson University, Clemson, SC 29634

Appropriate azide crosslinking agents have been synthesized and used as additives in polymer solutions that can be electrospun to produce nanofiber and microfiber substrates. These azides essentially react to crosslink, functionalize and covalently bind PET polymer chains. Electrospinning mixtures of PET with these additives enables fiber modification during or after the electrospinning process as heat can be used to initiate the crosslinking reaction. Modification of nanofiber/microfiber substrates was done by thermal post-spin treatments. This study demonstrates the effectiveness of the technique in modifying 100% PET electrospun substrates. The research shows how the process inherently changes the properties of the electrospun fibers including fiber Tg, melt temperature and how different azides over a series of concentrations affect the fundamental thermal properties of the fibers. The motivation for crosslinking PET electrospun materials was to examine methods that could assist with structure stability, thermal or mechanical.

Introduction

Polyester is the most widely used synthetic fiber. Many attempts have been made to modify fibers since modification offers the potential to create new products or new markets. Polymeric chemical finishes are applied to improve surface moisture transport and static behavior. Similar effects are achieved by conventional wet chemistry that utilizes alkali to hydrolyze the surface; enzymes and plazma or corona treatments have recently been suggested for this purpose. This wide range of research indicates the importance of polyester modification. In the past a large number of processes were traditionally based on large scale, wet processing techniques, the majority of which are extremely energy intensive, and invariably make use of chemical agents in their application. Due to the obvious economic and environmental implications of such processes, it may therefore be advantageous for certain applications to consider specialized technology alternatives.

In this respect our current research work builds on the rapid developments seen in electrospinning in recent years (1-4). We have designed and added appropriate azide additives to polymer solutions that can be electrospun. These azides can react, crosslink, functionalize and covalently bind polymer chains. Electrospinning mixtures of polymers with these additives can result in the covalent binding of synthetic polymers with natural polymers, e.g. proteins, in a single simple manufacturing step. By applying sufficient energy during the electrospinning process (i.e., heat or UV light) we can modify nanofiber substrates either during the fiber formation process or by post-spin treatments.

This study demonstrates the effectiveness of the technique in modifying 100% PET. We will show that the process inherently changes the properties of the electrospun fibers. The properties of nanoscale (less than 1 micron) fibrous materials made from standard polymers and modified polymers are clearly of importance to those interested in substrate biocompatibility, substrate adhesion, polymer/polymer adhesion, polymer compatibility as well as fiber stability and the general physical and chemical nature of such substrates. In our case the optimization or modification of electrospun materials /composites makes use of nitrene chemistry.

Azides on heating (or UV exposure) will form nitrenes that effectively insert into the C-H bonds of polymers and thus we can crosslink and functionalize electrospun fibers and covalently bind PET chains to each other. The synthesis and application of disulphonyl azides has recently been used to successfully crosslink and modify conventional synthetic fibers (5-8). The results from this prior work suggested that the degree of crosslinking was largely dependent upon the diffusional behavior of the azide into the fiber. However, in this study the azides were dissolved in the polymer solution prior to electrospinning and so no such fiber diffusional barriers existed. In addition, azides (on reaction) produce nitrogen and thus we believe opportunities are available to control or modify fiber porosity, polymer morphology and fiber/substrate properties.

Electrospinning

Formulation of solutions

Three crosslinking agents were synthesized (5); 1,3-benzenedisulfonyl azide (1,3-BDSA), 1,6-hexanedisulfonyl azide (1,6-HDSA), and 2,6-naphthalenedisulfonyl azide (2,6-NDSA).

The electrospinning solutions were made up as follows. PET (3.1g, Zimmer chip, supplied by ICI Synetix) was first dissolved in hexafluoroisopropanol (HFIP, 10ml, VWR). Intrinsic viscosity [IV] of the polymer was determined to be 0.63dl/g as measured in orthochlorophenol at 25°C. The azide crosslinking agents were then added to replicates of this solution so that each contained 0.0g (the "standard" containing no azide), 0.016g (15.27μmol/g (polymer); theoretical number of crosslinks per chain based on the assumption of 75% azide efficiency (X) is 0.25), 0.032g (30.54μmol/g; X is 0.5), 0.064g (61.08μmol/g; X is 1.00), 0.096g (91.62μmol/g; X is 1.5), 0.128g (122.16μmol/g; X is 2.00) or 0.256g (244.32μmol/g; X is 4.00) of 2,6-NDSA; 0.112g (122.16μmol/g; X is 2.00) of 1,6-HDSA or 0.109g (122.16μmol/g; X is 2.00) of 1,3-BDSA. The solutions were weighed, sealed and checked for solvent loss during the time taken for dissolution, and each was used only once for electrospinning.

Electrospinning of solutions

Each of the solutions was electrospun vertically at 40KV from a charged stainless steel fitting with a spherical spinneret head, having an opening of one millimeter, so that the fibers were deposited as a circular disk formed across a copper wire loop (shown in Figure 1, of 2.1cm diameter, 22.0cm from the spinneret). The level of each solution in the steel fitting was maintained so that the flow rate would be approximately the same for each sample under gravity. Many replicates were made of each sample. The replicated samples were then split into two groups. The first group was dried in air prior to thermal analysis, and these are referred to as the "as spun" samples. The second group was air-dried and then heat-treated at 220°C for 3 minutes in a Werner-Mathis stentor so that solvent removal and azide reaction would occur and are thus referred to as the "pre-reacted" samples.

Thermal analysis of electrospun samples

Thermogravimetric analysis (TGA) was performed using a Dupont TA 3000 TGA2950 instrument under N_2 with of flow rate 50ml/minute. Approximately 2.5mg of the "as spun" sample was heated from 25°C to 300°C at 5°C/min so that the solvent contents of the samples fibers could be calculated and DSC values adjusted accordingly. Replicated experiments were performed in order to examine variations within selected samples.

Differential scanning calorimetry (DSC) was performed using a Dupont TA 3000 DSC2920 instrument under N_2 of flow rate 50ml/minute. 2.5 to 2.7mg of sample was heated from 25°C to 300°C at 20°C/min (the first heating cycle) and held at 300°C for 2 min before being quenched on a liquid nitrogen cooled iron bar. The sample was then reheated (the second heating cycle) at 20°C/min up to 300°C. Replicated experiments were performed in order to examine variations within selected samples.

Analysis of Samples

Optimum Electrospinning Conditions

Many parameters were investigated to electrospin PET fibers from HFIP and a process window was found in the polymer concentration range from 30% to 33% (w./v.), the distance from the spinneret ranged from 20 to 28 cm and the voltage could be varied from 30 to 45KV. Deviation from these parameters gave samples that were either films or sputter coatings. The particular parameters chosen for this study, (31% w./v. PET, 40KV, 22.0cm), gave consistent fiber mats with fibers ranging in diameter from 500nm to 5μm as shown in Figure 1.

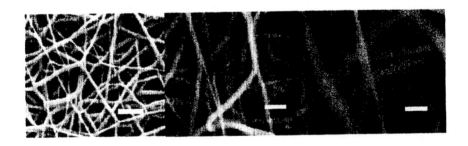

Figure 1. SEM micrographs of PET fibers electrospun from HFIP. Scale bar is 40μm, 10μm and 4μm from left to right.

Solvent Contents

All the "as spun" samples on TGA analysis provided thermograms that showed a mass loss between 50°C and 175°C corresponding to the loss of HFIP. For azide treated samples an additional mass loss, determined from derivative curves, overlapped and was observed at between 155-215°C corresponding to azide decomposition. The total mass loss was found to be approximately 13-15% for all samples and is typified by those samples shown in Figure 2. The correct sample mass for the DSC experiments was calculated and the data adjusted accordingly. In contrast to this, no mass loss was observed in the "pre-reacted" samples prior to decomposition because the heat-treatment had removed all of the solvent and the azide had already reacted.

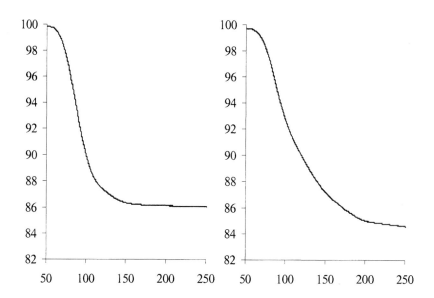

Figure 2. TGA thermograms of "as spun" samples containing (a) no azide and (b) 244.32 µmol/g 2,6-NDSA. Sample mass (%) versus Temperature (°C).

Thermal Behavior of "as spun" Standard Electrospun PET Fibers.

On the first heating cycle (Figure 3) the fibers gave a glass transition (Tg) on a sloping baseline, due to solvent loss, at 76.4°C. A crystallization exotherm was observed of 16.39 J/g with mid point 136.4°C and a melt endotherm of 31.01 J/g was observed with a mid point of 252.9°C. The difference between the enthalpies of crystallization and melting showed that the fiber was essentially amorphous as spun. On the second heating cycle (Figure 4) a Tg of 0.398 J/g/°C with mid point 81.5°C was recorded, followed by a crystallization exotherm of 29.45 J/g at a mid point of 159.7°C and a melt endotherm of 29.77 J/g at a mid point of 251.1°C.

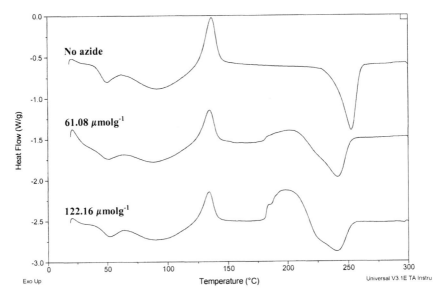

Figure 3. 1ˢᵗ run DSC traces of "as spun" PET fibers treated with differing amounts of 2,6-NDSA.

Thermal Behavior of "as spun" 2,6-NDSA Treated Electrospun PET Fibers.

Figure 3 shows that the first heating cycle was complex making numerical evaluation difficult. In addition to the thermal behavior observed in the standard sample, an exotherm was observed at a mid point of around 201°C corresponding to azide reaction. The magnitude of this exotherm varied from around 5 to 75 J/g depending on the quantity of azide used. The other features of the DSC thermograms were similar to that observed in the standard sample except for the melt endotherm which decreased in both temperature, from 252.9°C in the standard to 225.0°C in the sample with the highest level of treatment, and magnitude on increasing the concentration of the azide crosslinker. While the decrease in temperature was associated with crosslinking, the decrease in magnitude was thought to be due to both the increasing levels of crosslinking and the slight overlap between the melt endotherm and azide reaction exotherm making evaluation difficult.

Once the azide reaction had taken place and the sample was quenched, the second DSC heating cycle revealed definite trends in the fibers thermal properties depending on the level of crosslinking as shown in Figure 4. The first observable trend was in the Tg, shown in Figure 5, which increased in temperature from a mid point of 81.5°C in the standard sample to 86.2°C in the sample treated with the most azide. This was accompanied by a decrease in the magnitude of the transition from 0.398 J/g/°C to 0.223 J/g/°C respectively. The increase in Tg must be due to the progressively restricted molecular motion of the polymer chains while the decrease in energy is due to the increasing numbers of polymer chains unable to experience a Tg at all due to the increasing level of crosslinking.

On heating, the samples gave a crystallization exotherm that in general increases in crystallization temperature and decreases in enthalpy with increasing levels of crosslinking agent (Figures 4 & 6). Examination shows that while the trends have a general form, there is some variation in the crystallization mid point temperature.

In previous work on crosslinking Dacron PET fibers (5-8) this variation was analyzed as two competing effects. The first effect (A), noticed at lower levels of treatment, was thought to be due to the disruption of the crystallization process, so that chain mobility is more restricted. The second effect (B), is thought to be due to the encapsulation of crystallites during the crosslinking process.

During crosslinking the crystallites within the fibers are, we believe, surrounded by amorphous material that contains the latent crosslinking agent. If the amount of agent is not sufficient to form a crosslinked network then effect A predominates on melting, quenching and crystallization. Should the amount of crosslinking agent be sufficient that a crosslinked network arises (effect B), the crystallites become entrapped by the previously amorphous material, so that when these crystallites melt, they can be quenched and on heating will recrystallize with little interference from what was the previously amorphous material. However, at very high levels of crosslinking agent the material at the amorphous / crystalline (chain fold) interface becomes so restricted in motion that the crystallization temperature increases yet again whilst the enthalpy continues to decrease. Furthermore the competition between the two competitive processes is very specific to the heating method used in crosslinking and to the morphology and thermal histories of the PET fibers.

Additional evidence can be given in support of this explanation. The theoretical degree of crosslinking (X) was mentioned earlier during the formulation of electrospinning solutions. The first trend can be observed in those samples at lower levels of crosslinking where the number of crosslinks per chain is 0 to 0.5; insufficient to create a crosslinked network. The second trend (encapsulation) is observed at around 1 to 2 crosslinks per chain; sufficient to create a network in the amorphous material. The further reversal is seen at 4 crosslinks per chain.

The samples, on heating through the melt, gave a melt endotherm that

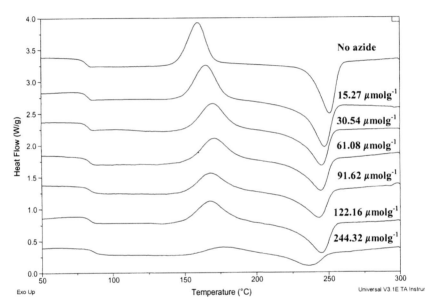

Figure 4. 2nd run DSC traces of "as spun" PET fibers treated with differing amounts of 2,6-NDSA.

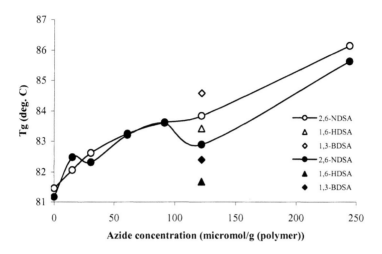

Figure 5. Dependence of Tg of "as spun" (hollow) and "pre-reacted" (solid) fiber samples on azide content.

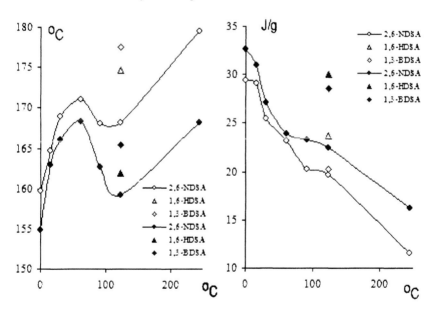

Figure 6. Dependence of crystallization temperature (left, °C) and enthalpy (right, J/g) on azide content of the "as spun" (hollow) and "pre-reacted" (solid) samples (micromol/g (polymer)).

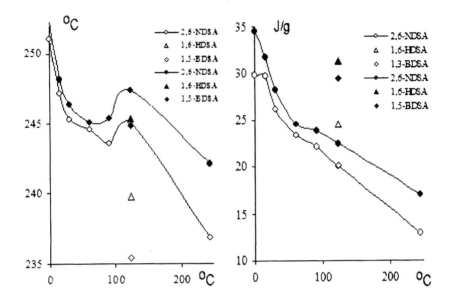

Figure 7. Dependence of melt temperature (left, °C) and enthalpy (right, J/g) on azide content of the "as spun" (hollow) and "pre-reacted" (solid) samples (micromol/g (polymer)).

generally decreased in temperature and enthalpy on increasing the degree of crosslinking (Figure 7) and thus reflects the crystallization properties.

Thermal behavior of the "pre-reacted" standard electrospun PET fibers.

On the first heating cycle the fibers gave a evidence of a small Tg step (i.e, change in heat capacity) at 76.4°C as shown in Figure 8. A small crystallization exotherm was observed of 1.26 J/g with mid point 125.5°C and a melt endotherm of 42.95 J/g was observed with a mid point of 251.4°C. The difference between the enthalpies of crystallization and melting showed that the fiber essentially crystallized during the heat treatment applied to induce the azide crosslinking reaction. On the second heating cycle a Tg of 0.390 J/g/°C with mid point 82.2°C was recorded, followed by a crystallization exotherm of 32.66 J/g at a mid point of 154.9°C and a melt endotherm of 34.58 J/g at a mid point of 252.4°C which can be observed in Figures 5 and 9.

Thermal behavior of "pre-reacted" 2,6-NDSA treated electrospun PET fibers.

The first heating cycle was simple in comparison to the "as spun" samples, so that the Tg, small crystallization exotherm and melt endotherm could be readily observed and compared as shown in Figure 8. The Tg of the samples increased slightly on increasing the extent of crosslinking, however an additional trend was observed in the crystallization data so that both the temperature and enthalpy of crystallization increased in comparison to the standard sample. This indicated that the crystallization process was restricted due to the crosslinking reaction during the heat treatment as would be expected. At higher temperatures, the samples experienced a melt endotherm that decreased in both enthalpy and temperature on increasing the degree of crosslinking.

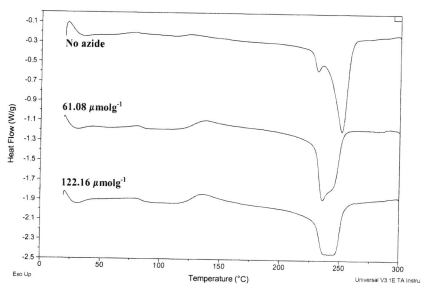

Figure 8. 1ˢᵗ run DSC traces of "pre-reacted" PET fibers treated with differing amounts of 2,6-NDSA.

Once the sample was quenched, the second DSC heating cycle revealed definite trends in the fibers thermal properties depending on the extent of crosslinking. The first observable trend was in the Tg, shown in Figure 5, which increased in temperature from a mid point of 82.2°C in the standard sample to 85.2°C in the sample treated with the most azide. This was accompanied by a decrease in the magnitude of the transition from 0.390 J/g/°C to around 0.270 J/g/°C respectively. This increase in Tg must be due to the incresingly restricted

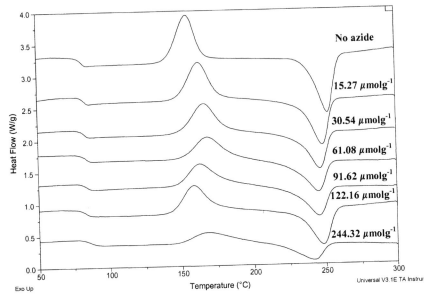

Figure 9. 2nd run DSC traces of "pre-reacted" PET fibers treated with differing amounts of 2,6-NDSA.

molecular motion of the polymer chains while the energy decrease is caused by the increasing numbers of polymer chains unable to experience a Tg at all due to the increasing level of crosslinking.

On heating, the samples gave a crystallization exotherm that in general increases in enthalpy and decreases in crystallization temperature with increasing levels of crosslinking agent (Figures 6, 7 and 9). Examination shows that while the trends have a general form, there is some variation in the crystallization mid point temperature. This variation is thought to be due to the same two competitive effects observed earlier in the "as spun" samples. The samples, on heating through the melt, gave a melt endotherm that generally decreased in temperature and enthalpy on increasing the degree of crosslinking (Figure 7).

Comparison of the two sets of samples reveals significant differences in their measured thermal properties. Though the trends are similar in form, their magnitude is different and this is due to the differing methods of heat treatment used to bring about the crosslinking processes. The trends in property change versus crosslinking level are most clearly demonstrated by the "pre-reacted" samples. This is thought to be because no melting of these samples had occurred during crosslinking, since the samples were held at 220°C for three minutes, whereas in the case of the "as spun" samples melting and crosslinking occurred simultaneously. This, in the latter case, would result in there being less crystalline material available for entrapment in comparison. The absolute differences, i.e. those unrelated to crosslinking, are thought to be due to the effects of differing thermal histories.

Comparison of the Effects Based on the Azide Used.

On comparing the effects of each azide, as shown in Figures 5, 6 and 7, we tentatively suggest that they had different impacts on the thermal properties of the crosslinked PET samples. It appears as though the effect of each azide on the reacted polymer properties was dependent on the azides molecular structure and insertion efficiency. Firstly, in the first heating cycle, the exotherm for the azide reaction for the 2,6-NDSA treated samples was notably larger than that found for the 1,6-HDSA or 1,3-BDSA treated samples. The temperature at which the reaction progressed most quickly was highest for 1,6-HDSA treated sample, at around 210°C, followed by the naphthalene derivative at 201°C and then the 1,3-BDSA treated sample, at 199°C. At equivalent levels of treatment, the Tg was least affected by the 1,6-HDSA and this might be a feature of its hydrocarbon flexibility so that molecular motion is less restricted than when the other derivatives are used.

Comparison of the crystallization properties showed that the magnitude of difference in the crystallization enthalpy was least for the sample treated with 1,6-HDSA and the greatest for both the aromatic azide treated samples. This again is thought to be due to the flexible structure of 1,6-HDSA causing the least restraint to crystallization. It should be noted though that the temperature of crystallization for this sample was higher than for the 2,6-NDSA treated sample. The melt properties of the samples treated with different azides reflected their crystallization properties. One particular feature of note is that on comparison of the graphs the order of the azides is maintained for each of the sets of samples. However, the systems involved are complex and any further discussion or correlation between the effects of the azides on the properties of the fibers can only be made with extensive additional study. The structures of the azides used are shown below in Figure 10 for comparison.

Summary

PET fibers have been electrospun from HFIP solution containing azide crosslinking agents. Subsequent heating of the electrospun substrates by two differing methods gave rise to reaction and crosslinking. The fibers were essentially amorphous as spun and were characterized by thermal analysis to determine the effects crosslinking had on their properties. Future work seeks to optimize the process conditions for the production of crosslinked nanofibers (during electrospinning and by post-spin treatments) and to further evaluate the effects of different azide structures in order to engineer permanent desirable

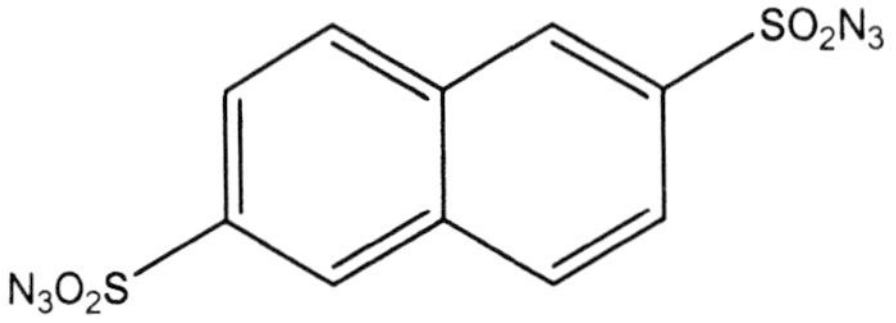

2,6-naphthalenedisulfonyl azide 2,6-NDSA.

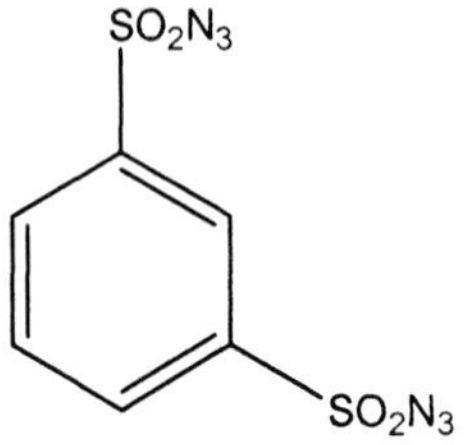

1,3-benzenedisulfonyl azide 1,3-BDSA

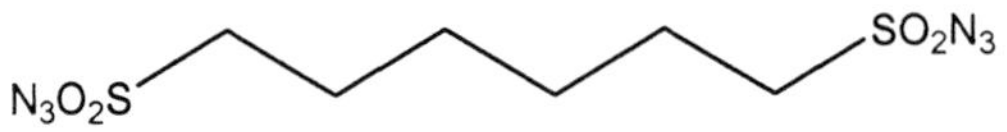

hexanedisulfonyl azide 1,6-HDSA

Figure 10. Azides synthesized and used in this study.

substrate characteristics. Electrospun materials are renowned for their high porosity and high surface area to volume ratio and our future work intends to examine how to utilize nitrogen release (by azide reaction) to enhance these micro morphological effects and to incorporate desirable functionality into these structures using the azide technology.

References

1. Doshi, J., Reneker, D.H. Electrospinning Process and Applications of Electrospun Fibers, *Journal of Electrostatics*, **1995**, 35, 151-160.
2. Fong, H., Chun, I., Reneker, D.H. Beaded nanofibers formed during electrospinning, *Polymer*, **1999**, 40, 4585-4592.
3. Shin, Y.M., Hohman, M.M., Brenner, M.P., & Rutledge, G.C. A whipping fluid jet generates submicron polymer fibers, *Applied Physics Letters*, **2001**, 78, 8, 1149-1151.
4. Hohman, M.M., Shin, M., Ruteledge, G., & Brenner, M.P., 2001, Electrospinning and electrically forced jets. 1. Stability theory, *Physics of Fluids*, **2001**, 13, 2201-2220.
5. Baker, D.A., East, G.C., Mukhopadhyay, S.K., Synthesis and Characterization of Some Disulfonyl Azides as Potential Crosslinking Agents for Textile Fibres., *Journal of Applied Polymer Science*, **2001**, 79, 6 1092-.1100.
6. Baker, D.A., East, G.C., Mukhopadhyay, S.K., The mechanical and thermal properties of polyester fibers crosslinked with disulfonyl azides. *J. Appl. Polym. Sci.*, **2002**, 83, 1517-1527.
7. Baker, D.A., East, G.C., Mukhopadhyay, S.K., The mechanical and thermal properties of acrylic fibers crosslinked with disulfonyl azides. *J. Appl. Polym. Sci.*, **2002**, 84, 1309-1319.
8. Baker, D.A., East, G.C., Mukhopadhyay, S.K., The thermomechanical and creep behaviour of polyester fibers crosslinked with disulfonyl azides. *J. Appl. Polym. Sci.*, **2003**, 88, 1556-1562.

Chapter 14

Electrospinning of Bioresorbable Polymers for Tissue Engineering Scaffolds

E. D. Boland[1], K. J. Pawlowski[1], C. P. Barnes[1], D. G. Simpson[2],
G. E. Wnek[3], and G. L. Bowlin[1]

Departments of [1]Biomedical Engineering and [2]Anatomy, Virginia
Commonwealth University, Richmond, VA 23298
[3]Department of Chemical Engineering, Case Western Reserve University,
Cleveland, OH 44106

Since the inception of the field of tissue engineering, there has been a considerable effort to develop an "Ideal" tissue engineering scaffold. To date, investigators have developed materials such as collagen, poly(glycolic acid) (PGA), poly(lactic acid) (PLA), and polycaprolactone (PCL) for use in matrix construction with frequently unacceptable results. The construction of an "Ideal" tissue engineering scaffold requires that multiple criteria are met. The first criterion is that the material be biocompatible and function without interrupting other physiological processes. This functionality includes an ability to promote normal cell growth and differentiation while maintaining three-dimensional orientation/space for the cells. Additionally, the scaffold should not induce any adverse tissue reaction. Another criterion involves the production of the scaffold. Scaffold production must include efficient material and construction parameters. The construction must also involve a process by which the scaffold can be easily reproduced to a wide range of shapes and sizes. Once implemented *in vitro* or *in vivo*, the material should be completely resorbable, leaving only native tissue. The method of electrospinning provides a simple method to achieve the goals of an "Ideal" tissue engineering scaffold. Electrospinning is the deliberate application of the phenomenon of electrostatic spraying which occurs when electrical forces at the surface of the polymer solution overcome the surface tension, creating a polymer solution jet. The jet produces a fiber with diameters in the micron to nano-scale range (down to 50 nm) as the solvent evaporates. The biodegradable polymer scaffold production (flexible, quick, and simple) utilizes the jet formation of the polymer-based solution from a nozzle to a grounded rotating mandrel. We have successfully electrospun PGA and PLA as pure polymers, co-polymers, and blends with each other as well as with PCL in order to develop novel tissue engineering scaffolds. The electrospinning process allows the production of PGA scaffolds comprised of 100 nm to 1.5 micron diameter fibers. For the PLA and PLA/PCL blends, the fiber diameter produced ranges from 8-10 microns. Obviously, with various composition and fiber dimensions, we can tailor both the mechanical properties and the degradation rates of the electrospun scaffoldings. The electrospinning process also allows us to control fiber orientation in the final fibrous scaffolds, ranging from completely random to predominately aligned. Thus, with these very promising results, it may now be possible to utilize this technology to create the "Ideal" tissue engineering scaffold (mimic tissue structure on demand).

Introduction

Tissue Engineering as an interdisciplinary field of engineering and life sciences is in its infancy, yet the principles are as old as interventional surgery. The desire to restore, maintain, or improve tissue function remains the end goal, but a new set of tools and techniques are emerging. To this end, the study of structure–function relationships in both normal and pathological tissues has been coupled with the development of biologically active substitutes or engineered materials. One such approach is to develop either cell-containing or cell-free implants that are capable of mimicking the native tissue. In order to duplicate all the essential intercellular reactions and promote native intracellular responses, the tissue structural component that must be mimicked is the extracellular matrix (ECM). These synthetic ECMs or scaffolds should be designed to conform to a specific set of requirements. The first requirement is that the material must be biocompatible and function without interrupting other physiological processes. This functionality includes an ability to promote normal cell growth and differentiation while maintaining three-dimensional orientation/space for the cells. Secondly, the scaffold should not promote or initiate any adverse tissue reaction (1). In addition, for clinical and commercial success, scaffold production should be simple yet versatile enough to produce a wide array of configurations to accommodate the size, shape, strength and other intricacies of the target tissue (1-5). Once implemented in vitro or in vivo, the material should either be removed though degradation and absorption or incorporated through native remodeling mechanisms, leaving only native tissue. Beyond these generalized requirements, we must look at the way a single native cell interacts with its immediate surroundings. It is no longer acceptable to view a cell as a self-contained unit sitting in a passive structural network (ECM). A dynamic three-dimensional inter-relation is constantly kept in balance and influenced by both internal and external stimuli. This model requires that any scaffold material must be able to interact with cells in three dimensions and facilitate this communication. In the native tissues, the structural ECM proteins are 1 to 2 orders of magnitude smaller than the cell itself, thus, allowing the cell to be in direct contact with many ECM fibers at once, and thereby defining its three dimensional orientation. This property may be a crucial factor in determining the success or failure of a tissue engineering scaffold.

We believe that electrospinning represents an advantageous processing method to meet both the general material requirements as well as the potential size issues raised above. In this overview, we will first discuss a brief history of the technique and then follow that with a simple electrospinning process that can and has been performed in the laboratory. Once the process is established, we will focus on the work done in our laboratories to merge this processing technique with the body of literature available on bioresorbable polymers to create a wide variety of scaffolds for tissue engineering.

Brief History of Electrospinning: Pioneering Work and Mechanism

Electrospinning can be documented as far back as Lord Rayleigh in the 19th Century. He theorized and subsequently determined through experiments that a sufficiently large electrical charge can cause a droplet at the tip of a nozzle to overcome surface tension and eject in a stream (*6,7*). Specifically, this work focused on a process known as electrospraying. This phenomenon and that of electrospinning are fundamentally identical. For very low viscosity liquids, the jet that is ejected dissociates into droplets due to overwhelming surface tension, resulting in a sprayed coating on the collection plate (*8*). In higher viscosity liquids, the jet does not break up, and the result is a continuous stream or, if dried, a fiber. Zeleny further documented the phenomenon of electrospraying in 1917 (*6,7*).

Many years later, Taylor studied and explained the events that occur at the nozzle as the electric field deforms the fluid and it forms a stream (*6,8*). When the voltage is initially applied to the solution, the droplet at the nozzle forms a hemispherical surface. As the electric field is increased, the surface undergoes a shape change from hemispherical to spherical and eventually to conical. These changes are due to the balancing of the increasing solution charge with its surface tension, and the final conical shape is known as the Taylor cone (*9*). Countless experiments since have shown that when the liquid has a finite conductivity (even very small values), the stable configuration known as the Taylor cone is actually a cone with a thin column or jet of liquid emerging from the tip. Such "cone-jet" configurations are commonly referred to as "Taylor cones," even though the conditions under which they are formed vary greatly from the assumptions underlying Taylor's analysis.

The first patent (U.S. Patent 1,975,504) in electrospinning was granted to Formhals in 1934 for a process that produced fine fibers from a cellulose acetate solution (*6,10*). He was later granted related patents (U.S. Patents 2,116,942; 2,160,962; and 2,187,306) in 1938, 1939, and 1940.

In the last decade, there has been increasing interest in the electrospinning process. A concise review of the process and history of electrospinning can be found in an article by Reneker and Chun (*6*). Some examples of particularly interesting work follow.

Srinivasan and Reneker dissolved Kevlar[®] (poly(p-phenylene terephthalamide)) in sulfuric acid and were able to obtain fibers with diameters of 1.3 μm that exhibited properties similar to that of commercial Kevlar[®] fibers spun from the liquid crystalline state (*11*). For these experiments, the fibers were collected in a grounded water bath, which allowed for removal of residual acid. Carbon nanofibers with diameters in the range of 100-500 nm have also been electrospun from polyacrylonitrile solution and molten mesophase pitch (*12, 13*).

Doshi and Reneker demonstrated the electrospinning of poly(ethylene oxide) (PEO) in water to obtain fibers with diameters in the range of 0.5-5.0 μm (*9*). Other groups have also succeeded in electrospinning fibers from PEO and biodegradable materials such as poly(lactic acid) (PLA) (*8, 14*). Bognitzki formed PLA fibers with an average diameter of one micron from a solution containing tetraethyl benzylammonium chloride (TEBAC). TEBAC, which is an organosoluble salt, facilitated a decrease in fiber diameter by increasing solution surface tension and electrical conductivity (*7*). PCL, PLA, PGA, and poly(D, L-lactide-co-glycolide) (PLGA) have been successfully electrospun into matrices with submicron fiber diameters and known degradation times for tissue engineering applications (15-18).

Until recently, few researchers have concentrated on critical analysis of the electrospinning process variables. The effects of polymer solution concentration and spinning voltage (electric field) on fiber morphology have been detailed by Deitzel et al. (8) and Sukigara et al. (19). Gibson et al. have studied transport properties of the fibers and fiber mats (15,16,20), and mathematical models describing the travel of the fiber jet have been recently detailed by Reneker et al. (17) and Fridrikh et al. (21).

From these bodies of work, we know that diameter, morphology, and orientation of the resultant fibers are a function of the many variables involved. The properties of the polymer solution, including its molecular weight, chemistry, and nature can greatly alter the ability of a fiber to be spun. The polymer to solvent ratio in solution affects viscosity, surface tension, and conductivity, all of which can influence fiber diameter. Voltage appears to play a role in determining fiber uniformity through jet stability. As voltage increases, the number of "beads on a string" defects increases. This is because the rate at which the solution is removed from the tip of the nozzle exceeds the rate at which the solution is delivered to the tip. Observationally, this can be seen as a jet initiating from the inside surface of the nozzle rather than from the Taylor cone (8, 14). These are not the only variables but are thought to be the most changeable (7, 9).

Additionally, if fibers are still wet when they contact the collecting plate, they will fuse at junctions through a process commonly referred to as solvent welding. The higher the polymer concentrations with respect to solvent volumes, the drier the fiber produced (8,14). Solution infusion rates into the system and distances between the initiation tip and collecting plate also affect fiber dryness (7,8,14). The principal variable in fiber dryness is solvent volatility, the computations of which must account for the temperature, environmental conditions, and distance to the target if trying to maximize fiber dryness and minimize the formation of fused fibers (9).

A Simple Laboratory Apparatus for Electrospinning

A simple apparatus (Figure 1) includes: a narrow conduit through which polymer solution is introduced to the system by a suitable pump or pressurized reservoir, a power supply capable of DC voltages in the 5 – 30 kilovolt range, and a conductive target. Because it is convenient to use stainless steel tubing from which hypodermic needles are produced, the injection conduit is frequently referred to as a "needle." Opposite the needle at some distance away is a grounded target that can take on virtually any configuration as long as it is conductive. For this illustration, consider a simple, static metal plate. Therefore, any potential applied to the needle (or to the entering liquid by means of a contact electrode) produces an electric field at the needle tip. The intensity of this field depends upon the potential difference between the needle and the target, the diameter of the needle, and the distance to the target. As previously described, the field induces flow of the emerging liquid toward the target in the form of a Taylor cone. When the applied voltage is sufficient to overcome surface tension, a tiny stream of fluid is ejected from the tip of the needle. If sufficiently high polymer concentrations are present, continuous fiber formation will occur rather than the droplets predicted by Rayleigh instability. The charges on the fibers eventually dissipate to the surrounding environment, and the final product of the process is a non-woven mat that is composed of fibers with diameters on the order of nanometers to microns (*11*). Polymer melts can also be electrospun (*12*). In actuality, the technique is even more versatile, as we have found that electrospun fibers can also be collected on a dielectric material interposed between the grounded target and the spinning solution. Thus, a wide variety of substrates other than metals can be coated.

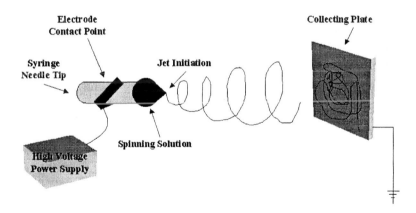

Figure 1. A schematic of a simple electrospinning system capable of producing micro- to nano-scale fibers which collect to become a biodegradable polymer tissue engineering scaffold.

Electrospinning of Bioresorbable Polymers

Rationale for Electrospinning Bioresorbable Polymers

The bioresorbable polymers under investigation in our laboratory are aliphatic polyesters of the poly(α-hydroxyl acids). These are some of the most commonly used biodegradable polymers and have the most clinical experience and initial acceptance. Their generalized formula is $-[-O-CH(R)-CO-]-$ where R equals H for glycolic acid, CH_3 side chain for lactic acid (both L and D isomers are possible), and CH_2CH_2 along the backbone for caprolactone. Degradation rates vary from as short as two weeks for some PGA constructs to as long as a few years for PCL. This degradation is believed to be a function of hydrolytic attack initially in amorphous regions and then more slowly in crystalline regions. This degradation is characterized by a sharp decrease in local pH and a decrease in crystallinity (22,23). As will be shown later in this text, careful combinations of polymers combined with the fiber diameters possible with electrospinning can create the flexibility needed to address the vast array of tissue engineering scaffold requirements.

Poly(glycolic acid)

Poly(glycolic acid) (PGA) possesses moderate crystallinity (30% - 50%), a high melting point (225°C), and low solubility in organic solvents. Through clinical experience with suture, PGA was found to have better than average tissue compatibility and reproducible mechanical properties such as strength, elongation, and knot retention (23).

PGA can be processed into fibers through the traditional extrusion methods. Due to the mechanics of the systems, fibers are restricted to diameters above 10 μm. Since electrospinning does not involve mechanical drawing, we have demonstrated the ability to produce scaffolds with individual fiber diameters from approximately 1.5 μm to less than 0.2 μm (24). This is accomplished by dissolving PGA in 1,1,1,3,3,3-hexafluoro-2-propanol (HFP) at various weight to volume ratios, the effect of which can be seen in the ability to control fiber diameter as illustrated in Figure 2.

Diametric control, while controlling orientation, is the key to tailoring the mechanical properties of a PGA scaffold, including tangential modulus of the construct, stress and strain to failure, and strength retention; smaller fibers have larger surface area to volume ratios and lose their strength faster during degradation. Testing has shown that PGA is a good material when an initially tough (high strength and elasticity) material is needed and when it is desirable to have the construct degrade as fast as possible. However, a possible negative attribute is also related to the degradation rate. Sharp increases in localized pH may cause unwanted tissue responses if the region does not have a high buffering capacity or sufficient mechanisms for the rapid removal of metabolic waste (*23*).

In terms of the *in vitro* and *in vivo* biocompatilbilty testing of electrospun PGA, the authors have recently published studies which evaluated PGA and acid treated PGA in cell culture and in an animal model (*25*). In general the cellular and *in vivo* responses were a function of fiber diameter and acid pretreatment. The acid pretreatment improved biocompatibility via a hypothesized mechanism of fiber polymer surface hydrolysis of ester bonds, exposing carboxylic acid and alcohol groups (*26*), which may improve vitronectin binding thereby improving the ability of cells to adhere to the surface.

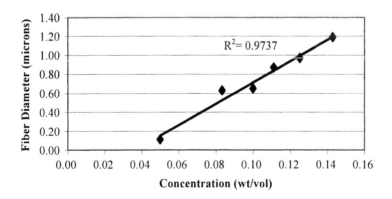

Figure 2. (Top) Electrospun PGA (0.143 g/ml in HFP) showing the random fiber orientation (1,600X magnification). (Bottom) This graph illustrates the ability to linearly control fiber diameter as a function of initial concentration of PGA in HFP. Fibers ranging from less than 0.2 microns to over 1.2 microns were produced (24).

196

As has been shown with PGA, electrospinning is an attractive approach for the production of sub-micron diameter fibers, which are of interest as tissue engineering scaffolds. The fiber diameter, as well as the fiber orientation, is important in controlling material properties of the scaffolds produced. In a published study (24), the authors have evaluated the mechanical properties of the different PGA electrospinning concentrations in both aligned and random fiber orientation scaffolds. Figure 3 summarizes the results of this study in terms of the elastic modulus and strain to failure of PGA in a uniaxial model. The overall results exhibit a correlation between the fiber diameter and orientation and the elastic modulus and strain to failure.

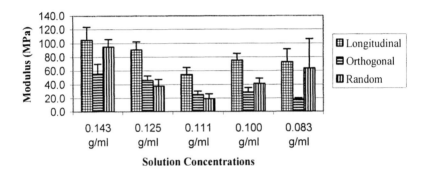

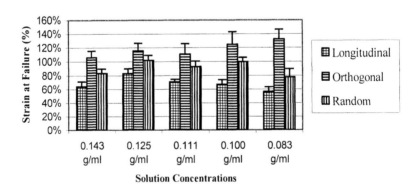

Figure 3. (Top) Modulus of elasticity and (Bottom) strain at failure measurements presented as a function of PGA concentration tested along the longitudinal and orthogonal axes (aligned fibers) and in a random mat (24).

Poly(lactic acid)

Poly(lactic acid) (PLA) is a very popular tissue engineering scaffold material and is available in either the L or D isomer. The L isomer is more commonly used, since it is derived from the only naturally occurring stereoisomer of lactic acid. The presence of the methyl group makes PLA more soluble in organic solvents and more hydrophobic than PGA. Despite a similar or higher degree of crystallinity compared to PGA, PLA commonly degrades in 30 to 50 weeks due to steric hindrance of the methyl group (23). Our laboratory testing has shown fiber formation by electrospinning PLA (27) in a range of 1/7 to 1/10 weight/volume ratios in both chloroform and methylene chloride. The decrease in concentration produces a similar decrease in fiber diameter, from about 10.5 μm to near 1 μm. If electrospun in HFP at similar concentrations, then one can create fibers down to 100 nm in diameter with PLA.

Mechanically, crystalline PLA exhibits relatively high strength and modulus and low strain to failure. Given that the mechanical properties of PLA present strong and non-compliant scaffolds, it becomes obvious why PLA has been extensively used in the tissue engineering of bone and cartilage.

Polydioxanone

Polydioxanone (PDS) is a highly crystalline degradable polyester originally developed for wound closure. The generalized formula of PDS is $-[O-CH_2-CH_2-O-CH_2-C=O]-$ and it has a degradation rate falling between that of PGA and PLA. PDS exhibits excellent flexibility due to the incorporation of an ester oxygen in the monomer backbone. PDS has been electrospun in HFP at concentrations between 42 and 167 mg/ml, producing fibers that ranged linearly from 0.18 μm to 1.4 μm, respectively, as seen in Figure 4. As a suture material, PDS is often criticized for shape memory. These sutures recoil to their original spooled shape which leads to kinking as fine sutures are tightened. While this is a negative trait for a suture material, it may provide rebound and kink resistance if formed into a vascular conduit. This unique trait coupled with excellent strength and elasticity may prove valuable for soft tissue engineering (28). The authors have evaluated the mechanical properties of the different PDS electrospinning concentrations in both aligned and random fiber orientation scaffolds (28). Figure 5 summarizes the results of this study in terms of the elastic modulus and peak stress of PDS in a uniaxial model. The overall results demonstrate a correlation between the fiber diameter and orientation and the elastic modulus and strain to failure.

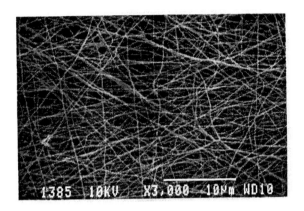

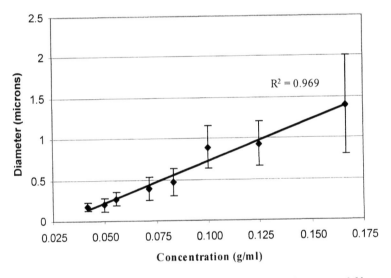

$R^2 = 0.969$

Figure 4. (Top) Micrograph of 0.18 μm diameter randomly oriented fibrous structures produced from 42 mg/ml PDS in HFP (3,000x magnification). (Bottom) Results of the fiber dimension analysis versus PDS concentration illustrating the linear relationship between electrospinning solution concentration and fiber diameter (28).

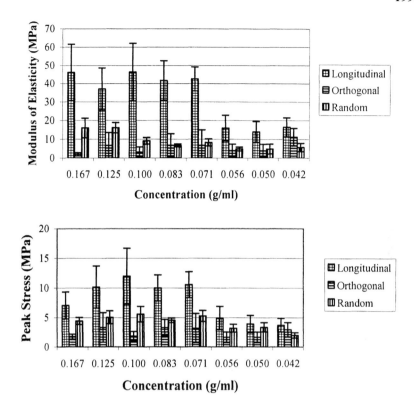

Figure 5. (Top) Elastic modulus and (Bottom) peak stress measurements presented as a function of PDS concentration tested along the longitudinal and orthogonal axes (aligned fibers) and in a random mat (28).

Polycaprolactone

Polycaprolactone (PCL) is an inexpensive, slow-degrading (1-2 years), highly elastic polyester that demonstrates a lack of toxicity (*23,29*). The ability to electrospin pure PCL was published by Reneker (*30*) as a solution in acetone at 14 to 18 weight percent. PCL fibers were shown to contact and merge within the electrospinning jet to form garland-like fibrous structures (*30*). Yashimoto et al. electrospun PCL at 10 weight percent from chloroform for a bone tissue engineering application (*29*), and Shin electrospun PCL from a 1:1 mixture of chloroform and methanol to form a tissue engineered cardiac graft (*31*). In both these tissue engineering applications, PCL was chosen for its good mechanical properties (modulus and elasticity) and slow degradation. After pre-coating with soluble collagen, the material exhibited acceptable cellular interaction.

Poly(glycolic acid) and Poly(lactic acid) Blends and Copolymers

Often, a single polymer does not have the exact properties desired of an engineering material. Tissue engineers must engage in organic chemistry to design and synthesize copolymers or simply mix the polymers in blends. Once again, electrospinning gives us this ability. Experimentally, our laboratory has tested 75%PLA - 25%PGA and 50%PLA - 50%PGA copolymers commonly referred to as PLGA. In addition to these true copolymers, we have blended PLA and PGA in HFP at these and other ratios. The properties of the copolymers and blends are different even at the same percentages. Observationally, both blended and co-polymerized fibers have similar diameters with uniform distributions. Further experimentation using techniques such as nuclear magnetic resonance (NMR) spectroscopy could elucidate the differences in fiber makeup (20). Li et al. have also demonstrated PLGA co-polymer electrospinning for tissue engineering applications (32).

While mechanical properties such as tangential modulus, peak stress, and strain do appear to be controlled by the composition, fiber diameters are comparable to PGA alone at the varying polymer to solvent ratios. This could be a function of the reduced solubility of PGA dictating surface tension and affecting the formation of the Taylor cone.

Poly(glycolic acid) and Polycaprolactone Blends

Addition of the highly elastic PCL to PGA leads to an increase in the extensibility of the electrospun blend matrix, as expected. Electrospinning of the blends appears to be limited to 1/3 PCL by weight. Experimentally, we have seen strain to failure in excess of 400% for these scaffolds. Like the PGA – PLA blends, we do not see a wide variance in fiber diameters and must assume that the individual polymers become entangled in a regular pattern dictated by their chemistry and concentration in HFP.

This blend seems well-suited to tissue engineering applications where high elasticity is the driving factor. In addition, due to the long degradation time of PCL (typically 1-2 years), it would be reasonable to expect that the blend would degrade slower than PGA alone. This has not yet been fully tested by our laboratory, but results so far support this prediction.

Poly(lactic acid) and Polycaprolactone Blends and Copolymers

The percent increase in elasticity for PLA – PCL blends and copolymers are even more pronounced than with PGA – PCL blends. The PLA – PCL blends and copolymers also afford the same solvent flexibility as pure PLA. Our preliminary experimentation was conducted in both chloroform and methylene chloride. With as little as 5% PCL by weight, the strain to failure of a scaffold increased from less than 25% to more than 200%. Typical strain to failure values for PLA – PCL blends are illustrated in Figure 6.

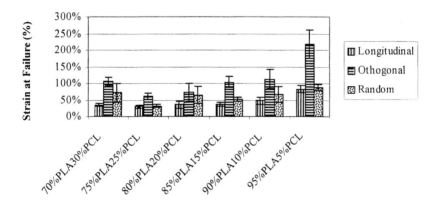

Figure 6. This figure illustrates that the strain to failure of a tissue matrix is a function of both composition (in this example, varying blend ratios of PLA and PCL) and fiber alignment. As predicted, stretch in the direction of the fibers is lowest, orthogonal to the fibers is greatest, and for a randomly oriented fiber matrix lies somewhere within those bounds.

With a strength comparable to PLA alone plus added elasticity provided by addition of PCL, these combinations show a great deal of promise for use as an arterial graft, which must be strong enough to accommodate a large pressure increase, elastic enough to passively expand to accept a large bolus of blood, and, finally, able to contract to push the blood downstream. Another important trait of arterial grafts is the fiber alignment, which adds radial and longitudinal strength to the construct. Once again, electrospinning can address the issue. It is routine to make seamless tubes from all of the previously mentioned materials in a controlled manner. The scanning electron micrograph in Figure 7 typifies the capacity to align fibers during electrospinning.

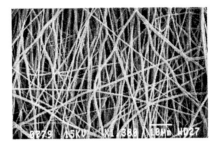

Figure 7. Scanning electron micrograph illustrating aligned fibers composed of a 85% PLA/15% PCL blend from a 0.143 g/ml solution in HFP.

Conclusion

In conclusion, it may be possible to construct true "biomimicking" fibrous scaffoldings for tissue engineering utilizing the process of electrospinning. With this process, scaffolds of various shapes and sizes can be constructed while precisely controlling the fiber orientation, composition (blended fibers), and dimensions. This flexibility may be the key needed to truly mimic the native extracellular architecture (multiple layers and orientation of natural fibers) during tissue engineering scaffold development. Another advantage is that any three-dimensional scaffold, for which a mold is fabricated, can be made seamless. By eliminating the seams, variations in scaffold properties can be minimized. Even with individual fiber sizes under 200 nm, all of the scaffolds produced exhibit substantial structural strength for ease of handling during scaffold sterilization, cell seeding, and transferring to the subsequent culture environment (static culture to complex bioreactors). In addition, the ability to co-electrospin polymers with various additives (e.g., growth factors, DNA) is being investigated. Their slow release (6,33) offers the possibility of enhancing positive cellular response while reducing negative responses. Thus, with the ease of processing and diversity afforded by the electrospinning of the polymers described (and other polymers including collagens and fibrinogen (34-38)), it may now be possible to create "Ideal" tissue engineering scaffolds that can be tailored to meet the specific needs (for growth and maintenance) of individual tissue.

Acknowledgments

The authors would like to thank the Whitaker Foundation (RG-98-0465) for the support of this research, as well as Alkermes, Inc., for the donation of PGA, PLA, and PCL polymers and copolymers. We would also like to thank Ms. Judy Williamson for assistance in obtaining the SEM micrographs.

References

1. Greisler, H. P.; Gosselin, C.; Ren, D.; Kang, S. S.; Kim, D. U. *Biomaterials* **1996**, *17*, 229-236.
2. How, T. V.; Guidoin, R.; Young, S. K. *Proc. Inst. Mech. Eng.* **1992**, *206*, 61-71.
3. Mooney, D. J.; Langer, R. S. In *The Biomedical Engineering Handbook*; Bronzino, J. D., Ed.; CRC Press: Boca Raton, FL, 1995; pp 1609-1618.
4. Starke, G. R.; Douglas, A. S.; Conway, D. J. In *Tissue Engineering of Prosthetic Vascular Grafts*; Zilla, P.; Greisler, H. P., Eds.; R.G. Landes Company: Austin, TX, 1999; pp 441-459.
5. Hsu, S.; Kambic, H. *Artif. Organs* **1997**, *21*, 1247-1254.
6. Reneker, D. H.; Chun, I. *Nanotechnology* **1996**, *7*, 216-223.
7. Bognitzki, M.; Frese, T.; Wendorff, J. H.; Greiner, A. *Proc. of the American Chemical Society, Division of Polymeric Materials: Science and Engineering* **2000**, *82*, 115-116.
8. Deitzel, J. M.; Kleinmeyer, J.; Harris, D.; Beck Tan, N. C. *Polymer*, **2001**, *42*, 261-272.
9. Doshi, J.; Reneker, D. H. *J. Electrostatics* **1995**, *35*, 151-160.
10. Kim, J.; Reneker, D. H. *Polymer Composites* **1999**, *20*, 124-131.
11. Srinivasan, G.; Reneker, D. H. *Polym. Intl.* **1995**, *36*, 195-201.
12. Chun, I.; Reneker, D. H.; Fong, H.; Fang, X.; Deitzel, J.; Tan, N. B.; Kearns, K. *J. Adv. Matls.* **1999**, *31*, 36-41.
13. Fong, H.; Chun, I.; Reneker, D. H. *24th Biennial Conference on Carbon* **1999**, *1*, 380-381.
14. Deitzel, J. M.; Tan, N. B.; Kleinmeyer, J. D.; Rehrmann, J.; Tevault, D. Army Research Laboratory Technical Report, ARL-TR-1989, 1999.
15. Gibson, P.; Schreuder-Gibson, H.; Pentheny, C. *J. Coated Fabrics* **1998**, *28*, 63-72.
16. Gibson, P.W.; Schreuder-Gibson, H.; Rivin, D. *AIChE J.* **1999**, *45*, 190-195.
17. Reneker, D. H.; Yarin, A. L.; Fong, H.; Koombhongse, S. *J. Appl. Phys.* **2000**, *87*, 4531-4547.
18. Chu, C. C.; Browning, A. *J. Biomed. Matr. Res.* **1998**, *22*, 699-712.
19. Sukigara, S.; Gandhi, M.; Ayutsede, J.; Micklus, M.; Ko, F. *Polymer* **2003**, *44*, 5721-5727.
20. Gibson, P.; Schreuder-Gibson, H.; Rivin, D. *Colloids and Surfaces A: Physicochemical and Engineering Aspects* **2001**, *187-188*, 469-481.
21. Fridrikh, S. V.; Yu, J. H.; Brenner, M. P.; Rutledge, G. C. *Phys. Rev. Letters* **2003**, *90*, 144502-1–144502-4.
22. Barrows, T. H. In *High Performance Biomaterials: A Comprehensive Guide to Medical and Pharmaceutical Applications*; Szycher, M., Ed.; Technomic Publishing Co.: Lancaster, PA, 1991; pp 243-257.

23. Wong, W. H.; Mooney, D. J. In *Synthetic Biodegradable Polymer Scaffolds*; Atala, A.; Mooney, D.J., Eds.; Birkhauser: Boston, MA, 1997; pp 50-82.

24. Boland, E. D.; Wnek, G. E.; Simpson, D. G.; Pawlowski, K.; Bowlin, G. L. *J. Macromol. Sci.* **2001**, 38, 1231-43.

25. Boland, E.D., Telemeco,T.; Simpson, D.G., Wnek, G.E.; Bowlin, G.L. *J. Biomed. Mat. Res. B, Appl. Biomater.* **2004**, 71B, 144-152.

26. Boland, E.D.; Wnek, G.E.; Bowlin, G.L. In *Encyclopedia of Biomaterials and Biomedical Engineering.* Editors, G.E. Wnek and G.L. Bowlin, Marcel Dekker, New York, NY, **2004**, 1246-53.

27. Stitzel, J. D.; Pawlowski, K.; Wnek, G. E.; Simpson, D. G.; Bowlin, G. L. *J. Biomaterials Applications* **2001**, *15*, 1-12.

28. Boland, E.D.; Coleman, B.D.; Barnes, C.P.; Simpson, D.G., Wnek, G.E.; Bowlin, G.L. *Acta Biomaterialia,* In Press.

29. Yoshimoto, H.; Shin, Y. M.; Terai, H; Vacanti, J. P. *Biomaterials* **2003**, *24*, 2077-2082.

30. Reneker, D. H.; Kataphinan, W.; Theron, A.; Zussman, E.; Yarin, A. L. *Polymer* **2002**, *43*, 6785-6794.

31. Shin, M.; Ishii, O.; Sueda, T.; Vacanti, J. P. *Biomaterials,* **2004**, 25, 3717-3723.

32. Li, W.-J.; Laurencin, C. T.; Caterson, E. J.; Tuan, R. S.; Ko, F. K. *J. Biomed. Mater. Res.* **2002**, *60*, 613-621.

33. Kenawy, E. R.; Bowlin, G. L.; Mansfield, K.; Layman, J.; Simpson, D. G.; Sanders, E. H.; Wnek, G. E. *J. Contr. Release* **2002**, *81*, 57-64.

34. Matthews, J.A.; Simpson, D.G.; Wnek, G.E.; Bowlin, G.L. *Biomacromolecules* **2002**, 3, 232-238.

35. Matthews, J.A.; Boland, E.D.; Wnek, G.E.; Simpson, D.G.; Bowlin, G.L. *J. Bioactive and Compatible Polymers* **2003**, 18, 125-34.

36. Wnek, G.E.; Carr, M.E.; Simpson, D.G.; Bowlin, G.L. *Nano Letters* **2003**, 3: 213-16.

37. Shields, K.J.; Beckman, M.J.; Bowlin, G.L.; Wayne, J.S. *Tissue Engineering* **2004**, 10, 1510-1517.

38. Boland, E.D.; Matthews, J.A.; Pawlowski, K.J., Simpson, D.G., Wnek, G.E.; Bowlin, G.L. *Frontiers in Biosciences* **2004**, 9, 1422-1432.

Chapter 15

Understanding the Effects of Processing Parameters on Electrospun Fibers and Applications in Tissue Engineering

Cheryl L. Casper[1], Weidong Yang[2], Mary C. Farach-Carson[2], and John F. Rabolt[1,*]

Departments of [1]Materials Science and Engineering and [2]Biological Sciences, University of Delaware, Newark, DE 19716

The effects of molecular weight and atmospheric conditions on the electrospinning process have been investigated. Electrospun polymer microfibers with a nanoporous surface texture have been produced in the presence of humidity. The density of pores, their depth, and their shape have been shown to vary with relative humidity, molecular weight of the polymer, and solvent volatility. Although polystyrene (PS) was investigated in detail, other commodity polymers were also shown to exhibit the nanoporous surface structure under a judicious choice of spinning conditions. The effect of molecular weight on electrospun fiber formation was also studied. It was determined that sufficient chain entanglements are necessary for fiber formation. Molecular weight was also found to affect fiber diameters. Understanding the effects of molecular weight and humidity on electrsospun fibers allows for a greater understanding of how to control the electrospinning process to meet specific applications needs. Utilizing this knowledge, we have investigated the use of electrospun collagen and gelatin fibers as tissue engineering scaffolds. This study shows that cells readily attach to this unique fiber morphology.

Introduction

Electrospinning is a fiber processing technique that involves applying a high voltage to a polymeric solution. The diameters of electrospun fibers are in the nanometer to micron range making them applicable in a wide variety of applications. One challenge in the electrospinning field is the ability to understand and control processing parameters. A multitude of parameters directly affect the electrospinning process. The majority of the research has focused on the effects of solution concentration (1), applied voltage (1), electric field (2), and the effect of the polymer/solvent system (2, 3). This initial work focuses on providing some insight into how molecular weight and humidity affect this fiber formation technique. A fundamental understanding of how to control the formation, shape, texture, and morphology of these fibers is essential. Determining the link between electrospinning parameters and electrospun fiber morphology will allow for the design of polymeric fibers to meet specific application needs.

Electrospun fibers have been investigated for use in a variety of applications such as filtration (4, 5), electronic and fuel cell applications (6-8). However, much of the current research has focused on using electrospun fibers for biomedical applications such as wound dressings (9), drug delivery vehicles (10), and tissue engineering scaffolds (11-13). Properties such as an interconnected porous network, small fiber diameters, controllable degradation rate, and mechanical integrity make electrospun membranes ideal candidates for tissue engineering constructs (14). The nanometer diameter of electrospun fibers mimics the size scale of fibrous proteins found in the extracellular matrix (ECM) of the body (15). The porous nature of the membrane is vital for cells to move throughout the membrane and transport fluids and waste materials through the membrane. Studies have been done in our laboratory to introduce nanoporous features on the surface of electrospun fibers to eventually aid in fabrication of electrospun tissue engineering scaffolds (3, 16). The goal of this work is to investigate how simple processing parameters, such as humidity and molecular weight, affect the electrospinning process. This knowledge will then be used to study electrospun fibers in applications such as tissue engineering scaffolds.

The effect of molecular weight on fiber formation

One important issue in the electrospinning process is the effect of polymer molecular weight on fiber formation. Literature shows that electrospun fibers can vary in diameter from nanometer to micron (1, 3, 17-21), but little is known about what causes these diameter differences. Studies also show that electrospinning equipment can be used to produce freestanding beads, which would make this an electrospray process *(22)*. We believe that molecular weight plays a major role in controlling fiber formation and fiber diameter. To investigate this theory, the molecular weight of polystyrene (PS) was varied while keeping other processing parameters constant..

PS molecular weights from 31,600 g/mol to 1,800,00 g/mol were studied as outlined in Table I. The solutions were electrospun under the following experimental conditions: 35 wt% PS dissovled in tetrahydrofuran (THF), 10 kV, 25% relative humidity, and working distance of 30 cm. The needle diameter increased with molecular weight due to changes in viscosity. Field Emission Scanning Electron Microscopy (FESEM) was used to characterize the samples. Figures 1a and 1b show that the lower molecular weights PS yielded beads, thus electrospraying had occured. Fiber formation began at the 75,700 g/mol molecular weight and was apparent by the appearance of fiber 'tails' on the end of the beads, Figure 1c. However, it was not until a sufficient molecular weight of 171,000 that typical electrospun fibers began to form, Figure 1d. The normal ribbon-like shape of PS electrospun fibers was not observed when electrospinning the 560,900 and 1,800,000 molecular weight solutions. Instead, these fibers appeared to have a rippled surface as seen in Figures 1e and 1f. Thus, molecular weight affects both fiber formation and surface morphology.

We believe that fiber formation is dependent upon the presence of chain entanglements and solution viscosity. Recently, Koski et al. showed that solution concentration plays a major role in the diameter of electrospun polyvinyl alcohol (PVA) fibers *(18)*. Fibers were formed when the solution was in the semi-dilute entangled regime. This semi-dilute region can be noted by $[\eta]C>4$, where C is the solution concentration and $[\eta]$ is the intrinsic viscosity calculated by the Mark-Houwink relationship: $[\eta]= KM_w^a$. For PS in THF, K= 11×10^{-3} and a= 0.725. Applying this equation to the various molecular weights of PS studied, the $[\eta]C$ values are as follows in Table II.

Table I. Electrospinning Various Molecular Weight PS in THF

M_W (g/mol)	Needle Inner Diameter (mm)	Observation
31,600	0.26	Beads only (15-20 μm)
44,100	0.34	Beads only (15-20 μm)
75,700	0.51	Beads (15-20 μm), fiber 'tails' (0.5 μm)
171,000	0.51	No beads, fibers (3-10 μm)
560,900	1.60	No beads, fibers (5-40 μm)
1,800,000	1.60	No beads, fibers (5-30 μm)

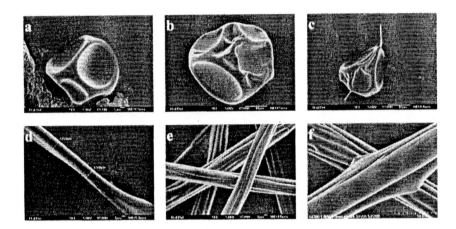

Figure 1. FESEM micrographs of (a)31,600 g/mol, (b) 44,100 g/mol, (c) 75,700 g/mol, (d) 171,000 g/mol, (e) 560,900 g/mol, and (f) 1,800,000 g/mol PS/THF electrospun fibers.

Table II. Applying the Mark-Houwink Equation to Various Molecular Weight PS Studied

PS Molecular Weight (g/mol)	$[\eta]C$
31,600	7
44,100	9
75,700	13
171,000	24
560,900	57
1,800,000	132

These results show that in order for PS/THF fibers to be formed, the $[\eta]C$ must be greater than 13. These results agree with the findings of Koski et al. for PVA systems *(18)*. To further test this idea, the 31,600 g/mol PS/THF was electrospun again, but this time the concentration was increased to 80 wt%. At this concentration and molecular weight, $[\eta]C= 16$ which is in the entangled regime that should produce fibers. Figure 2 shows that increasing the concentration of this low molecular weight sample results in micron diameter fibers. This leads us to believe that sufficient chain entanglements ($[\eta]C>13$) are necessary for fiber formation.

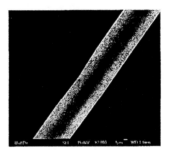

Figure 2. FESEM micrograph of 31,600 g/mol PS electrospun from an 80 wt % solution of PS/THF.

The effect of humidity on surface morphology

Electrospinning was used to produce PS fibers that exhibit a nanoporous surface texture whose morphology is highly dependent upon processing parameters. Previous work has suggested that humidity plays a critical role in the formation of surface features in electrospun fibers *(3, 16)*. The presence of nanopores on the surface of electrospun fibers increases the surface area and provides sites for incorporating drugs, nanoparticles, or enzymes. Learning how to control the surface properties of these fibers is of prime importance for the fabrication of tissue engineering constructs. The goal of this work is to study the effect of increasing humidity and molecular weight on the surface morphology of electrospun PS fibers. Understanding this link between humidity, molecular weight, and surface morphology will allow for tailoring of polymeric fibers to meet specific application needs.

The effect of humidity on the electrospinning process was studied using a 35 wt% PS/THF electrospun under varying humidity ranges *(16)*. The humidity was varied using a humidifier (Holmes HM-1700) placed inside of an enclosed electrospinning box. Five different humidity ranges were studied: <25%, 31-38%, 40-45%, 50-59%, and 60-72%.

The surface of PS fibers appeared to be smooth and featureless, as shown in Figure 3a, when electrospinning in a humidity of <25%. Electrospinning under a humidity range of 31-38% produced small, circular pores randomly distributed on the fiber surface. Increasing the humidity to 40% resulted in more pores being present, Figure 3b. Further increases in humidity, above 50%, caused an obvious increase in the size and number of porous features found on the fiber surface, Figure 3c. At these high humidity ranges, the fiber surface was rich with pores leaving little space between adjacent pores. The pores began to lose their uniform shape due to coalescence of smaller pores into larger, nonuniform shaped features.

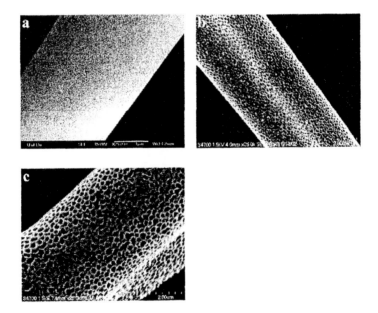

Figure 3. FESEM images of 190,000 g/mol PS/THF fibers electrospun under varying humidity: (a) <25%, (b) 40-45%, (c) 50-59%.

Scion Image Software was used to determine the size and distribution of the pores. It was found that at a humidity range of 31-38%, pores ranged in size from 60-190 nm with the most frequent pore diameter being 85 nm. The 40-45% humidity range yielded a range of pore diameters of 90-230 nm with most pores being approximately 115 nm. When increasing the humidity to the 50-59% humidity range, the most frequent pore diameter of 115 nm remained constant but the distribution broadened to 50-270 nm diameters. The largest pores were found in the 66-72% humidity range with pores ranging from 50 nm to 280 nm, the most frequent being 135 nm in diameter. This shows that humidity plays a direct role in the formation and growth of these nanoporous structures.

Further examinations into humidity and molecular weight allowed for a more comprehensive look at the role of humidity in the electrospinning process. Polystyrene with a molecular weight of 560,900 g/mol was studied to determine if molecular weight affected pore formation. The solution (35 wt% PS/THF) was electrospun under the same conditions as described previously. The only difference is that a larger needle (1.60 mm instead of 0.51 mm) was used due to the high viscosity of the solution.

Similar to the previous experiments, a humidity of over 30% was needed for the formation of pores on the surface of the high molecular weight (560,900 g/mol) fibers as well. Electrospinning the high molecular weight PS solution under 31-38% humidity produced a fiber with pores ranging in size from 150-650 nm, the most frequent being 250 nm in diameter. This is significantly larger than the pores produced from the same humidity using the lower molecular weight solution (190,000 g/mol). Increasing the humidity to 40-45%, Figure 4a, humidity yielded the most frequent pore diameter being 350 nm with a range between 150-600 nm. The 50-59% humidity range broadened the size distribution to 100-850 nm in diameter with most pores being approximately 300 nm, Figure 4b. The largest pore distribution was observed in the 60-72% humidity range. At this humidity, the pores ranged in size from 200-1800 nm in diameter with the most frequent pore size being 350 nm.

It was found that humidity impacts the width of the nanoporous features on electrospun PS fibers but does humidity affect the depth of the pores? Atomic force microscopy (AFM) was used to answer this question. AFM measurements were taken for the higher humidity samples due to the fact that the pores were not large enough in some of the lower humidity samples to allow for probing by the AFM tip. AFM studies revealed that the 190,000 g/mol fibers electrospun under 60-72% humidity had a fiber depth of 48 nm. The 560,900 g/mol fibers had an average pore depth of 171 nm at 50-59% humidity and 220 nm at 60-72% humidity. Upon comparison of the pore depths between the 190,000 and 560,900 g/mol fibers, it is evident that the higher molecular weight sample resulted in significantly deeper pores.

212

Figure 4. FESEM micrographs of 560,900 g/mol PS/THF fibers electrospun in (a) 41-45% humidity and (b) 50-59% humidity.

The exact mechanism of pore formation is intricate. We believe the driving force for pore formation is due to breath figures. Breath figures form due to evaporative cooling that occurs as the solvent evaporates upon leaving the tip of the needle and traveling towards the target. The surface of the jet cools and water from the air condenses on the surface of the fiber. The water droplet leaves an imprint behind after evaporating. This phenomenon has been observed in thin films. Srinivasarao et al. introduced pores on thin films by evaporating PS from a volatile CS_2 solution. Passing humid air across the surface resulted in the formation of hexagonally ordered arrays of pores *(23)*. Peng et al. made thin films of PS evaporated out of toluene, carbon disulfide, THF, and chloroform *(24)*. Peng found a linear relationship between atmospheric humidity and pore size suggesting that breath figures is a dominating mechanism *(24)*.

The differences in pore size upon changing molecular weight is most likely due to reduced diffusion times. Diffusion is proportional to M_w^{-2}. Therefore when increasing the molecular weight, diffusion times are significantly reduced. Also, the diameter of the fibers increases with increasing molecular weight, which also slows diffusion of the solvent out of the fiber.

These results show that humidity has the same effect, independent of molecular weight, in that increasing the humidity causes an increase in the pore diameter, depth, and distribution of pores on the surface of the fiber. However, higher molecular weight polymers resulted in larger pores with a more broad pore size distribution. This information is critically important for eventually customizing the surface of the fiber.

Pores have been introduced on thin films to be used as cell culture substrates by methods such as photolithography, micro-contact printing, and self-organization of polymers *(25)*. Studies show that stretching these porous films allows for cardiac myocytes cells to form fibrous tissue along the axis of the elongated pores *(26)*. Therefore by monitoring the humidity, pores can be introduced onto electrospun fibers in a relatively quick, one-step method that may provide useful as cell culture substrates.

The use of electrospun fibers in tissue engineering applications

After discovering how to control certain processing parameters, the electrospinning process was studied to determine if electrospun fibers could be used in tissue engineering applications. The previous results on controlling fiber diameters and surface morphology via molecular weight were employed to ensure that fibers were produced.

We chose to use collagen (type I from calf skin) and gelatin as materials to electrospin due to the fact that these components are found in the ECM. The eventual goal of this work is to produce an electrospun scaffold that will promote the repair and regeneration of new bone tissue.

Collagen is a major constituent of natural ECM in skeletal tissue. Collagen alone promotes cellular recognition *(27)* and exhibits a high affinity for proteins such as other ECM proteins and growth factors *(28)*. As previously mentioned, the characteristics of electrospun membranes make them ideal candidates for biomaterial applications because electrospun fibers structurally mimic the ECM. The choice of electrospinning parameters for the fabrication of collagen fibers is the first step in producing electrospun fibers for tissue engineering applications. Collagen can be electrospun from a variety of solvents including formic acid, acetic acid, glycolic acid, and 1,1,1,3,3,3-hexafluoro-2-propanol (HFIP) *(29, 30)*. The fibers shown in Figure 5a were electrospun using an 18 wt% collagen in a 2:1 ratio of HFIP:acetic acid. The fibers in Figure 5b were electrospun using 15 wt% gelatin/HFIP. The collagen membranes are exposed to a glutaraldehyde vapor for 19 hours to crosslink the collagen. The fibers were exposed to UV radiation, ethanol solution, and an antibiotic/antimycotic treatment to sterilize the mats prior to cell culture.

Initial cell culture studies were completed on gelatin to determine the success of the crosslinking and sterilization methods. Electrospun scaffolds were seeded for 2 hours with a human osteoblastic cell line (MG63) and imaged using laser scanning confocal microscopy (LSCM). The cells were stained with a syto-13 nuclear stain. Plate 1 reveals the presence of cells on the scaffold. A cell proliferation assay was also completed to ensure the proliferation of viable cells in the scaffold (data not shown). These results indicate that the environment provided by the electrospun fibers is hospitable for cell infiltration and proliferation. Further studies will focus on attaching biologically relevant molecules to the electrospun matrix in order to further increase the bioactivity of the scaffold.

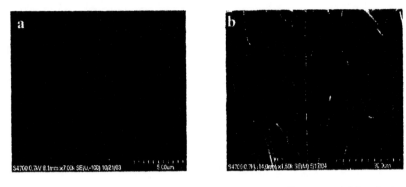

Figure 5. FESEM micrographs of (a) electrospun collagen and (b) gelatin.

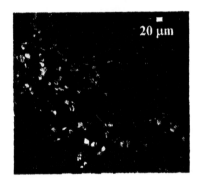

Plate 1. LSCM image of gelatin scaffold seeded with MG63 osteoblast cells.
(See page 12 of color inserts.)

Conclusions

Molecular weight and humidity has proven to have a significant impact on the electrospinning process. The results of these studies show how to control fiber formation and surface morphology through monitoring molecular weight of the polymer and atmospheric conditions. The findings allow for better control over the electrospinning process in order to customize fibers.

Electrospinning of polymeric fibrous membranes with and without surface texture has shown considerable promise for use in tissue engineering scaffolds. Preliminary studies on collagen and gelatin have proven these membranes to be effective in initial cell attachment studies. Future studies will involve the incorporation of biologically relevant molecules (heparan sulfate modified perlecan domain I) to increase the bioactivity of the electrospun matrix in order to further promote tissue regeneration.

Acknowledgements

The authors would like to acknowledge many helpful discussions and contributions from Dr. Bruce Chase and Dr. Jeannie Stephens. We thank Nancy Tassi for her assistance with the AFM data. J. Rabolt would like to acknowledge NSF (DMR0210223) and NASA (NA68-01923) for partial support of this work. W. Yang and M.C. Farach-Carson acknowledge the NIH/NIDCR (DE013542) grant. In addition, C. Casper would like to thank the University of Delaware IGERT program for support during part of this project.

References

1. J. M. Deitzel, J. Kleinmeyer, D. Harris, and N. C. B. Tan, *Polymer*, **2001**, *42*, 261.
2. S. Koombhongse, W. X. Liu, and D. H. Reneker, *Journal of Polymer Science Part B-Polymer Physics*, **2001**, *39*, 2598.
3. S. Megelski, J. S. Stephens, D. B. Chase, and J. F. Rabolt, *Macromolecules*, **2002**, *35*, 8456.
4. H. Schreuder-Gibson, P. Gibson, K. Senecal, M. Sennett, J. Walker, W. Yeomans, D. Ziegler, and P. P. Tsai, *Journal of Advanced Materials*, **2002**, *34*, 44.
5. P. Gibson, H. Schreuder-Gibson, and D. Rivin, *Colloids and Surfaces a-Physicochemical and Engineering Aspects*, **2001**, *187*, 469.
6. J. Y. Lu, C. Norman, K. A. Abboud, and A. Ison, *Inorganic Chemistry Communications*, **2001**, *4*, 459.
7. C. Drew, X. Y. Wang, L. A. Samuelson, and J. Kumar, *Journal of Macromolecular Science-Pure and Applied Chemistry*, **2003**, *A40*, 1415.
8. H. Q. Dai, J. Gong, H. Kim, and D. Lee, *Nanotechnology*, **2002**, *13*, 674.
9. E. R. Kenawy, J. M. Layman, J. R. Watkins, G. L. Bowlin, J. A. Matthews, D. G. Simpson, and G. E. Wnek, *Biomaterials*, **2003**, *24*, 907.
10. E. R. Kenawy, G. L. Bowlin, K. Mansfield, J. Layman, D. G. Simpson, E. H. Sanders, and G. E. Wnek, *Journal of Controlled Release*, **2002**, *81*, 57.

216

11. K. Nagapudi, W. T. Brinkman, J. E. Leisen, L. Huang, R. A. McMillan, R. P. Apkarian, V. P. Conticello, and E. L. Chaikof, *Macromolecules*, **2002**, *35*, 1730.
12. H. Yoshimoto, Y. M. Shin, H. Terai, and J. P. Vacanti, *Biomaterials*, **2003**, *24*, 2077.
13. C. Y. Xu, R. Inai, M. Kotaki, and S. Ramakrishna, *Biomaterials*, **2004**, *25*, 877.
14. L. C. Lu and A. G. Mikos, *Mrs Bulletin*, **1996**, *21*, 28.
15. W. J. Li, C. T. Laurencin, E. J. Caterson, R. S. Tuan, and F. K. Ko, *Journal of Biomedical Materials Research*, **2002**, *60*, 613.
16. C. L. Casper, J. S. Stephens, N. G. Tassi, D. B. Chase, and J. F. Rabolt, *Macromolecules*, **2004**, *37*, 573.
17. D. H. Reneker and I. Chun, *Nanotechnology*, **1996**, *7*, 216.
18. A. Koski, K. Yim, and S. Shivkumar, *Materials Letters*, **2004**, *58*, 493.
19. Y. M. Shin, M. M. Hohman, M. P. Brenner, and G. C. Rutledge, *Applied Physics Letters*, **2001**, *78*, 1149.
20. G. Srinivasan and D. H. Reneker, *Polymer International*, **1995**, *36*, 195.
21. G. E. Wnek, M. E. Carr, D. G. Simpson, and G. L. Bowlin, *Nano Letters*, **2003**, *3*, 213.
22. L. Reyderman and S. Stavchansky, *International Journal of Pharmaceutics*, **1995**, *124*, 75.
23. M. Srinivasarao, D. Collings, A. Philips, and S. Patel, *Science*, **2001**, *292*, 79.
24. J. Peng, Y. C. Han, Y. M. Yang, and B. Y. Li, *Polymer*, **2004**, *45*, 447.
25. T. Nishikawa, J. Nishida, R. Ookura, S. I. Nishimura, S. Wada, T. Karino, and M. Shimomura, *Materials Science & Engineering C-Biomimetic and Supramolecular Systems*, **1999**, *10*, 141.
26. T. Nishikawa, M. Nonomura, K. Arai, J. Hayashi, T. Sawadaishi, Y. Nishiura, M. Hara, and M. Shimomura, *Langmuir*, **2003**, *19*, 6193.
27. P. X. Ma and R. Y. Zhang, *Journal of Biomedical Materials Research*, **1999**, *46*, 60.
28. S. A. Gittens and H. Uludag, *Journal of Drug Targeting*, **2001**, *9*, 407.
29. J.S. Stephens, Ph.D. Thesis, University of Delaware, Newark, DE, 2003.
30. J. A. Matthews, G. E. Wnek, D. G. Simpson, and G. L. Bowlin, *Biomacromolecules*, **2002**, 3, 232.

Chapter 16

Nanofibers from Polylactic Acid Nanocomposites: Effect of Nanoclays on Molecular Structures

Huajun Zhou[1,2], Kyoung-Woo Kim[2], Emmanuel Giannelis[2], and Yong Lak Joo[1,*]

Schools of [1]Chemical and Biomolecular Engineering and [2]Materials Science and Engineering, Cornell University, Ithaca, NY 14853

Poly(Lactic) acid (PLA) and its nanocomposite nanofibers have been successfully obtained via electrospinning from both solution and melt. Our results show that molecular structures of electrospun nanofibers are significantly influenced by both electrospinning conditions and the presence of nanoclays. Highly oriented structures of PLA, which give rise to the formation of a fibrillar β crystal form, are observed in electrospun fibers, and the formation of β crystal is enhanced by strongly aligned nanoclays. It is also found that due to rapid solidification, cold crystallization occurs between 100°C and 130°C, and the degree of cold crystallization becomes more prominent in electrospun nanofibers from melt. Annealing effect on fiber structures is also investigated by DSC and XRD.

Poly(Lactic) acid (PLA) which is derived from renewable resources such as corn has received much attention as a biodegradable and biocompatible replacement for various polymers in packaging and biomedical applications. Despite its promises, the widespread use of PLA is limited by its thermo-mechanical properties. To improve the mechanical and barrier properties, and thermal stability of PLA, researchers have been developing its composites with nanoclays. Blends of PLA with organically modified layered silicate (OMLS) were first prepared by Ogata *et al.* using a solution casting method (*1*). Their results demonstrated that only tactoids which consists of stacked silicate monolayers were observed in the composites. Even with this level of poor dispersion, however, mechanical properties are shown to be slightly improved. Due to the advances in nanocomposite technology, OMLS nanoclays with a few nanometer of layer thickness and very high aspect ratio (*e.g.*, 10-1000) can be homogeneously dispersed in a polymer matrix by various processing schemes including solution casting (*1*), melt extrution (*2*) and in-situ polymerization (*3*). It has been reported that mechanical and barrier properties, and even biodegradability of PLA can remarkably be improved by the inclusion of nanoclays (*4,5,6*). Such improvements in material properties are evidently caused by the enhanced interfacial interactions between clays and polymer molecules in nanocomposites via either intercalated structures or exfoliated morphologies (*2*).

In the meantime, electrospinning where a droplet of a polymer solution or melt is elongated by a strong electrical field has become a useful scheme to fabricate sub-micron scale fibers. Although the proof-of-concept of the electrospinning process was instituted much earlier, it is quite recent that electrospinning has been revisited and thoroughly studied by Reneker and others. According to the review by Huang *et al.*, more than fifty different polymers have been successfully electrospun mostly via solution electrospinning (*7*). Electrospinning typically produces nonwoven fiber mats with extremely high surface area to volume ratios. Nonwoven fiber mats obtained by electrospinning have promising applications in the areas of filtration (*8*), sensor, tissue scaffolding (*9,10*), drug delivery (*11,12*) and composite materials.

Although people noticed that polymer solution or melt driven by the electrostatic force undergoes strong elongational deformation during electrospinning, thorough studies on microstructural changes caused by this deformation are very rare. Dersch *et al.* studied the internal structure of electrospun nylon-6 and PLA fibers (*13*). Their electron diffraction results showed only inhomogeous, weak elongation of nylon crystals resulted from electrospinning. As to PLA fibers, however, there is only a negligible amount of crystals formed during electrospinning, and thus no (even local) orientation of crystals was observed. DSC studies on electrspun PLA fibers demonstrated that aligned structures may form during the electrospinning which caused the development of the cold crystallization peak at lower temperature (*14,15*).

Studies of electrospinning have just begun to scratch at the surface of the crystalline structures of nanocomposite fibers. The effects of nanoclays on the degree of crystallinity in electrospun fibers were investigated by a few researchers and they agreed on one point that nanclays greatly inhibited the growth of crystals in the fibers *(16,17)*. Fong *et al.* studied in detail the alignment of the exfoliated clays along the nylon-6/montmorillonite nancomposite fibers *(18)*. α crystal structure dominates nylon-6 casting film from solution and γ crystal structure is adopted into the electrospun fibers with crystal layer normal parallel to the fiber axis. However, no studies on the structural study of PLA nanocomposites nanofibers have been done till now. The structural study of electrospun PLA nanocomposite fibers is particulary important because PLA exhibit two types of crystal structures: α structure with a lamellar folded-chain morphology and β structure with a fibrillar morphology *(19)*. The pseudoorthorhombic α structure is found at relatively low drawing temperatures and/or low hot-draw ratios. A second so-called β structure appears only at higher hot-draw ratios, and thus probing the β crystal structure can provide useful information associated with elongational deformation during the electrospinning process. We will portray how the inclusion of nanoclays influences the molecular structures of the electrospun fibers from both its solutions and melts.

PLA and its Nanocomposite Fibers from Solutions

Experimental Procedures

Poly(Lactic) acid pellets which majorly consist of poly(L-lactic) acid (>98%) are supplied by Cargill-Dow. Molecular weight is about 186,000 and polydispersity of 1.76. To prepare PLA nanocomposites, PLA pellets were first ground into fine powders with an averge size of 0.5 mm using a Retsch ZM 100 ultra-mill (Glen Mills, Inc.). A polymer/nanoclay premix was prepared by combining appropriate amounts of organically modified montmonrillonite (Closite 30B, Southern Clay Products, Inc.) and polymer powders in a DAC 150FV speed-mixer (Flacktek, Inc.). This premixed powders were then dried in vacuum oven at 80°C for 12 hours followed by melt-extrusion at 220°C in a nitrogen atmosphere, with a screw speed of 100 r.p.m. and a residence time of 3 minutes using a bench-top micro twin extruder. Two different compositions of nanocomposites (3 wt% and 5 wt% of nanoclays) are used in the study.

Solutions of PLA/nanoclay composites were prepared in two different ways. The first scheme is to dissolve 10 to 25 wt% of these compounded PLA/nanoclay nanocomposites in chloroform (Sigma-Aldrich). The second method is a simple blending scheme in which nanoclay particles are simply stirred and blended with chloroform until nanoclay particles are dispersed homogeneously in chloroform, and then pre-dried PLA powders are directly added to nanoclay/chloroform suspension. Hence, compounding PLA and nanoclay in a twin screw extruder is by-passed in the second scheme. Solutions made by these two schemes are named as follows; for example, in PLA-NC3, NC denotes the first scheme with twin-screw extrusion compounding and 3 denotes the content of nanoclay in PLA composites. PLA-BL5 represents PLA composites with 5 wt% clay prepared by the second blending scheme.

Electrospinning experiments were conducted in a horizontally placed electrospinning setup. A high voltage source (ES30P, Gamma High Voltage Research, Inc.) was used to apply 10 to 30 kV to a 24 gauged needle on a syringe through copper wiring. A precisely-controlled syringe pump (PHD2000, Harvard Apparatus) was used to continually renew the droplet at the syringe tip. The volumetric flow rate ranged from 0.005 to 0.025 ml/min. Grounded aluminum foil on a metal sheet was placed 6 to 12 inches away from the syringe tip acting as a collector. As the applied voltage is increased, a droplet at the needle tip deforms into a conical shape and then an electrically charged jet is formed. The jet solidifies due to the evaporation of solvent, as it goes through a vigorous whipping motion. As a result, non-woven fiber mats are formed on the surface of the collector.

These fiber fabrics were then characterized by various methods. Thermal analysis was conducted by DSC (Seiko, DSC 220C). Approximately 5 mg of samples were loaded and heated in a nitrogen atmosphere at a rate of 10°C/min until the temperature reached 200°C. Morphology of electrospun fibers was examined by SEM (Leica 440). Structrual study was performed through XRD (Scintag, Inc. Theta-Theta Diffractometer) and TEM (JEOL 1200EX). XRD data were collected in the 2θ range of 1-40°, in steps of 0.02° and a scanning rate of 4 sec per point. Tensile experiments were carried out on the Instron 1125 test system by cutting fiber mats into rectangular strips. For comparison purposes, casting films were also prepared. First, powders of PLA or nanocomposite were dissolved in chloroform. After removing the solvent in air at room temperature, PLA cast films were put into a vacuum oven for 24 hours. These films were then characterized under the same conditions as electrospun fibers.

Fiber Morphology

Fibers with submicron diameter were successfully spun from chloroform solutions of neat PLA, and PLA composites prepared by two different schemes described above. Figure 1 shows the electrospun fibers obtained from PLA-

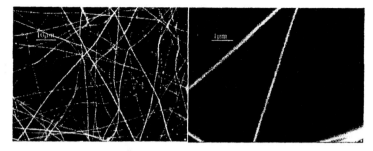

*Figure 1 SEM micrographs of electrospun fibers obtained from
PLA/Chloroform solutions with 3 wt % nanoclay (PLA-NC3)*

aspects compared with electrospun neat PLA fibers collected under the same condition.

Although the nanoclay does not affect the surface morphology of fibers, it does affect the size of fibers. Our measurements show that adding nanoclay (up to 5 wt%) can decrease fiber diameter under the same processing conditions. This is possibly due to the fact that inclusion of nanoclays can affect the charge density of solutions since nanoclay can be ionized and carry more charges.

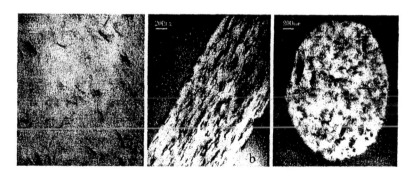

*Figure 2 TEM micrographs of a) PLA-NC5 cast film, b) PLA-NC3 fiber
microtomed along its axis and c) cross section of PLA-NC3 fiber*

The dispersion of the clay layers in electrospun fibers is exaimined using TEM studies. TEM micrographs taken from a cast film and an electrospun nanofiber of a PLA nanocomposite are compared in Figure 2. The micrograph of the cast film, Figure 2a, shows that the organoclay is well dispersed in the PLA matrix, although some parts of agglomerated layers still exist. As expected, no significant orientation effect could be observed for this bulk sample. In contrast, the TEM images of an electrospun fiber microtomed along and perpendicular to its axis (Figure 2b and 2c) show the clay layers aligned along the fiber axis throughout the entire fiber, suggesting strong prossessing-structure correlations. Highly aligned mont-morillonite layers have also been reported in electrospun fibers of nylon nanocomposite (18). The TEM study also reveals that intercalation is preserved during electrospinning. This is in agreement with the XRD study which shows a peak at a low value of 2θ around 1.86° in Figure 4.

Effect of Naniclays on Molecular Structures of Electrospun Fibers

An electrically charged jet of polymer solution can significantly be elongated during electrospinning. The microstructural state resulting from interaction between the dynamics of the polymer molecules and the deformation during electrospinning process are not well understood. In particular, we are interested in the effect of nanoclays on the microstructural development. Figure 3 shows the first run of the DSC for various electrospun PLA and its nanocomposite fibers. It should be noted here that the residual solvent was removed by drying in all the samples before testing. Compared to cast films, all electrospun fibers have two distinct melting peaks which can be deconvoluted by Gaussian fitting (dotted lines). The melting peak at lower temperature is for fibrillar β crystal, created by elongational deformation (19-21), whereas the other peak at higher temperture is for that of α crystal. The presence of the melting peak for β crystal becomes more prominent in nanocomposite fibers (PLA-NC3).

We also note that due to highly aligned molecular structures, the cold crystallization occurs between 100°C and 130°C, and the degree of cold crystallization becomes more prominent in nanocomposite fibers. The cold crystallization peak for nanocomposite fibers (PLA-NC3) is much sharper and stronger than those of neat PLA fibers and simply blended composite fibers (PLA-BL3). The cold crystallization temperature is also shifted to lower temperature, around 110°C. These results may suggest that the alignment of intercalated (partially exfoliated) nanoclays caused by elongational deformation during electrospinning enhance the formation of oriented PLA molecules which transform into crystals upon heating.

The characteristics of these curves are summarized in Table 1 including glass transition temperature T_g (°C), cold crystallization temperature T_{cold}^* (°C), β crystal melting temperaturue T_m^β (°C), α crystal melting temperature T_m^α (°C) and the degree of crystallinity χ (%). The fraction of β crystal is also calculated by the ratio of β crystal peak area to the total peak area. The degree of crystallinity χ is determined using the following equation:

$$\chi = \frac{\Delta H_f - \Delta H_c}{\Delta H_f^0} \times 100\% \tag{1}$$

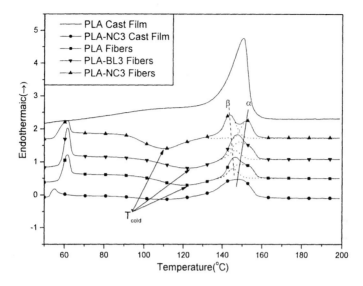

Figure 3 DSC thermograms of various electrospun PLA fibers. Cast films of PLA and PLA/3 wt%nanoclay are included for comparison.

where ΔH_f is the heat of fusion, ΔH_c is the heat of cold crystallization and $\Delta H_f^{\,0}$ is the heat of fusion for 100% crystalline PLA, 93 J/g (*22*). It is observed that the crystallinity of PLA-NC3 cast film is significantly smaller than that of neat PLA cast film. Despite the fact that nanoclays can act as nucleates (*23*), recent studies (*24*) demonstrated that nanoclays may impede the mobility of polymer molecules thus slow crystallization kinetics (*25*). We note that the degree of crystallinity for electrospun fibers is remarkably smaller than that of cast films. This decrease in crystallinity may be associated with the rapid solidification and short residence time during the electrospinning process. Contrary to casting films, the inclusion of nanlclays of 3 wt% in electrospun fibers greatly enhances crystallization, indicating that the presence of nanoclays under the deformation may induce crystallization. Crystallization, however, appears to be suppressed at higher nanoclay contents (5 wt%). It should also be noted that electrospun PLA fibers exhibits significant β fractions, which indicates that substantial deformation during electrospinning. The fraction of the β crystal is greatly increased in electrospun PLA nanocomposite fibers.

Table 1 Summary of DSC Curves for Different PLA Films/Fibers

Film/Fiber	T_g	T_c^*	T_m^β	T_m^α	$\chi(\%)$	$F(\beta)$
Neat PLA Cast Film	N/A	N/A	N/A	150.6	32.4	0
PLA-NC3 Cast Film	52.3	N/A	N/A	148.6	15.7	0
Neat PLA Fibers	57.8	119.2	145.3	149.3	2.2	0.27
PLA-BL3 Fibers	58.6	121.4	146.3	148.7	1.8	0.23
PLA-NC3 Fibers	55.1	109.9	143.8	152.9	4.4	0.59
PLA-BL5 Fibers	60.5	113.5	143.4	148.9	0.8	0.20
PLA-NC5 Fibers	59.9	108.9	144.2	153.2	1.4	0.59

Note: Data for PLA-BL5 and PLA-NC5 are also obtained from DSC curves which are not shown in Figure 3.

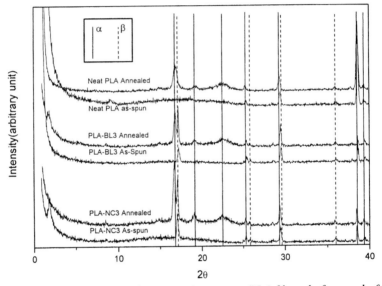

Figure 4 XRD patterns of various electrospun PLA fibers before and after annealing at 120 °C for 3hours.

The existence of the β crystal can also be verified by the XRD study. The electrospun PLA and PLA nanocomposite fibers are annealed at 120°C for 3 hours and the XRD patterns of samples before and after the annealing are shown in Figure 4. Peaks marked by dotted lines belong to the β crystal structure (*19*). It is observed that electrospun fibers exhibit a peak associated with the β structure at $2\theta = 29.5°$ (orthorhombic (330) or (003) reflection) and nanocomposite fibers exhibit significant β peak at $2\theta = 17.2°$ (orthorhombic (200) reflection). Meanwhile, the XRD patterns of annealed fibers show a significant increase in both α and β crystals. The intensity of the α peak at 2θ = 16.8° (pseudoorthorhombic (200) or (110) reflection) and that of the β crystal peak at 2θ = 17.2° are increased significantly after annealing. Finally, we note that after annealing the β peak at 2θ = 29.5° shifts to 2θ =29° which can be for the α structure (pseudoorthorhombic (216) reflection). This may indicate the chain relaxation of the β crystal and its transformation into the α crystal, since the α form is the more stable phase. Similar transformation has been observed the nylon fibers (26). The transformation of crystal structures will be discussed using temperature-dependent XRD studies in the melt electrosping section.

The DSC thermograms and XRD patterns of PLA and PLA/nanoclay nanofibers suggest that the β structure of PLA is formed during the electrospinning process and that the presence of highly aligned nanoclays enhances the formation of the β structure.

Mechanical Properties and Biodegradability of Electrospun Fiber Mats

Since the intercalation in nanocomposite fibers has been preserved during electrospinning, we would expect enhancement in the mechanical properties of nonwoven fiber mats (see Table 2). The inclusion of nanclays via two different schemes improves the yield strength, but slightly decreases the flexibility of fiber mats. We note that nanofibers from compounded composite of 3wt% nanoclay exhibit significant increases in both modulus and yield strength. This improvement is partly due to the effective alignment of nanoclays in electrospun fibers. Similar increases are obtained at the higher contents of nanoclays (5%).

Table 2 Mechanical Properties of Electrospun Nonwoven Fiber Mats

	Modulus(MPa)	Yield Strain(%)	Yield Stress(MPa)
Neat PLA Fibers	18.29	27.79	0.54
PLA-BL5 Fibers	17.56	24.11	0.73
PLA-NC3 Fibers	36.36	19.11	1.84
PLA-NC5 Fibers	39.03	18.25	1.89

Biodegradability studies in a buffer solution of $KCl/NaCl/Na_2HPO_4/KH_2PO_4$ at pH = 7.38 and 37°C show that electrospun PLA fiber mats exhibit a faster biodegradation rate than films. The high surface area to volume ratio of the fiber mats and their low degree of crystallinity may contribute to this improvement. However, the inclusion of nanoclays appears to slow down the degradation partly due to the increase in crystallinity.

PLA and its Nanocomposite Fibers from Melts

Despite its promises, very few studies on electrospinning from polymer melts have been reported till now (*27,28*). Main difficulties underlying this melt electrospinning (ME) process are precise temperature control, higher viscosity and poor conductivity of polymer melts. In this section we illustrate our ME scheme and present the structural study of PLA nanocomposite fibers directly from melts.

Experimental Setup

The major difference between ME and conventional SE setups is the heating element. In ME, a heating chamber needs to be incorporated to melt the sample. In our design, we also add a heated guiding chamber and a temperature controlled collection plate to control the solidification and annealing process during electrospinning. Figure 5 depicts the schematic of the current design which includes the basic SE elements: syringe pump, high voltage supply and grounded collector as well as ME components: shielded heating unit, temperature controller, heated guiding chamber, in which needle temperature can be adjusted independently, and temperature controllable collector. This design results in four different temperatures: syringe temperature (T_1), needle temperature (T_2), guiding chamber temperature (T_3) and collection temperature (T_4). These temperatures are controlled separately in the setup.

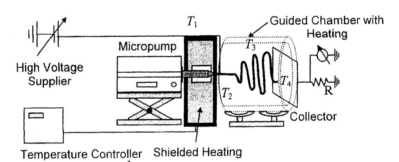

Figure 5 Schematic Sketch for the melt electrospinning setup

Fiber Morphology

Solvent evaporation in SE gives rise to additional decrease in fiber diameter while thicker fibers are expected under the same flow rate since no solvent is involved in ME. As seen in Figure 6, micron-sized fibers are collected through

ME using a nozzle diameter of 0.84mm. Some of these fibers scale down to 150 nm as seen in the insert of Figure 6b) with a smaller nozzle. It should be noted that contrary to previous studies on melt electrospinning (28), the size distribution of melt electrospun fibers in the current study is relatively uniform possibly due to the precise temperature control in various regions in the process.

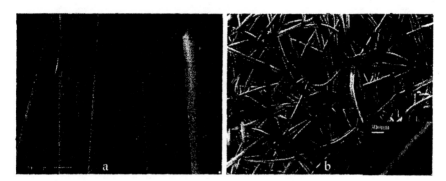

Figure 6 SEM micrographs for neat PLA fibers obtained a) without heating the guiding chamber $T_3=25^0C$ and b)with guiding chamber at $T_3=100^0C$

As described earlier, the control of polymer melt viscosity before the ejection by the electric field is done by heating the syringe and the needle separately (T_1 and T_2). The effect of temperature of spinning region T_3 on the fiber morphology is shown in Figure 6. The common processing conditions are as follows: flow rate at 0.005 ml/min, voltage at 20 kV, collecting distance at 6 cm and $T_1 = 200°C$, $T_2 = 220°C$, $T_4 = 23°C$. If the temperature in the spinning region T_3 is kept at room temperature and thus the ejected polymer melt jet is almost quenched, thicker fibers are formed (Figure 6a). The average diameter of fibers spun at room temperature is 15 μm which is about three times as thick as that of fibers at T_3 at 100°C under the same processing condition. It is also observed that the fiber diameter steadily increases with increasing flow rate and/or nozzle diameter in the range investigated in the current study.

It is observed that the internal structure of melt electrospun PLA nanocomposite fibers is quite different from that of solution electrospun fibers. More exfoliated structures are observed in the TEM mircograph of melt electrospun nanocomposite fibers (Figure 7a). It also shows a skin-core morphology of the fibers, in which nanoclays are aligned along the fiber axis but the extent of their alignment varies along the radial direction. The skin area exhibits a higher degree of alignment than the core region. In addition, the dispersion of the nanoclays is quite nonuniform radially. The absence of solvent and large radial thermal gradient in melt electrospinning together with larger fiber dimension can induce such inhomogeniety. This skin-core differentiation can also be manifested by the SEM micrograph of partially peeled fibers (Figure 7b).

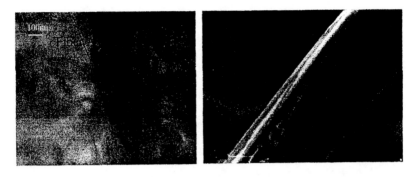

Figure 7a) TEM micrograph of PLA-NC5 fibers microtomed along its axis and b) SEM micrograph of partially peeled PLA-NC5 fibers

Effects of Nanoclays and Spinning Temperature on Molecular Structures

Figure 8a shows DSC diagrams for neat PLA fibers and PLA-NC5 fibers collected under the same processing condition other than different T_3. It should be noted that the substantial degree of cold crystallization is observed for fibers from both neat PLA and PLA nanocomposite melts, manifesting the great deal of aligned structures in electrospun fibers. We also note that both presence of nanoclay as well as temperature in the spinning region has a strong influence on the degree of cold crystallization. The peaks associated with cold crystallization and crystal structure become more distinct and are shifted to lower temperature as the nanoclay content increases or spinning temperature decreases. As a result, the degree of cold crystallization and presence of crystal structure is most prominent when PLA nanocomposite melt is electrospun at room temperature. DSC data show that the crystallinity of PLA-NC5 fibers spun at $T_3 = 100°C$ is $5.8 \pm 1.2\%$ which is higher than that of PLA-NC5 fibers spun at $T_3 = 23°C$ ($4.1 \pm 0.8\%$). These results indicate that the degree of cold crystallization and thus the crystallinity can easily be controlled by changing spinning temperature in melt electrospinning. We surmise that the slow crystallizing nature of PLA together with substantial elogational deformation during electrospinning can give rise to such significant cold crystallization. Figure 8b illustrates how molecular structure changes during the heat up using temperature-dependent XRD analysis. After the occurrence of cold crystallization starting at around 100°C, crystal peak at $2\theta = 28.2°$ (orthorhombic (311) reflection) shifts to a lower $2\theta = 27.5°$ (pseudoorthorhombic (117) reflection) and eventually disappears as temperature increases further. In the later communication, more rigorous studies on crystallization kinetics of PLA during melt electrospinning in the presence of nanoclays will be presented.

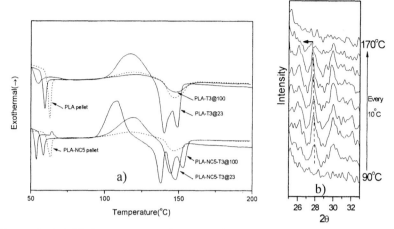

Figure 8 a) DSC thermograms of various electrospun PLA fibers. PLA and PLA-NC5 pellets are used for comparison. b) Temperature-dependent XRD spectrum of PLA-NC5 fibers

Conclusions

PLA is a biocompatible material which has various applications including biomedical fields. Nanofibers of PLA and its nanocomposites from solution and melt have been obtained through electrospinning. We studied the effects of nanoclay and processing scheme on molecular structures of electrospun fibers. We have demonstrated that inclusion of nanoclay enhances the formation of oriented structures in electrospun fibers, which gives rise to substantial cold crystallization between 100°C and 130°C. Our results also reveal that electrospinning induces β PLA crystal structure with a fibrillar morphology and the addition of nanoclay enhances the formation of the β crystal structure. This change in molecular structures due to inclusion of nanoclays during electrospinning is shown to significantly increase the mechanical properties of the electrospun fiber mats and influence the biodegradability.

References

1. Ogata, N.; Jimenez, G.; Kawai, H.; Ogihara, T. *J. Polym. Sci. Part B: Polym. Phys.* 1997, 35, 389-396.
2. Ray, S.S.; Maiti, P.; Okamoto, M.; Yamada, K.; Ueda, K. *Macromolecules* 2002, 35, 3104-3110.

230

3. Yang, F.; Ou Y.C.; Yu, Z.Z. *J APPL POLYM SCI* 1998, 69, 355-361.
4. Ray, S.S.; Yamada, K.; Okamoto, M.; Ueda, K. *Polymer* 2003, 44, 857-866.
5. Ray, S.S.; Okamoto, M. *Macromol. Rapid Commun.* 2003, 24, 815-840.
6. Ray, S.S.; Yamada, K.; Okamoto, M.; Ueda, K. *Nano Lett.* 2002, 2, 1093-1096.
7. Huang, Z.M.; Zhang, Y.Z.; Kotaki, M.; Ramakrishna, S. *Compos. Sci. Technol.* 2003, 63, 2223-2253.
8. Groitzsch, D.; Fahrbach, E. US patent 4,618,524, 1986
9. Luu, Y. K.; Kim, K.; Hsiao B. S.; Chu, B.; Hadjiargyrou, M. *J. Controlled Release* 2003, 89, 341-353.
10. Kim, K.; Yu, M.K. et. al. *Biomaterials* 2003, 24, 4977-4985.
11. Zeng, J.; Xu, X.Y.; Jing, X.B. et. al. *J. Controlled Release* 2003, 92, 227-231.
12. Kenawy, E.R.; Bowlin, G.L.; Wnek, G.E. et. al. *J. Controlled Release* 2002, 81, 57-64.
13. Dersch, R.; Liu, T.Q.; Schaper, A.K.; Greiner, A.; Wendroff, J.H. *J. Polym. Sci. Part A: Polym. Chem.* 2003, 41, 545-553.
14. Zong, X.H.; Kim, K.S.; Chu, B. et. al. *Polymer* 2002, 43, 4403-4412.
15. Bognitzki, M.; Czado, W.; Frese, T. et. al. *Adv. Mater.* 2001, 13, 70-72.
16. Shao, C.L.; Kim, H.Y.; Park, S.J. et. al. *Mater. Lett.* 2003, 57, 1579-1584.
17. Gong, J.; Li, X.D.; Kim, H.Y. et. al. *J. Appl. Polym. Sci.* 2003 ,89, 1573-1578.
18. Fong, H.; Liu, W.D.; Wang, C.S.; Vaia, R.A. *Polymer* 2002, 43, 775-780.
19. Hoosteen, W.; Postema, A.R.; Penning, A.J.; Brinke, G.T.; Zugenmaier, P. *Macromolecules* 1990, 23, 634-642.
20. Eling, B.; Gogolewski, S.; Pennings, A. J.; *Polymer* 1982, 23, 1587-1593.
21. Sawai, D.; Takahashi, K.; Imamura, T.; Nakamura, K.; Kanamoto, T.; Hyon, S. H.; *J. Polym. Sci. Part B: Polym. Phys.* 2002, 40, 95-104.
22. Migliaresi, C. D.; Cohn, D.; De Lollis, A.; Fambri, L. *J. Appl. Polym. Sci.* 1991, 43, 83-95.
23. Nam, J.Y.; Ray, S.S.; Okamoto, M. *Macromolecules* 2003, 36, 7126-7131.
24. Ke, Y. C.; Yang, Z. B.; Zhu, C. F.; *J. Appl. Polym. Sci.* 2002, 85, 2677-2691.
25. Pluta, M.; Galeski, A.; Alexandre, M.; Paul, M. A.; Dubois, P. *J. Appl. Polym. Sci.* 2002, 86, 1497-1506.
26. Stephens, J.S.; Chase, D.B.; Rabolt, J.F. *Macromolecules* 2004, 37, 877-881.
27. Kim, J. S.; Lee, D. S.; *Polym. J.* 2000, 32, 616-618.
28. Larrondo; Manley, R.; St, J. *J. Polym. Sci.: Polym. Phys. Ed.* 1981, 19, 909-940.

Chapter 17

Coelectrospinning of Carbon Nanotube Reinforced Nanocomposite Fibrils

Frank K. Ko, Hoa Lam, Nick Titchenal, Haihui Ye,
and Yury Gogotsi

Department of Materials Science and Engineering, Drexel University,
Philadelphia, PA 19104

In order to assess the potential of carbon nanotube (CNT) for
structural composite applications, a method was proposed to
convert CNT to continuous fibrils and fibrous assemblies
using the co-electrospinning process. Preliminary experiments
demonstrated the feasibility for the formation of composite
CNT fibrils. These fibrils provide a convenient material form
to carry the CNT and facilitate the formation of macro
composite structures. The level and the nature of CNT and
fiber alignments, and the inclusion of CNT in the nanofiber
were elucidated through SEM, TEM observations and Raman
spectra analysis. Mechanical testing of the nanofibril,
nanofibril spunbonded mats and composite yarns were carried
out to assess the effect of fibril alignment and verify the
nancomposite fibril concept.

Carbon nanotubes (CNTs) (*1*) are seamless graphene tubule structures with nanometer-size diameters and high aspect ratios. This new class of one-dimensional material is shown to have exceptional mechanical, thermal and novel electronic properties. The elastic moduli of the CNTs are in the range of 1-5 TPa (*2-4*) and fracture strains of 6 to 30%, both are about an order of magnitude better than those of the commercial carbon fibers, which typically have 0.1-0.5 TPa elastic moduli and 0.1-2% fracture strains (*5*). The factor of 10 enhancement in strength implies that, for the same performance, replacing the commercial carbon fibers with CNTs will lead to significant reduction in the volume and weight of the structural composites currently used in space applications.

Based on a NASA study by Harris et al (*6*) using micromechanics computation as shown in Table I, it was predicted that an order of magnitude increase in specific modulus can be achieved with CNT composites. However, it was also recognized that it would be significantly more challenging in the conversion of CNT to useful structures.

Table I. Properties of SWNT and their Composites (*6*)

Properties	Aluminum 2219-T87	IM7/8552 Quasi-isotropic Composite	CNT/Polymer Quasi-isotropic Composite*	CNT SWNT Crystal
Tensile strength (GPa)	0.46	1.3	2.5	180
Tensile Modulus (GPa)	73	58	240	1200
Rupture Elongation (%)	10	1.6	6	15
Density (g/cc)	2.83	1.59	0.98	1.2
Specific Strength	0.16	0.80	2.5	170
Specific Modulus	26	36	240	1000
Thermal Conductivity (W/mK)	121	5	5	5000
Manufacturability**	9	6-9	1	1

* based on 60% fiber volume fraction in a quasi-isotropic laminate, with strength at 1% strain

**manufacturability rating, in the range of 0 to10, with 10 being the best.

In addition to potential applications as high performance reinforcing fibers, carbon nanotubes are shown to have promising materials properties for applications as hydrogen storage materials (*7, 8*) high energy capacity battery electrodes (*9*), and cold-cathode electron emitters (*10, 11*). Depending on their chiralities and diameters, CNTs can be either semiconducting or metallic (*12*). The electrical conductivity of the metallic CNTs is (6000 S/cm) significantly higher than the best commercial carbon fibers. Because of their high degree of graphitization, CNTs are expected to have higher thermal conductivity (2000 W/m-K) than the best carbon fibers.

Despite their promises, most of the current studies are limited to the physics and chemistry of individual CNTs. There is limited knowledge on the properties of macroscopic materials comprising CNTs as the basic building blocks for macroscopic structures. It is still not clear whether the superb properties observed at the individual molecular level can be translated to the macroscopic structures. For example, most of the studies on nanotube - polymer composites have been on electron microscopy investigation of deformation of the individual CNTs embedded in polymeric matrices (*13-14*). No significant enhancement in the mechanical strength has been achieved in nanotube-polymer composites (*15*), presumably due to the weak interface between CNTs and composites. In order to realize the exciting potential of CNT, there is a need for processing methodologies to convert the CNT to macroscopic structures. Accordingly, it is the objective of this paper to introduce a concept that converts CNT into nano scale filamentous composites by the co-electrospinning process.

The Concept of CNT Nanocomposite Fibrils

It is well known that the translation of reinforcement properties to the composite depends on the alignment or orientation θ, of the reinforcement for a given volume fraction of the reinforcement, V_f, with θ and V_f functions of fiber architecture as illustrated in Figure 1 (*16*). Figure 1 shows the range of obtainable elastic moduli for various composites normalized by the fiber modulus E_f, versus the appropriate fiber volume for the fiber architecture indicated. It can be seen that the aligned fibers in discrete or continuous form have the most efficient translation of the material property of the reinforcing fibers to the composite.

CNTs are known to have highly anisotropic mechanical, thermal, and electrical properties. To measure and utilize these anisotropic properties, many attempts have been made to fabricate materials with controllable degree of CNT alignment. These methods include:

1. Mechanical stretching - nanotubes can be aligned inside polymeric matrices by mechanical stretching and developed procedures to determine the direction and the degree of alignment (*17*).

2. Roll-cast-membranes CNTs embedded in thermoplastic matrices by solution cast and produced composites with uni-axially aligned SWNTs by mechanical shearing (*15, 17*).

3. Magnetic alignment - thick film of SWNT and ropes are aligned by filtration/deposition from suspension in strong magnetic fields (*18*).

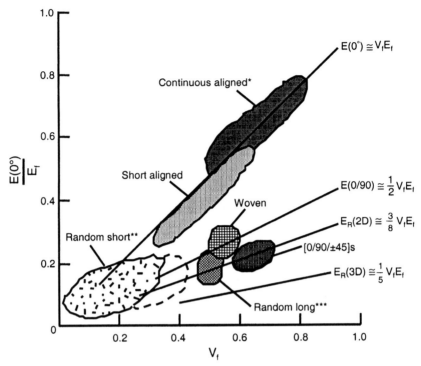

Figure 1. Effect of fiber architecture on material property translation in fiber reinforced composites (Reproduced with permission from reference 16. Copyright 1997 John Wiley & Sons.)

In this study the alignment of CNT and fibrils is demonstrated by co-electrospinning of mixtures of CNT and polymer solution to form aligned nanocomposite fibrils. In this electrostatic induced self-assembly process, ultrafine fibers down to the nanoscale are produced. In the electrospinning process, a high voltage electric field is generated between an oppositely charged polymer fluid contained in a glass syringe with a capillary tip and a metallic collection screen (*19*). Once the voltage reaches a critical value, the electric field overcomes the surface tension of the suspended polymer with cone formed on the capillary tip of the syringe (spinneret or glass pipette) and a jet of ultrafine fibers is produced. As the charged polymer jets are spun, the solvent quickly evaporates and the fibrils are accumulated on the surface of the collecting screen. This results in a nonwoven mesh of nano to micron scale fibers. A nanoscale fiber or nanofiber is also referred to as a fibril. Varying the electric field strength, polymer solution concentration and the duration of electrospinning can control the fiber diameter and mesh thickness. A schematic illustration and an example of a composite formed by the process are shown in Figure 2a. The concept of CNT nanocomposites (CNTNC) can be illustrated in Figure 2b, showing the orientation of the CNT in a polymer matrix through the electrospinning process by flow and charge induced orientation as well as confinement of the CNT in a nanocomposite filament (*20, 21*).

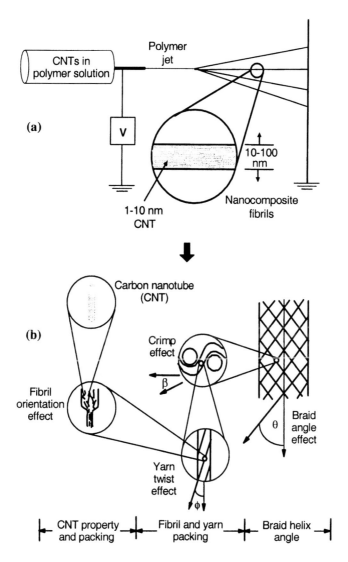

Figure 2. (a) Co-electrospinning of CNT; and (b) Concept of CNTNC.

The nanofiber composite can also be subsequently deposited as a spunbonded nanofibril mat for subsequent processing into composites or for use as a nonwoven mat as illustrated in a previous study (22). Various techniques of electrospinning have successfully produced aligned nanofibers (23-25). Alternately, as shown in Figure 3, by proper manipulation, the CNTNC filaments can be aligned as a flat composite filament bundle or twisted to further enhance handling and/or tailoring of properties in higher order textile preforms for structural composites.

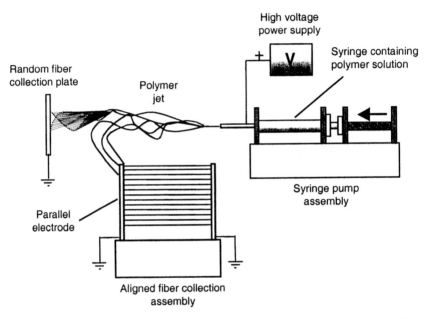

Figure 3. Schematic setup of the electrospinning of random and aligned fiber assemblies.

To demonstrate the co-electrospun CNTNC concept and to study the reinforcement and alignment effects, 1 wt. % SWNT-loaded PAN aligned nanofibers were produced and characterized. The produced aligned nanofibers of pristine PAN and SWNT-loaded PAN were characterized by a field emission environmental electron scanning microscope (ESEM) for fiber morphology and diameter measurement. The incorporation of SWNT and their orientation in the nanofibers was verified by Raman microspectroscopy and HRTEM. The effectiveness of the reinforcement and fibrils alignment was investigated through tensile testing.

Materials and Methods

Spinning Dope Preparation

SWNTs (produced by HiPco method) and MWNTs used for the study were produced at Rice University and Duke University, respectively. Polyacrylonitrile Dimethylformamide (DMF) and polyvinylpyrrolidone (PVP, $M_w \sim 10,000$) were purchased from Aldrich Chemical Co. For the PAN/SWNT spinning dope, 1 wt. % SWNT (weight percent relative to polymer) was added to DMF and magnetically stirred at ambient temperature for 24 hours. The SWNT/DMF suspension was then sonicated for 3 hours. Upon completion, 1 wt. % PVP (relative to PAN) was added to the dispersion to wrap the SWNT. Sonication continued for an additional hour before the mixture was transferred to a heating/stirring plate. PAN powder (11 wt. % relative to DMF) was incrementally added to the mixture and allowed to stir at 90° C for 3 hours.

Co-electrospinning of CNT/PAN Nanofibers

The homogenously mixed solution was transferred to a 3 ml capacity syringe with a 16G stainless steel needle attached. A syringe pump (KD Scientific, KDS-220) was used to maintain a consistent flow rate of the solution during spinning. The syringe pump assembly was mounted horizontally with the needle perpendicular to the collection plate, approximately 17 cm apart. A positive electrode was connected to the needle while the collection plate was grounded. The fiber collection plate consisted of a 12 cm x 12 cm copper plate covered with aluminum foil for ease of removal of the fiber mat. Modification was made to the traditional electrospinning process to allow simultaneous fabrication of random and aligned fiber configurations. A set of parallel electrodes was mounted adjacent to the polymer jet to collect aligned fibers as shown in Figure 3. A flow rate of 5 µl per minute and a potential of 10 kV were used in the electrospinning process. During the electrospinning process, some of the polymer jets traveled toward the front target and form a random mat, while the rest bent 90° and deposited between the parallel electrodes forming aligned fibers. The deposited fibers were collected in regular intervals of 2 – 3 minutes and carefully transferred to a holder.

Characterization

The collected fiber aligned fiber bundles were characterized by various techniques including scanning electron microscopy, Raman microspectroscopy, transmission electron microscopy, and tensile testing. A field emission environmental scanning electron microscope (ESEM, Phillips XL30) was used to measure the fiber diameter. To avoid over heating of the fiber sample, an accelerating voltage of 22 kV and a spot size of 3 were used to analyze the fiber samples. Thirty readings were taken at various locations within a 1cm x 1cm fibers mat. In the case of the aligned fiber bundles, same number of readings was taken along a 1 cm length specimen.

The inclusion of SWNT in the electrospun nanofibers was verified by Raman microspectroscopy (Renishaw 1000). For the random fiber sample, a small and thin sample was removed from the nonwoven mat and laid flat on a clean glass slide. Similarly, a small sample was removed from the aligned fiber bundle and laid flat on a glass slide. A low energy excitation wavelength (diode laser, $\lambda = 780$ nm) operating at 10 % power was used to analyze the pristine PAN and SWNT-loaded PAN in both random and aligned fibers.

Mechanical properties of the random and aligned fiber assemblies were characterized using a micro-tensile tester (Kawabata KES-G1). Strips of 5mm x 40mm were cut from the random fiber mat and the aluminum foil backing was carefully removed. The fiber specimen was then mounted on a paper frame with a gauge length of 30mm. Double-sided adhesive tape was used to mount the fiber strips to the paper frame for testing. For the aligned fiber assembly, yarns of approximately 300 denier and 50 mm in length were slightly twisted at 1 twist per centimeter. A gauge length of 30 mm was used for the aligned fiber bundles. Super glue was used to fix the yarn ends between the paper tabs. Silicon rubber was also used at the inner edges of the tabs to prevent stress concentration imposes during gripping. Tensile testing was carried out at an extension rate of 0.2 mm/sec. Five specimens were tested from each fiber samples.

In order to gain an understanding of the failure mechanism of the nanofibers, ESEM and HRTEM analysis were performed on the fracture surfaces of the tested specimens. The samples were prepared by drawing a small amount of the fibers at the failure surface using a pair of tweezers and then placed on the lacey carbon coated copper grids. HRTEM was performed using JEOL 2010F TEM with a relatively low accelerating voltage of 100 kV to minimize SWNT damage by the electron beam. Further verification of the inclusion of SWNT and their alignment in the fibers were also carried out by HRTEM.

Results and Discussion

Shown in Figure 4 are ESEM images of the electrospun aligned (Fig. 4a-c) and random (Fig. 4d) PAN nanofibers with 1 wt. % SWNT-loading. Although some misalignment of fibers are present but the degree of fibrils alignment in the yarn axis is considerably higher than that of the random fiber mats. Fiber diameters were measured by ESEM and the average diameter was determined to be approximately 370nm and 430nm for the pristine PAN and 1 wt. % SWNT/PAN, respectively. A large distribution of fiber diameter is observed in all cases. The fibers are relatively uniform in cross section and are free of beads and surface defects. A strong dependency of fiber diameter on polymer concentration was observed in the SWNT/PAN fibers. Incorporation of SWNT increases the solution viscosity resulting in thicker fibers. The presence of SWNT also changes the spinning dope properties leading to a narrower window for processing. Therefore, the polymer concentration must be adjusted accordingly in order to maintain spinnablility of the spinning dope. The degree of difficulty in spinning became greater as the SWNT loading increases, partly due to the significant change in solution viscosity resulting in higher concentration of SWNT aggregates. The inclusion of SWNT in the aligned composite nanofibrils was verified by Raman microspectroscopy. Similar characterization was performed on the aligned fiber assembly to ensure the inclusion of SWNT in the fibrils. Characteristic peaks (1561 and 1589 cm^{-1}) in the tangential and the radial breathing (RBM, 166 – 268 cm^{-1}) modes confirmed the incorporation of SWNT in the aligned fibers as well. As can be seen in

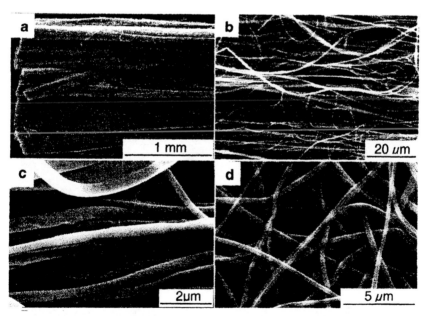

Figure 4. ESEM images of electrospun nanofibers. (a-c) aligned 1 wt. % SWNT/PAN fibers at various magnifications; and (d) random 1 wt. % SWNT/PAN fibers.

Figure 5, the Raman spectrum of pristine PAN fiber (top curve) contains no distinctive peak. The lower spectrum is of purified SWNT with characteristic peaks in the tangential mode (1561 cm^{-1} and 1589 cm^{-1}) and the radial breathing mode (140 – 256 cm^{-1}). The spectrum of the composite aligned nanofibril containing 1 wt. % SWNT (middle spectrum) appears similar to that of purified SWNT except that the peaks in the RBM are up shifted by 20 – 30 cm^{-1}. This up shift in wave number indicates the interaction between the SWNT and the PAN matrix. The electrospun SWNT/PAN random fibrils [13] possess similar peaks as those seen in the aligned fibers. Thus, confirming the success in incorporating the SWNT in the both the random and aligned electrospun composite nanofibers.

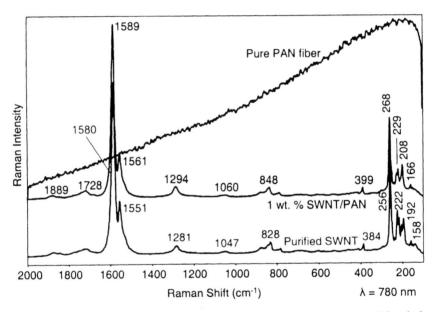

Figure 5. Raman spectra of purified SWNT, pristine PAN and SWNT-loaded PAN composite fibrils.

Results from Raman microspectroscopy confirmed the presence of SWNT in the samples. However, whether the SWNT are included inside or resided on the surface of the fibers could not be determined by Raman spectroscopy. Further confirmation by HRTEM is necessary in order to conclude that the SWNT are indeed incorporated in the fibers. Shown in Figure 6 are HRTEM images evidencing the incorporation of SWNT in the PAN fibers. Figure 6a is a featureless image of an unfilled PAN fiber whereas, Figure 6b shows the HRTEM image of the nanocomposite fibril that contains 1 wt. % of SWNT. The alignment of SWNT along the fiber axis is clearly visible. The alignment of SWNT along the fiber axis could be attributed to the following three mechanisms: (1) shear flow induced alignment; (2) charge induced alignment; and (3) fiber diameter confinement induced orientation of SWNT along the fiber direction.

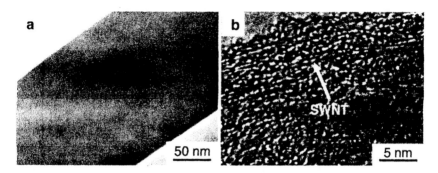

Figure 6. HRTEM of electrospun nanofibers. a) Pristine PAN; b) 1 wt. % SWNT/PAN.

In order to assess the translation efficiency of properties, tensile test of the aligned fiber bundles were performed. For comparison purposes, the random mat specimens were also tested under the same conditions. Figure 7 shows the stress-strain properties of electrospun pristine PAN and 1 wt. % SWNT/PAN nanofibers. The CNT reinforcement effect can be observed by comparing the stress-strain properties of the pristine PAN samples with that of the random fiber mat with 1wt% CNT. It can be seen that over two folds increase in tensile strength and elastic modulus was obtained with the addition of 1 wt. % SWNT. The increase of strength and modulus was off-set by an over 50% reduction of breaking elongation.

By aligning the CNT composite fibrils a significant increase in strength and toughness were realized. Specifically, the aligned 1 wt. % SWNT/PAN has a tensile strength of 30 MPa and over 15% elongation at break. The nearly two-fold increase in failure stress and strain compared to that of the random fiber mat indicates that the reinforcement effect of CNT can be significantly enhanced by proper alignment of the composite fibrils. Of particular interest is the large increase in the area under the stress-strain curve of the aligned SWNT/PAN specimen, indicating the effectiveness of SWNT in toughening and strengthening the fibers.

In order to understand the deformation and failure behaviors of the nanocomposite fibrils, ESEM and HRTEM examination of the failure surfaces

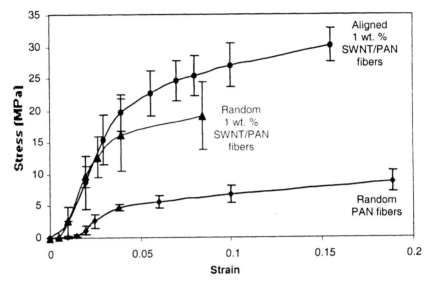

*Figure 7. Tensile properties of electrospun pristine PAN and 1 wt. %
SWNT/PAN fibers in random and aligned configuration.*

were performed (Figure 8a). As shown in Figure 8, a substantial deformation in the form of necking prior to failure was observed. Such extensive multiple necking (Figure 8b-c) are characteristic behavior of materials with high toughness and ductility. Figure 8d-e show ESEM and HRTEM images of a bundle of SWNT protrudes from the end of a rupture fiber. It appears that the fibril bundle was pulled out during the necking process. SWNT pulled out suggested poor bonding between SWNT and PAN matrix. However, pull out mechanism consumes extra energy which is responsible for the increase in toughness. Proper tailoring of interfacial bonding will allow better load transferring therefore would further increase the toughness and strength of the fiber.

As previously reported (26), further increase in SWNT content did not increase the mechanical properties of the nanofiber mat. This was due to the non uniform dispersion of SWNT in the solution since aggregates act as inclusion that weakened the fiber. The level of dispersion is a function of the dispersing medium, the dispersion method, the type of polymer, and the CNT and polymer concentrations. Increase SWNT content creates more challenge in obtaining a high degree of dispersion. As the SWNT concentration increase the fibers become more brittle resulting in inconsistency in sample preparation. A combination of defect inclusion due to poor SWNT dispersion and defect introduced to the fiber during sample preparation could be the contributing factors. An improved dispersion method has been developed and preliminary results indicate significant improvement in the degree of CNT dispersion.

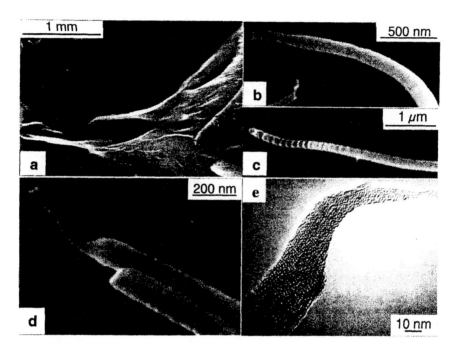

Figure 8. ESEM images of ruptured fiber specimens. (a-d) rupture surface of random fiber strip; (b, c) rupture ends of 1 and 10 wt. % SWNT/PAN nanofibers showing multiple necking prior to failure; (d) Pullout of SWNT from tensile testing; and (e) HRTEM image showing ductile failure behavior of SWNT/PAN nanofiber.

Conclusions

In order to assess the potential of CNT for structural composite applications, a method was proposed to convert CNT to continuous fibrils and fibrous assemblies using the electrospinning process. Preliminary experiments demonstrated the feasibility for the formation of composite CNT fibrils. These fibrils provide a convenient material form to carry the CNT and facilitate the formation of macro composite structures. The level and the nature of CNT alignment have been elucidated through TEM/AFM observations and spectra analysis. Mechanical testing of the nanofibril, nanofibril spunbonded mats and composite yarns were carried out to assess the level of alignment and verify the nanocomposite fibril concept.

Our preliminary results have successfully demonstrated that CNTs reinforced polymer nanofibers can be produced by the co-electrospinning process. Up to 10 wt. % SWNT reinforcing PAN can be electrospun in a range of diameters ranging from 50 – 400 nm by controlling the SWNT/PAN concentration in the solution. It is demonstrated in this study that orientation of SWNTs in the fiber direction as well as fibrillar alignment in the yarn assembly can greatly enhance the properties of the fibers. Our continuing effort in improving the dispersion of CNT, interfacial bonding between CNT and matrix through post-treatment of the fiber promises to maximize the translation of the superior properties of CNT to larger scale composite structures. The feasibility in producing continuous aligned composite nanofibers that allows them to be integrated into textile structure for advanced composite applications has been demonstrated. Furthermore, the ability to co-electrospin with various polymer/reinforcement systems and reinforcement concentrations allow a wide range of multifunctional composite nanofibers to be produced thus suitable for many applications.

It should be further noted that none of the electrospun specimen has undergone any post treatment such as drawing, annealing and further heat treatment at various temperature regimes. Future study will also explore higher level of CNT loading and optimization of dispersion and spinning processes. Accordingly, with proper tailoring of CNT-matrix interface and post treatment, it is anticipated that further improvement of the mechanical properties and translation efficiency of the CNT properties to the nanocomposite assemblies is expected.

Acknowledgements

This work was supported by the National Science Foundation through the Integrative Graduate Education and Research Traineeship (NSF-IGERT) program. Support from NASA and the Pennsylvania Nanotechnology Institute are also gratefully appreciated.

References

1. Ijimi, S. *Nature.* **199,** 354, 53-58.
2. Treacy, M. M.; Ebbesen, T.W.; Gibson, J.M. *Nature.* **1996,** 381, 67-80.
3. Wang, E.W.; Sheehan, P.E.; Lieber, C.M. *Science.* **1997,** 277(26), 1971-1975.
4. Lu, J.P. *Phys. Rev. Lett.* **1997,** 179:1297.

5. Peebles, L. H. *Carbon Fibers: Formation, Structure, and* Properties; CRC Press, Inc.: Boca Raton, 1995.
6. Harris, C.E.; Starnes, Jr., J.H.; Shuart, M.J. *An Assessment of the State of the Art in the Design and Manufacturing of Large Composite Structures for Aerospace vehicles*; April 2001, NASA/TM-2001-210844.
7. Dillon, A.C.; Jones, K.M.; Belledahl, T.A.; Kiang, C.H.; Bethune, D.S.; Heben, M.J. *Nature.* **1997**, 386,377-379.
8. Ye, Y.; Ahn, C.C.; Witham, C.; Fultz, B.; Liu, J.; Rinzler, A.G.; Colbert, D.; Smith, K.A.; Smalley, R.E. *Appl. Phys. Lett.* **1999**, 74(16), 2307-2309.
9. Gao, B.; Kelinhammes, A.; Tang, X.P.; Bower, C.; Wu, Y.; Zhou, O. *Chem. Phys. Lett.* **1999**, 307, 153-157.
10. de Heer, W.A.; Chatelain, A.; Ugarte, D. *Science.* **1995**, 270, 1171-1180.
11. Zhu, W.; Bower, C.; Zhou, O.; Kochanski, G.P.; Jin, S. *Appl. Phys. Lett.* , **1999**, 75(6), 873-875.
12. Saito, R.; Dresselhaus, G.; Dresselhaus, M.S. *Physical properties of Carbon Nanotubes;* Imperial College Press: 1998.
13. Laurie, O.; Cox, D.E.; Wagner, H.D. *Appl. Phys. Lett.* **1998**, 81(8), 1638-1641.
14. Bower, C.; Rosen, R.; Jin, L.; Han, J.; Zhou; O. *Phys. Lett.* **1999**, 74(22), 3317-3319.
15. Schadler, L.S.; Giannaris, S.C.; Ajayan, P.M. *Appl. Phys. Lett.* **1999**, 73, 3842-3844.
16. *Advanced Composites Manufacturing;* Gutowski, T.G. ed.; John Wiley & Sons: New York, 1997, pp 1-41.
17. Jin, L.; Bower, C.; Zhou, O. *Appl. Phys. Lett* **1999**,.73(9), 1197-1199.
18. Smith, Z.; Benes, Z.; Luzzi, D. E.; Fischer, J. E.; Walters, D. E.; Casavant, M. J.; Schmidt, J.; Smalley, R. E. *Appl. Phys. Lett.* **2000**, 77, 663.
19. Reneker, D.; Chun, I. Nanotech**1996**, 7, 216-233.
20. Han, W., Ko, F. K.; Rosen, R.; Zhou, O. *Mater. Res. Soc. Symp. Proc.* Fall Meeting, 2000.
21. Ko, F.; *In Advanced Composites Manufacturing, Processing of Preforms for Composites*; Gutowski, T.G. Ed.; Wiley Interscience: 1997
22. Ko, F.; Gogotsi, Y.; Ali, A.; Naguib, N., Ye, H.; Yang, G.; Li, C.; Willis, P. *Adv. Mater.* **2003**, 15, 1161-1165.
23. Li, D.; Wang, Y.; Xia, Y. *Adv. Mater.* **2004**, 16, 4, 361-366.
24. Fennessey, S. F.; Farris, R. J. *Polymer.* **2004**, 45, 4217-4225.
25. Katta, P.; Alessandro, M.; Ramsier, R. D.; Chase, G. G. Nano Lett. **2004,** in press.
26. Lam, H.; Titchenal, N.; Naguib, N.; Ye, H.; Gogotsi, Y.; Ko, F. *Mater. Res. Soc. Symp. Proc.* **2003**. Q10.5.1.

Chapter 18

Polymer Nanofibers and Polymer Sheathed Carbon Nanotubes for Sensors

D. Aussawasathien[1], P. He[1,2], and L. Dai[1,3,*]

[1]Department of Polymer Engineering, College of Polymer Science and Polymer Engineering, The University of Akron, Akron, OH 44235
[2]Chemistry Department, East China Normal University, Shanghai, China
[3]Department of Chemical and Materials Engineering, School of Engineering, University of Dayton, Dayton, OH 45469

Conducting polymers, carbon nanotubes, and their hybrid structures are promising for various sensing applications. While the electrospun lithium perchlorate ($LiClO_4$)-doped polyethylene oxide (PEO) nanofibers have been shown to be good humidity sensing materials, camphosulfonic acid (HCSA)-doped polyaniline (PANI)/polystyrene (PS) electrospun nanofibers, after having electrochemically immobilized with glucose oxidase (GOX), exhibited a good sensitivity for glucose sensing due to their large specific surface area and good electrical properties. We have also demonstrated that aligned carbon nanotubes electrochemically coated with appropriate conducting polymers can be used for making new highly sensitive glucose sensors, and that aligned carbon nanotubes chemically grafted with single-strand DNA chains of specific base sequences could be used for sensing complementary DNA and/or target DNA chains with a high sensitivity and selectivity.

Introduction

Measurement represents one of the oldest methods used by human beings to better understand and control the world. Many measurement systems are primarily physical sensors, which measure time, temperature, weight, distance, and various other physical parameters. The need for cheaper, faster and more accurate measurements has been a driving force for the development of new systems and technologies for measurements of materials, both chemical and biological. In fact, chemical and biological sensors (or biosensors) are the evolved products of physical measurement technologies. Chemical sensors are measurement devices that convert a chemical or physical change of a specific analyte into a measurable signal, whose magnitude is normally proportional to the concentration of the analyte. On the other hand, biosensors are a subset of chemical sensors that employ a biological sensing element connected to a transducer to recognize the physicochemical change and to produce the measurable signal from particular analytes, which are not necessary to be biological materials themselves although sometimes they are (1).

Depending on the basis of the transduction principle, chemical and biological sensors can be classified into three major classes with different transducers: sensors with electrical transducers, sensors with optical transducers, and sensors with other transducers (e.g. mass change). The unusual electronic and/or optical properties of conducting polymers (both electronically and ionically conducting polymers) have made them very attractive as transducer-active materials in various sensing devices. Indeed, electronically conducting polymers based on conjugated structures with alternating single and double bonds have been demonstrated to possess interesting optoelectronic properties, arising from the highly delocalized electrons over the extended π-orbital along the polymer backbone (2). Their electrical conductivities depend strongly, among many other factors, on the charge-transfer interaction with various electron-accepting/donating dopants (3). Ionically conducting polymers are, however, normally prepared by dissolving salt species in solid polymer hosts and their conductivities result from the ion migration between coordination sites repeatedly generated by the local motion of polymer chain segments (2). Consequently, the conductivity of ionically conducting polymers may change with any phenomenon that changes the mobility of the polymer chain segments and/or the solvation state of the ionic charge carriers. Electronically and ionically conducting polymers are thus promising for sensing applications. The early work on the conductivity measurements of polyacetylene films upon doping with iodine, bromine or AsF_5, and subsequent compensation with NH_3, thus constitutes the simplest conducting polymer gas sensors (4). Similarly, various humidity sensors have been devised from ionically conducting polymers by measuring conductivity changes associated with the dissolution of counter ions by the adsorbed water *(5, 6, 7)*. However, the rather low surface to volume ratio of these film-type sensors strongly limited their device performance.

With the recent significant advances in nanoscience and nanotechnology, the development of micro- or nano-structured polymers of large surface/interface areas has opened up novel fundamental and applied frontiers, which has attracted tremendous interest in recent years *(8)*. In this regard, conducting polymer micro- or nano-structures (e.g. conducting polymer nanofibers and polymer sheathed carbon nanotubes) provide ideal transducer-active systems for sensing applications. In particular, electrospinning has recently been studied as an economic and effective method for large scale fabrication of non-woven mats of nanofibers *(9)*. The electrospinning process produces continuous polymer fibers with nano- or micro-scale diameters through the action of an external electric field imposed on a polymer solution or melt. Non-woven mats of electrospun nanofibers thus produced possess a large specific surface area, low bulk density, and high mechanical flexibility. The conductivity of an individual electrospun conducting nanofiber composite (e.g. 72 wt% HCSA-doped PANI/PEO) has been measured, which is almost 300 times higher than the corresponding value of a cast film due to the highly aligned nature of the PANI chains in the electrospun fiber *(10)*. The high surface to volume ratio and good electrical properties associated with electrospun conducting polymer nanofibers make them very attractive for a wide range of potential applications *(9)*. Of particular interest, we present here the use of the $LiClO_4$-doped PEO and HCSA-doped PANI/PS electrospun nanofibers for humidity and glucose sensing applications, respectively.

Having conjugated structures with a unique molecular symmetry, carbon nanotubes also show interesting electronic, photonic, mechanical and thermal properties *(11)*. Like conjugated conducting polymers, the principles for carbon nanotube sensors to detect the nature of gases are based on change in electrical properties induced by charge transfer with the gas molecules (e.g. O_2, H_2) *(4)*. For these sensors, therefore, large electrode surfaces of the sensor array and good conductivity of the electrode will enhance significantly their performance. Thus, the high surface area and good electronic property provided by carbon nanotubes is an attractive feature in the advancement of a chemical or biosensor. However, the number of analytes that can be determined using a carbon nanotube based sensor is limited by the very few transduction mechanisms employed, mainly the conductivity measurements. Aligned carbon nanotubes not only provide a well-defined large surface area for a sensor but also facilitate surface modification of the carbon nanotubes with various transduction materials for broadening the scope of analytes to be detected by the nanotube sensor *(4, 12)*. Indeed, we have prepared novel aligned conducting polymer-carbon nanotube coaxial nanowires by electrochemically depositing a concentric layer of an appropriate conducting polymer onto the individual aligned carbon nanotubes *(13)*. We have also successfully developed novel glucose sensors by encapsulating glucose oxidase onto the conducting polymer-coated aligned carbon nanotubes.

In this article, we will summarize our work on the development of polymer nanofiber humidity and glucose sensors and polymer sheathed aligned carbon nanotube glucose and DNA biosensors.

Experimental

PEO (M.W. = 400,000), emeraldine base PANI (M.W. = 65,000) and PS (M.W. = 200,000) were supplied by Aldrich, as were LiClO$_4$, HCSA, and pyrrole. Glucose oxdase (GOX, 15,500 units/g) and dextrose glucose were purchased from Sigma. Buffer solution (pH = 7.0) was received from Fisher Scientific. A 0.1 M buffer solution (pH = 7.45) was prepared from sodium phosphate in Milli-Q water. The aligned carbon nanotubes were prepared by pyrolyzing iron (II) phthalocyanine under Ar/H$_2$ at 900°C according to our previously published procedures *(12)*. To construct a nanotube electrode for electrochemical work, the *as-synthesized* aligned carbon nanotube film was transferred onto a thin layer of gold *(12)*.

For DNA sensing, oligonucleotides with an amino group at the 5'-phosphate end (i.e. [AmC6]TTGACACCAGACCAACTGGT-3', I); the complementary oligonucleotide pre-labeled with ferrocenecarboxaldehyde, FCA, (designated as: [FCA-C6]ACCAGTTGGTCTGGTGTCAA-3', II); the FCA-labeled non-complementary oligonucleotide: [FCA-C6]CTCCAGGAGTCGTCGCCACC-3' (III); and the target oligonucleotide: 5'-GAGGTCCTCAGCAGCGGTGGACCAGTTGGTCTGGTGTCAA-3' (IV) were purchased from QIAGEN Operon. Ferrocennecarboxaldehyde (FCA), 1-[3-(Dimethylamino)propyl]-3-ethylcarbodiimide hydrochloride (EDC), tris (hydroxymethyl) aminomethane hydrochloride (Tris), and dodecyl sulfate sodium salt (SDS) were purchased from Aldrich. The 2×SSC buffer (0.3 M NaCl + 0.03 M sodium citrate, pH = 7.0) and TE buffer (10 mM Tris-HCl + 1mM EDTA, pH = 7.0) were used. All other chemicals were obtained from commercial sources and used without further purification.

Fabrication and characterization of LiClO$_4$-doped PEO nanofiber humidity sensors

A predetermined amount of PEO sample was dissolved in a mixture solvent of ethanol and distilled water (EtOH/H$_2$O = 0.7 w/w) to produce 5 wt% PEO solution, to which 1.0 wt% LiClO$_4$ was added at room temperature under stirring. To construct a conductometric sensor for humidity measurements, the homogenous PEO/LiClO$_4$ solution was then electrospun onto an interdigited comb-shaped aluminum electrode supported by a glass substrate (1.25 cm^2). Electrospinning was performed at room temperature in air under a voltage of ca.30 kV and at a distance of 30 cm between the polymer droplet and nanofiber collector (vide infra). As a reference, the corresponding film-type humidity sensor was prepared by spin-casting the PEO/LiClO$_4$ solution onto the comb-shaped aluminum electrode under the same conditions. Both the nanofiber- and film-type sensors were dried in a vacuum oven at 30°C for 1 hr, followed by measuring their humidity-resistance characteristics over a humidity range from 25 to 65% controlled by a humidifier (Lasko) at 25°C. Scanning electron microscopy (SEM, Hitachi S-2150 SEM unit) was used to characterize the electrospun fibers in the LiClO$_4$-doped PEO nanofiber sensor before and after the humidity-resistivity measurements.

Fabrication and characterization of HCSA-doped PANI/PS nanofiber glucose sensors

Prior to electrospining, a mixture of 2 wt% PANI, 4 wt% HCSA, and 7.5 wt% PS in chloroform was magnetically stirred over night to produce a homogeneous solution for electrospining under an applied voltage of 30 kV at room temperature in air. The gap distance between the tip of the pipette and the collector (aluminum foil) was 30 cm. The resulting HCSA doped-PANI/PS non-woven mat nanofibers were transferred from the Al collector onto a redox inactive substrate (0.5 cm x 0.5 cm) for the electrochemical measurements. For comparison, an HCSA-doped PANI/PS film sensor was also prepared by spin-coating the same solution under the same conditions. Both the HCSA-doped PANI/PS nanofiber and film sensors were dried in a vacuum oven at 60°C for 2 hr.

Thereafter, GOX was immobilized onto the HCSA-doped PANI/PS nanofiber and film electrodes by electrodeposition from a solution containing 2.5 mg/ml GOX in acetate buffer (pH = 5.2) under a potential of 0.4 V at a scan rate of 100 mV/s for 1 hr *(14)*. The resulting GOX-containing HCSA-doped PANI/PS nanofiber and film sensors were then used to monitor concentration changes of H$_2$O$_2$ generated during the glucose oxidation reaction by measuring redox current at the oxidative potential of H$_2$O$_2$ (i.e. the amperometric method) by cyclic voltammetry (CV) performed on an AD Instruments Power Lab (4SP electrochemical unit). Prior to the glucose sensing measurements, however, the HCSA-doped PANI/PS nanofiber and film sensors were tested with the pristine H$_2$O$_2$. In both cases, a single compartment cell with a working volume of 2 ml phosphate buffer (pH = 7) was used. The three-electrode system consisted of the HCSA-doped PANI/PS working electrode, platinum counter electrode, and

Ag/AgCl reference electrode. Scanning electron microscopic (SEM) images were taken on a Hitachi S-2150 SEM unit under 20 kV for the HCSA-doped PANI/PS nanofiber sensor before and after the H_2O_2 and glucose measurements.

Fabrication and characterization of aligned carbon nanotube-conducting polymer coaxial nanowire (CP-NT) glucose sensors

To construct the aligned carbon nanotube electrode, a piece of gold-supported aligned carbon nanotube film was cut into 1.5 mm × 1.0 mm sheets *(12)*, and then connected with a thin platinum wire using the commercially available conductive silver epoxy glue (Chemtronics®). The nanotube electrodes thus prepared were then used for electropolymerizing conducting polymers around individual aligned carbon nanotubes with Ag/AgCl and platinum foil as the reference and counter electrode, respectively. In particular, GOX immobilization was performed by oxidation of pyrrole (0.1 M) in a solution containing 0.1 M $NaClO_4$ and 2 mg/ml GOX in a phosphate buffer solution (pH = 7.4) at 10°C. A constant potential of 1.0 V was employed for the electropolymerization for 1 min. Amperometric response of the prepared electrodes to glucose was examined in 0.1 M phosphate buffer solution (pH = 7.4) by measuring the oxidation current at a potential of ca. 0.25 V. The redox potential of H_2O_2 has been previously demonstrated to shift towards lower voltages with nanostructured electrodes *(15)*. The background current was allowed to stabilize for at least 30 min before injection of glucose. Figure 1 shows the procedure for fabricating an aligned carbon nanotube-DNA electrochemical sensor, which involves plasma activation of the Au-supported aligned carbon nanotubes with acetic acid for chemical immobilization of H_2N-oligonucleatide chains *(16)*.

Briefly, the plasma polymerization was carried out at 20 W, 200 kHz under acetic acid vapor pressure of 0.1 torr for less than 1 min. The plasma-activated aligned carbon nanotubes were then reacted with H_2N-oligonucleatide chains in the presence of EDC, followed by washing with TE buffer for 5 min. The oligonucleotide-immobilized aligned carbon nanotube electrode was then immersed into the hybridization buffer (2×SSC) containing the FCA labeled DNA probe. The solution was incubated in a water bath at 42°C for 2 hr. After hybridization, the electrode was washed for more than three times with 0.4 M NaOH and 0.25% SDS solution to remove the physically absorbed DNA chains, if any. The electrochemical response from the hybridized FCA-labeled DNA chains was recorded in a 1 ml electrochemical cell using the hybridized aligned carbon nanotubes as a working electrode, Ag/AgCl as a reference electrode, and platinum wire as a counter electrode. The CV measurements were performed in a 0.1 M H_2SO_4 electrolyte solution, and the scan potential window was from 0.0 V to +0.8 V.

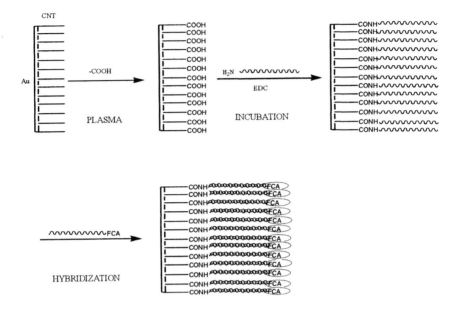

Figure 1. Schematic illustration of the procedure for fabricating an aligned carbon nanotube electrochemical DNA sensors (Reproduced with permission from reference 16. Copyright 2003 John Wiley & Sons, Inc.).

Results and discussion

LiClO₄-doped PEO nanofiber humidity sensors

Polymer nanowires and naofibers are attractive for a large variety of potential applications. The large-scale fabrication of polymer nanowires/nanofibers at a reasonably low cost has been a big challenge. Electrospinning, a technique that was first patented in 1934, *(17)* has recently been re-examined, notably by Reneker's group, *(18, 19, 20)* as an effective method for large-scale fabrication of ultrafine polymer fibers. As schematically shown in Figure 2, the electrospinning process involves the application of a high electric field between a droplet of polymer fluid and a metallic collection screen at a distance *h* from the polymer droplet. As electrically charged jet flows from the polymer droplet toward the collection screen when the voltage reaches a critical value (typically, ca.20 kV for *h* = 0.2 m) at which the electrical forces overcome the surface tension of the polymer droplet. The electrical forces from the charge carried by the jet further cause a series of electrically driven bending instabilities to occur as the fluid jet moves toward the collection screen. The repulsion of charge on adjacent segments of the fluid jet causes the jet to elongate continuously to form ultrafine polymer fibers. The solvent evaporates and the stretched polymer fiber that remains is then collected on screen as a non-woven sheet composed of one fiber many kilometers in length. The electrospun polymer fibers thus produced have diameters ranging from several microns down to 50 nm or less. This range of diameters overlaps conventional synthetic textiles and extends to diameters two or three orders of magnitude smaller *(21)*.

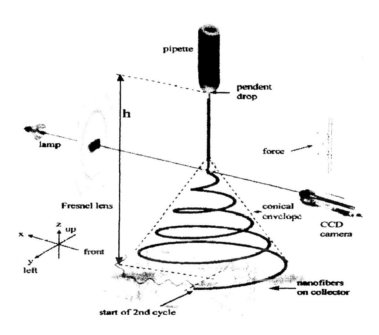

Figure 2. Schematic drawing of the electrospinning process (Reproduced with permission from reference 20. Copyright 2000 American Institute of Physics).

The electrospinning process schematically shown in Figure 2 was used for spinning the LiClO$_4$-doped PEO nanofibers over the comb-shaped aluminum electrodes on the glass substrate. A typical SEM micrograph of the electrospun LiClO$_4$-doped PEO fibers thus prepared is given in Figure 3 (A), which shows fiber diameters in the ranges of ca.250-600 nm. Electrospinning conditions, including the applied voltage, the concentration of polymer solution, and the nature of solvent used, will affect the fiber diameter. Nanofibers with desirable fiber diameters could be obtained by optimizing the electrospinning conditions. As we shall see later, the porous structures, together with the high surface-to-volume ratio, allowed a fast humidification and desiccation. After carrying out humidity measurements on the nanofiber sensor, however, some deformation of the nanofibers was observed due to the moisture-induced change in the fiber structure (Figure 3 (B)), having approximate fiber diameters in the ranges of ca.250-800 nm. Consequently, the as-prepared LiClO$_4$-doped PEO nanofiber sensors might be utilized only as a disposable humidity sensor, though further modification of the material/device design could circumvent this problem.

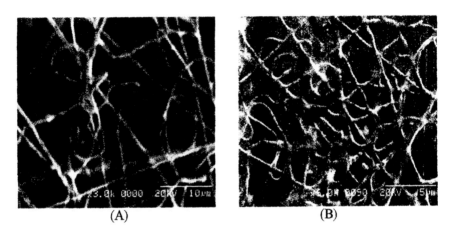

(A) (B)

Figure 3. SEM images of electrospun LiClO$_4$-doped PEO: (A) before, (B) after humidity measurements.

Figure 4 shows changes in the resistance (in the logarithm scale) as a function of % humidity for LiClO$_4$-doped PEO nanofiber and film sensors, respectively. A linearly inverse proportional relationship between the resistance and the % humidity was observed in both cases. However, the rate of resistance reduction with the % humidity is much higher for the LiClO$_4$-doped PEO nanofiber sensor than that of the corresponding LiClO$_4$-doped PEO film sensor, as reflected by the different values of the slopes for the two lines. The greater value of the slope for the nanofiber sensor (ca. 0.06 vs. 0.01) indicates the higher humidity sensitivity, most probably, due to its high specific surface area, as mentioned earlier. As also seen in Figure 4, the initial resistance of the electrospun fiber mat was higher than that for a cast film, though they have been made from the same starting solution. The lower conductivity values for the electrospun fibers than those of cast films can be attributed to the porous nature of the non-woven electrospun fiber mat as the present method measures the volume resistivity rather than the conductivity of an individual fiber. Although the measured conductivity for the electrospun mat of conducting polymer nanofibers is relatively low, their porous structures, together with the high surface-to-volume ratio and good electrical properties, have been demonstrated to be important advantages for the development of advanced humidity sensors with a high sensitivity.

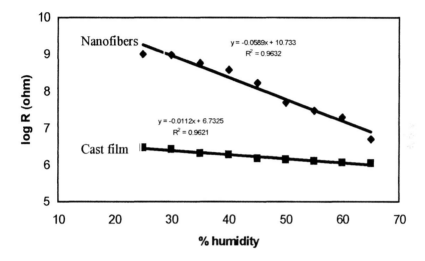

Figure 4. Humidity dependence of resistivity: LiClO$_4$-doped PEO nanofibers and film.

HCSA-doped PANI/PS nanofiber glucose sensors

Since the discovery of the first electronically conducting polymer, namely polyacetylene, by Shirakawa et. al. in 1977 *(22)*. Many conjugated conducting polymers, including PANI, have been prepared for a wide range of potential applications. Examples include electrochromic devices *(23, 24)*, batteries *(25)*, capacitors *(26)*, photoelectrochemical cells, light-emitting diodes, anti-static coatings *(27)*, sensors *(28)*, and conducting composites with conventional polymers *(29, 30, 31, 32, 33)*.

The relatively high environmental stability and excellent electrical properties, together with the fact that PANI can be made from inexpensive raw materials, make PANI an ideal electronically conducting polymer for sensing applications. In the case of glucose sensing, GOX was immobilized onto the conducting polymer layer. Upon reaction with glucose, gluconolactone and hydrogen peroxide were produced as shown in Figure 5.

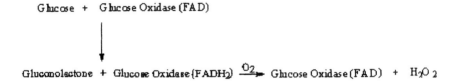

Figure 5. Schematic illustration of the reaction during glucose sensing measurement.

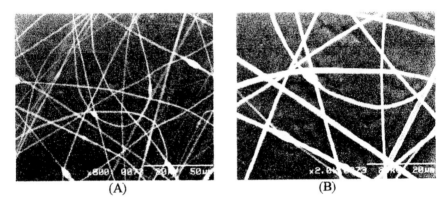

Figure 6. (A) Scanning electron micrographs of HCSA doped-PANI/PS nanofibers; (B) as for (A), under a higher magnification.

Thus, the detection of H_2O_2 generated by the enzymatic reaction could serve as the basis of signal generation for the glucose sensing *(34)*. Owing to the clinical significance of measuring glucose levels in blood, numerous attempts have been made to construct low-cost glucose sensors with a high sensitivity and selectivity. For instance, electrodeposited conducting polymers, including polypyrrole, poly(N-methylpyrrole), poly(o-phenylenediamine) and PANI, have been used as matrices for the immobilization of glucose oxidase in the preparation of amperometric glucose biosensors *(35, 36)*. Since the rate of electrochemical reaction is proportional to the surface area of an electrode and diffusion rate of the electrolyte, the surface area of the electrode plays an important role in many electrochemical sensors, including the glucose sensor. The high surface area and low bulk density characteristic of the electrospun fiber mats (vide supra) make conducting polymer electrospun fibers very attractive for electrochemical sensing applications. Of particular interest, we report here the use of HCSA-doped PANI/PS electrospun nanofibers, after having been electrochemically immobilized with GOX, for glucose sensing applications. Corresponding results from the HCSA-doped PANI/PS film will also be discussed as appropriate. Figure 6 (A) and (B) show typical SEM images of the as-spun HCSA-doped PANI/PS nanofibers, in which bead fiber-like structures of diameters in the range of 600-1500 nm are clearly evident.

The CV spectra measured at various H_2O_2 concentrations for the HCSA-doped PANI/PS nanofiber electrode are given in Figure 7 (A). The corresponding CV spectra for a spin-cast HCSA-doped PANI/PS film sensor were also measured under the same condition. As expected, the film sensor shows much weaker current signals than those from the nanofiber sensor over a large range of the redox potential. Although a linear response of the peak redox current to the H_2O_2 concentration was observed for both the HCSA-doped PANI/PS nanofiber and film sensors, the nanofiber sensor shows a much higher sensitivity, as indicated by a much larger value for the slope (Figure 7 (B)).

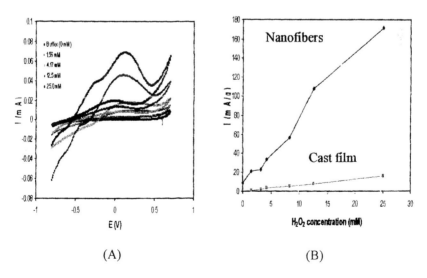

(A)   (B)

Figure 7. (A) CVs of HCSA doped-PANI/PS nanofiber sensors at different H_2O_2 concentrations. (B) The current response of HCSA-doped PANI/PS nanofiber and spin-cast sensors at various H_2O_2 concentrations. Note that the current response has been scaled by the weight of the polymer material deposited on the electrodes. *(See page 13 of color inserts.)*

Just as the measurements on pure H_2O_2 has shown a higher sensitivity for the HSCA-doped PANI/PS nanofiber sensor than the corresponding film sensor, the amperometric response from the GOX-immobilized HCSA-doped PANI/PS nanofiber sensor to glucose was also found to be much higher than that of the HCSA-doped PANI/PS film sensor. As can be seen in Figure 8, the redox current at the oxidative potential of H_2O_2 increased with increasing glucose concentrations. Therefore, the HCSA-doped PANI/PS electrospun nanofibers could be promising for making new glucose sensors with a higher sensitivity.

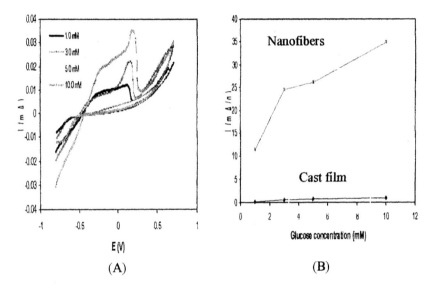

(A)   (B)

Figure 8. (A) CVs showing the H_2O_2 peaks with varying concentration of glucose solution for HCS-doped PANI/PS nanofiber sensor. (B) The current response of HCSA-PANI/PS nanofiber and spin-cast film sensors at various glucose concentrations. Note that the current response has been scaled by the weight of the polymer material deposited on the electrodes (scan rate = 100 mV/s). *(See page 13 of color inserts.)*

Unlike the $LiClO_4$-doped PEO nanofiber sensor, there was no observable deformation for the H_2O insoluble HCSA-doped PANI/PS nanofiber sensors, the fiber diameters (ca.400-1000 nm) remain unchanged (see Figure 9 (A)-(C)), suggesting reusability of the PANI nanofiber sensor

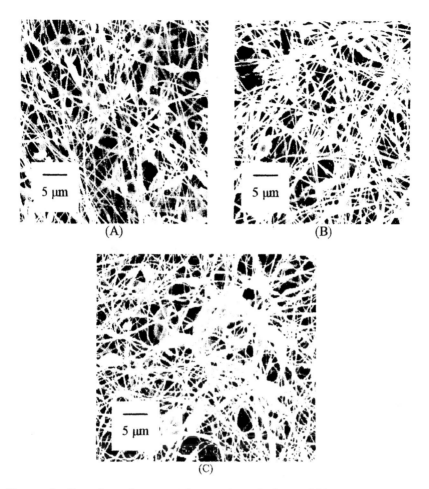

Figure 9. Scanning electron micrographs of the HCSA-doped PANI/PS nanofibers; (A) before and (B) after H_2O_2 measurements and (C) after glucose sensing.

Polymer sheathed carbon nanotube glucose sensors

As briefly mentioned earlier, we have used the aligned carbon nanotubes produced from FePc to make novel aligned carbon nanotube-conducting polymer (CP-NT) coaxial nanowires by electrochemically depositing a concentric layer of an appropriate conducting polymer onto each of the constituent aligned nanotubes as shown in Figure 10 *(37)*.

Figure 10. A typical SEM image of the CP-NT coaxial nanowires produced by cyclic voltammetry on the aligned carbon nanotube electrode, showing a thin layer of conducting polymer (polypyrrole) coating surrounding each of the constituent aligned carbon nanotubes. Scale bar is 1 μm. (Reproduced with permission from reference 37. Copyright 2000 Wiley-VCH Publishers, Inc.).

The electrochemical performance of the aligned CP-NT coaxial nanowires was evaluated by carrying out CV measurements. As for PANI films electrochemically deposited on conventional electrodes, the CV response of the PANI-coated nanotube array in an aqueous solution of 1 M H_2SO_4 (Figure 11A (a)) shows oxidation peaks at 0.33 and 0.52 V (but with much higher current densities) *(38)*. This indicates that PANI films thus prepared are highly electroactive. As a control, the CV measurement was also carried out on the bare aligned nanotubes under the same conditions (Figure 11A (b)). In the control experiment, only capacitive current was observed with no peak attributable to the presence of any redox-active species. The coaxial structure allows the nanotube framework to provide mechanical stability *(39, 40)* and efficient thermal/electrical conduction *(41, 42)* to and from the conducting polymer layer. The large surface-interface area obtained for the nanotube-supported conducting polymer layer is an additional advantage for using them in sensing applications. In this context, we have immobilized GOX onto the aligned carbon nanotube substrate by electrooxidation of pyrrole (0.1 M) in the presence of GOX (2 mg/ml) and $NaClO_4$ (0.1 M) in a pH 7.45 buffer solution *(43)*. Then, the GOX-containing polypyrrole-carbon nanotube coaxial nanowires were used to monitor concentration change of H_2O_2 during the glucose oxidation reaction by measuring the increase in the electrooxidation current at the oxidative potential of H_2O_2 (i.e. the amperometric method).

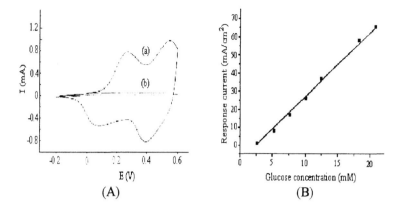

Figure 11. (A) Cyclic voltammograms of (a) the polyaniline-coated CP-NT coaxial nanowires and (b) the bare aligned carbon nanotubes. Measured in an aqueous solution of 1 M H_2SO_4 with a scan rate of 50 mV/s (Reproduced with permission from reference 37. Copyright 2000 Wiley-VCH Publishers, Inc.). (B) The dependence of electrooxidation current at the oxidative potential of H_2O_2 on the glucose concentration for the CP-NT coaxial nanowire sensor (Reproduced with permission from reference 43. Copyright 2003 Wiley-VCH Publishers, Inc.).

As shown in Figure 11 (B), a linear response of the electrooxidation current to the glucose concentration was obtained for the CP-NT nanowire sensor *(43)*. The linear relationship extends to a glucose concentration as high as 20 mM, which is higher than the 15 mM typical limit used for the detection of blood glucose in practice *(44)*. Furthermore, the amperiometric response was found to be about ten orders of magnitude higher than that of more conventional flat electrodes coated with glucose oxidase-containing polypyrrole films under the same conditions *(44)*. The CP-NT nanowire sensors were also demonstrated to be highly selective, with their amperiometric responses being almost unchanged even in the presence of some interference species, including ascorbic acid, urea and D-fructose. Therefore, the CP-NT nanowires could be used for making new glucose sensors with a high sensitivity, selectivity and reliability.

Aligned carbon nanotube-DNA electrochemical sensors

A major feature of the Watson-Crick model of DNA is that it provides a vision of how a base sequence of one strand of the double helix can precisely determine the base sequence of the partner strand for passing the genetic information in all living species. The principle learned from this breakthrough

262

has been applied to the development of biosensors for DNA analysis and diagnosis through the very specific DNA pairing interaction *(45)*.

Owing to their high sensitivity, low cost, and good compatibility with optical detection technologies, DNA biosensors with optical transducers are currently under intense investigation *(47, 48)*. Comparing with those optical transduction methods, the electronic transduction has many advantages including easier data processing, greater simplicity, and broader applicability. Consequently, a few of DNA electrochemical biosensors have recently been reported *(46)*. Due to their unique electronic properties *(11)* and large surface area, aligned carbon nanotubes provide additional advantages for making electrochemical biosensors. In this regard, Meyyappan and co-workers *(49)* have recently developed a very sensitive DNA sensor by immobilizing DNA chains onto aligned carbon nanotubes protected with a spin-on glass (SOG) layer *(50)*. In a somewhat related, but independent study, we have developed a simple, but effective, method for preparing aligned carbon nanotube-DNA sensors by chemically coupling DNA probes on both the tip and wall of plasma-activated aligned carbon nanotubes *(51)*.

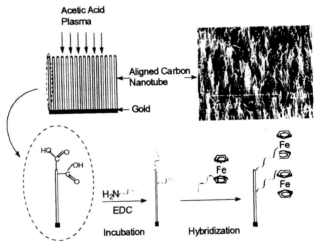

Figure 12. A schematic illustration of the aligned nanotube-DNA electrochemical sensor. The up-right SEM image shows the aligned carbon nanotubes after having been transferred onto a gold foil. For reasons of clarity, only one of the many carboxyl groups is shown at the nanotube tip and wall, respectively (Reproduced with permission from reference 51. Copyright 2004 The Royal Society of Chemistry).

In particular, we carried out acetic-acid-plasma treatment *(51)* on gold-supported aligned carbon nanotubes generated from pyrolysis of iron (II) phthalocyanine *(12)* , followed by grafting single-strand DNA (ssDNA) chains with an amino group at the 5'-phosphate end (*i.e.* [AmC6]TTGACACCAGACCAACTGGT-3', I) onto the plasma-induced COOH group through the amide formation in the presence of EDC coupling reagent. Complementary DNA (cDNA) chains pre-labeled with ferrocennecarboxaldehyde, FCA, ([FCA-C6]ACCAGTTGGTCTGGTGTCAA-3', II) were then used for hybridizing with the surface-immobilzed oligonucleotides to form double strand DNA (dsDNA) helices on the aligned carbon nanotube electrodes (Figure 12).

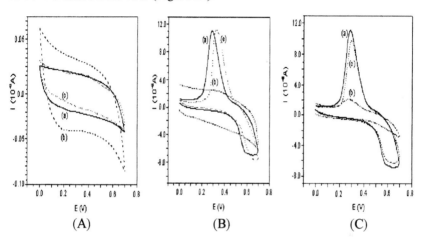

(A)   (B)   (C)

Figure 13. CVs of (A) the pristine aligned carbon nanotube electrode (a), the nanotube electrode after the acetic-acid-plasma treatment (b), and the plasma-treated nanotube electrode after the immobilization of ssDNA (c); (B) the ssDNA (I)-immobilized aligned carbon nanotube electrode after hybridization with FCA-labeled complementary DNA (II) chains (a), in the presence of FCA-labeled noncomplementary DNA (III) chains (b), and after hybridization with target DNA (IV) chains in the presence of the FCA-labeled noncomplementary DNA (III) chains (c); and (C) the ssDNA (I)-immobilized aligned carbon nanotube electrode after the 1st hybridization with FCA-labeled complementary DNA (II) chains (a), after being denatured (b), and after the 2nd hybridization with FCA-labeled complementary DNA (II) chains (c). All the cyclic voltammograms were recorded in 0.1 M H_2SO_4 solution with a scan rate of 0.1V/s. The concentration of the FCA-labeled DNA probes is 0.05 µg/ml. Thermal denaturing was achieved by heating the combined cDNA probes on the aligned carbon nanotube electrode in 2xSSC buffer solution (i.e. 0.3 M NaCl + 0.03 M sodium citrate, pH=7.0) at 100°C for 6 min, followed by subsequent rapid cooling in a salt ice bath for 10 min. (Reproduced with permission from reference 51. Copyright 2004 The Royal Society of Chemistry).

Figure 13 (A) shows the electrochemical characteristics for the pristine aligned carbon nanotubes before (curve a) and after (curve b) the acetic-acid-plasma treatment, along with the ssDNA-immobilized nanotubes (curve c), in 0.1 M electrolyte H_2SO_4 solution. Only capacitive current was observed for the pristine aligned carbon nanotubes (curve a of Figure 13 (A)). The capacitive current increased after treating the nanotube electrode with the acetic acid plasma (curve b of Figure 13 (A)), presumably because the plasma-induced carboxyl groups facilitated the charge-transfer between the nanotube electrode and H_2SO_4 electrolyte through the enhanced hydrophilic-hydrophilic interaction (51). Upon grafting the ssDNA chains onto the plasma-induced surface carboxyl groups, a significant decrease in the capacitive current was observed (curve c in Figure 13 (A)), indicating the replacement of carboxyl groups by a thin layer of the covalently-grafted DNA chains. The performance of the surface-bound ssDNA(I) chains on the plasma-treated nanotube electrode for sequence-specific DNA diagnoses was demonstrated in Figure 13 (B). The strong reduction peak seen at 0.29 V in curve a of Figure 13 (B) can be attributable to ferrocene (53, 54) and indicates the occurrence of hybridization of FCA-labeled cDNA(II) chains with the nanotube-supported ssDNA(I) chains, leading to a long-range electron transfer from the FCA probe to the nanotube electrode through the DNA duplex (55, 56). In contrast, the addition of FCA-labeled non-complementary DNA chains (i.e. [FCA-C6]CTCCAGGAGTCGTCGCCACC-3', III) under the same conditions did not show any redox response of FCA (curve b of Figure 13 (B)). This indicates that, as expected, there was no specific DNA pairing interaction with the non-complementary DNA chains, and that physical adsorption of the FCA-labeled DNA chains, if any, was insignificant in this particular case. Subsequent addition of target DNA chains (i.e. 5'-GAGGTCCTCAGCAGCGGTGGACCAGTTGGTCTGGTGTCAA-3', IV) into the above solution, however, led to a strong redox response from the FCA-labeled DNA (III) chains (curve c of Figure 13 (B)) because the target DNA (IV) contains complementary sequences for both DNA (I) and DNA (III) chains.

More interestingly, the electrochemical responses seen in Figure 13 (B) were revealed to be highly reversible. As can be seen in Figure 13 (C), the electrochemical response of the FCA-labeled cDNA(II) chains (curve a of Figure 13 (C)) diminished almost completely after being thermally denatured from the aligned carbon nanotube electrode (curve b of Figure 13 (C)). Re-hybridization with fresh FCA-labeled cDNA(II) chains, however, led to a rapid recovery of the electrochemical response characteristic of FCA (curve c of Figure 13 (C)).

The above results suggest that the ssDNA immobilized aligned carbon nanotubes can be repeatedly used as a highly-selective electrochemical sensor for sequence-specific DNA diagnoses. Furthermore, the amperiometric response from the aligned carbon nanotube-DNA sensors (curve a of Figure 14 (A)) was found to be much higher (ca.20 times) than that of more conventional flat electrodes immobilized with the ssDNA(I) chains under the same conditions (curve b of Figure 14 (A)). The linear dependence of the redox current on the complementary DNA concentration shown in Figure 14 (B) further ensures the use of the aligned carbon nanotube-DNA sensors for DNA sensing and/or sequence-specific diagnoses over a wide range of the cDNA concentrations.

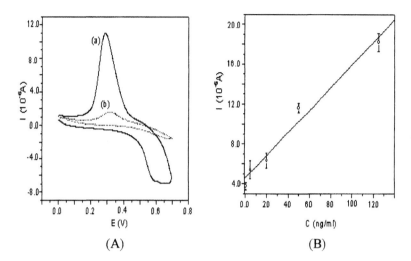

(A)   (B)

Figure 14. (A) Cyclic voltammograms of the aligned carbon nanotube electrode immobilized with ssDNA(I) chains followed by hybridization with the FCA labeled cDNA(II) probes (a) and an Au electrode immobilized with ssDNA(I) chains followed by hybridization with the FCA-labeled cDNA(II) probes under the same conditions (b). Note, the geometric area of the aligned carbon nanotube electrode is 1.5 mm × 1.0 mm and the area of the gold is 2.0 mm × 1.5 mm. The electrochemical measurements were carried out in an aqueous solution of 0.1 M H_2SO_4 v.s. Ag/AgCl at a scan rate of 0.1V/s. The concentration of the FCA labeled cDNA(II) probe is 0.05 µg/ml. (B) The dependence of redox current at the reduction potential of FCA (0.29 V) on the cDNA(II) concentration for the aligned carbon nanotube-DNA sensor. (Reproduced with permission from reference 51. Copyright 2004 The Royal Society of Chemistry).

Conclusion

In summary, we have demonstrated that conducting polymers, carbon nanotubes, and their hybrid structures are promising for various sensing applications. While the electrospun $PEO/NaClO_4$ nanofibers have been shown to be good humidity sensing materials, HCSA-doped PANI/PS electrospun nanofibers, after having electrochemically immobilized with glucose oxidase, exhibited a good sensitivity for glucose sensing due to their large specific surface area and good electrical properties. The size of the electrospun fibers may affect the sensitivity of the particular kind of sensors. Generally speaking, the smaller diameter (higher surface area) would yield better sensing signals.

We have also demonstrated that aligned carbon nanotubes coated with appropriate conducting polymers (e.g. CP-NT coaxial nanowires) can be used for making new highly sensitive glucose sensors. Furthermore, single-strand DNA chains of specific base sequences could be covalently immobilized onto plasma-activated aligned carbon nanotubes for sensing complementary DNA and/or target DNA chains of specific sequences with a high sensitivity and selectivity.

The above results should have an important implication to the use of polymer and carbon nanostructures for a wide range of sensing applications. This, coupled with so many micro-/nano-fabrication and surface modification methods already reported and more to be developed, will surely lead to the development of sensors and sensor arrays based on the polymer nanofibers and carbon nanotubes (e.g. sensor arrays from micropatterns of aligned carbon nanotubes and their derivatives with conducting polymer coatings) of practical significance. Clearly, a promising potential for future research and development exists in this area.

Acknowledgment

We thank our colleagues for contributions to the research reviewed in this article. In particular, we are grateful to Darrell H. Reneker, Woraphon Kataphinan, Gordon G. Wallace and Mei Gao. We are also grateful for the financial support from the American Chemical Society (ACS-PRF 39060-AC5M), National Science Foundation (NSF-CCF-0403130), and the Materials and Manufacturing Directorate of the Air Force Research Laboratory, University of Dayton, The Dayton Development Coaltition, and Wright Brothers Institute for supporting the WBI Endowed Chair Professorship in Nanomaterials to LD.

References

1. *Polymer Sensors and Actuators;* Osada, Y.; De Rossi, D. E., Eds.; Springer-Verlag, Berlin, 2000.
2. Dai, L. *Intelligent Macromolecules for Smart Devices: from Materials Synthesis to Device Applications;* Springer-Verlag, London, 2004.

267

3. Dai, L.; Winkler, B.; Huang, S.; Mau, A. W. H. In *Semiconducting Polymers: Applications, Properties, and Synthesis;* Hsieh, B. R.; Wei, Y., Eds.; American Chemical Society: Washington, DC, 1999.
4. Dai, L.; Soundarrajan, P.; Kim, T. *Pure Appl. Chem.* **2002**, *74*, 1753.
5. Sadaoka, Y.; Sakai, Y. *J. Mater. Sci.* **1986**, *21*, 235.
6. Xin, Y.; Hirata, M.; Yosomiya, R. *Proc. Transducers* **1987**, *87*, 669.
7. Huang, H.; Dasgupta, P. K. *Anal. Chem.* **1991**, *63*, 1570.
8. *Encyclopedia of Nanoscience and Nanotechnology;* Nalwa, H. S., Ed.; American Scientific Publisher, CA, 2004.
9. Doshi, J.; Srinivasan, G.; Reneker, D. H. *Polym. News* **1995**, *20*, 206.
10. MacDiarmid, A. G.; Jones, W. E. Jr.; Norris, I. D.; Gao, J.; Johnson, A. T. Jr.; Pinto, N. J.; Hone, J.; Han, B.; Ko, F. K.; Okuzaki, H.; Llaguno, M. *Synth. Met.* **2001**, *119*, 27.
11. Peter J. F. Harris; *Carbon Nanotubes and Related Structures-New Materials for the Twenty-first Century;* Cambridge University Press: Cambridge, 2001.
12. Dai, L.; Patil, A.; Gong, X.; Guo, Z.; Liu, L.; Liu, Y.; Zhu, D. *ChemPhysChem* **2003**, *4*, 1150.
13. Gao, M.; Huang, S.; Dai, L.; Wallace, G.; Gao, R.; Wang, Z. *Angew. Chem. Int. Ed.* **2000**, *39*, 3664.
14. Garjonyte, R.; Malinauskas, A. *Biosensors and Bioelectronics* **2000**, *15*, 445.
15. Gao, M.; Dai, L.; Wallace, G. *Electranaly.* **2003**, *15*, 1089.
16. Dai, L.; He, P.; Li, S. *Nanotechnology* **2003**, *14*, 1081.
17. Formhals, A. U.S. Patent 1,975,504, 1934.
18. Doshi, J.; Reneker, D. H. *J. Electrost.* **1995**, *35*, 151.
19. Reneker, D. H.; Chun, I. *Nanotechnology* **1996**, *7*, 216.
20. Reneker, D. H.; Yarin, A. L.; Fong, H.; Koombhongse, S. *J. Appl. Phys.* **2000**, *87*, 4531.
21. *Nanowires and Nanobelts Vol. II: Materials, Properties and Devices;* Wang, Z. L., Ed.; Kluwer Academic Publishers: Dordrecht, 2003.
22. Shirakawa, H.; Louis, E. J.; MacDiarmid, A. G.; Chiang, C. K.; Heeger, A. J. *J. Chem. Soc., Chem. Chem. Commun.* **1977**, 578.
23. Genies, E. M.; Boyle, A.; Lapkowisky, M.; Tsivantis, C. *Synth Met.* **1990**, *36*, 139.
24. Cao, Y.; Smith, P.; Heeger, A. *J. Synth. Met.* **1992**, *48*, 91.
25. Somasiri, N. L. D.; MacDiarmid, A. G. *J. Appl. Electrochem.* **1998**, *18*, 92.
26. *Handbook of Organic Conductive Molecules and Polymers;* Arbizzani, C.; Mastragostino, M.; Scrosati, B.; Nalwa, H. S., Eds.; Wiley and Sons: New York, 1997.
27. *Metallized Plastics 5 & 6: Fundamental and Applied Aspects;* Fornari, C. C. M. Jr.; Heilmann, C.; Mittal, K. L., Eds.; VSP BV: The Netherlands, 1998.
28. Bartlet, P. N.; Birkin, P. N. *Synth. Met.* **1993**, *61*, 15.

268

29. Genies, E. M.; Boyle, A.; Lapkowisky, M.; Tsivantis, C. *Synth. Met.* **1990,** *36,* 139.
30. Mattoso, L. H. C. *Quimica Nova* **1996,** *19,* 388.
31. Ziberman, M.; Narkis, M. *J. Appl. Polym. Sci.* **1997,** *66,* 243.
32. Paul, E. W.; Ricco, A. J.; Wrighton, M. S. *J. Phys. Chem.* **1985,** *89,* 1441.
33. Tsutumi, H.; Yamashita, S.; Oishi, T. *J. Appl. Electrochem.* **1997,** *27,* 477.
34. Adeljou, S. B.; Shaw, S. J.; Wallace, G. *Anal. Chem. Acta.* **1997,** *341,* 155.
35. Trojanowicz, M.; Krawczy Ski Vel, T.; Geschke, O.; Cammann, K. *Sensors and Actuators B: Chem.* **1995,** *28,* 191.
36. Garjonyte, R.; Malinauskas, A. *Biosensors and Bioelectronics* **2000,** *15,* 445.
37. Gao, M.; Huang, S.; Dai, L.; Wallace, G.; Gao, R.; Wang, Z. *Angew. Chem. Int. Ed.* **2000,** *39,* 3664.
38. Sazou, D.; Georgolios, C. *J. Electroanal. Chem.* **1997,** *429,* 81.
39. Poncharal, P.; Wang, Z. L.; Ugarte, D.; de Heer, W. A. *Science* **1999,** *283,* 1513.
40. Gao, R.; Wang, Z. L.; Bai, Z.; de Heer, W. A.; Dai, L.; Gao, M. *Phys. Rev. Lett.* **2000,** *85,* 622.
41. Frank, S.; Poncharal, P.; Wang, Z. L.; de Heer, W. A. *Science* **1998,** *280,* 1744.
42. Odom, T. W.; Huang, J. L.; Kim, P.; Lieber, C. M. *J. Phys. Chem. B* **2000,** *104,* 2794.
43. Gao, M.; Dai, L.; Wallace, G. *Electranaly.* **2003,** *15,* 1089.
44. Yasuzawa, M.; Kunugi, A. *Electrochem. Commun.* **1999,** *1,* 459.
45. Gooding, J. J. *Electroanalysis* **2002,** *14,* 1149.
46. Dai, L. *J. Aust. Chem.* **2001,** *54,* 11.
47. Li, J.; Lu, Y. *J. Am. Chem. Soc.* **2000,** *122,* 10466.
48. Biggins, J. B.; Prudent, J. R.; Marshall, D. J.; Ruppen, M.; Thorson, J. S. *Proc. Natl. Acad. Sci., U. S. A* **2000,** *97,* 13537.
49. Nguyen, C. V.; Delzeit, L.; Cassell, A. M.; Li, J.; Han, J.; Meyyappan, M. *Nano Lett.* **2002,** *2,* 1079.
50. Li, J.; Ng, H. T.; Cassell, A.; Fan, W.; Chen, H.; Ye, Q.; Koehne, J.; Han, J.; Meyyappan, M. *Nano Lett.* **2003,** *3,* 597.
51. He, P.; Dai, L. *Chem. Commun.* **2004,** 348.
52. Chen, Q.; Dai, L.; Gao, M.; Huang, S.; Mau, A. *J. Phys. Chem. B* **2001,** *105,* 618.
53. Xu, C.; He, P.; Fang, Y. *Anal. Chim. Acta* **2000,** *411,* 31.
54. Cai, H.; Wang, Y.; He, P.; Fang, Y. *Anal. Chim. Acta* **2002,** *469,* 165.
55. Giese, B.; Amaudrut, J.; Köhler, A. K.; Spormann, M.; Wessely, S. *Nature* **2001,** *412,* 318.
56. Wong, E. L. S.; Gooding, J. *J. Anal. Chem.* **2003,** *75,* 3845.

Chapter 19

Effect of Carbon Black Loading on Electrospun Butyl Rubber Nonwoven Mats

Nantiya Viriyabanthorn[1], Ross G. Stacer[1], Changmo Sung[2], and Joey L. Mead[1]

Departments of [1]Plastics Engineering and [2]Chemical Engineering, University of Massachusetts at Lowell, Lowell, MA 01854

The focus of this chapter is the effect of carbon black loading on the resulting fiber structure and tensile properties of thermoset butyl rubber membranes prepared using the electrospinning process. These materials offer potential for use in stretchable, selectively permeable membranes for chemical protective applications. The effect of carbon black loading (0, 25, 50, 75 per hundred parts rubber) on dimensional stability, fiber structure, and mechanical properties was studied. The addition of carbon black filler was found to decrease the fiber diameter and number of beads. Increasing carbon black content decreased the density and tensile modulus. When the tensile properties were corrected for the differences in density, the behavior of the electrospun butyl rubber membranes showed the expected increase with carbon black loading.

Part of this chapter is based on Nantiya Viriyabanthorn's thesis. She did the work at University of Massachusetts at Lowell in 2003. Doctor Viriyabanthorn holds copyright to the thesis. Those portions of the chapter are used with permission of the copyright holder.

Introduction

The electrospinning process produces submicron size fibers through the use of a strong electric field and a polymer solution (*1*). One application for the electrospinning process is the development of selectively permeable membranes for chemical protection. Current materials for protective clothing include polytetrafluoroethylene (PTFE) or flash-spun polyethylene membranes (*2, 3*). Although these two materials provide breathability, they lack the flexibility and stretch provided by elastomeric materials. Butyl rubber offers excellent resistance to a wide range of chemicals, as well as the flexibility of elastomers, but provides no breathability. Both breathability and stretch are important to providing comfort to the wearer. The ability of the electrospinning process to control porosity and surface area offers promise for the use of electrospun membranes in chemical protective applications. Research has been directed toward the use of the electrospinning process for the preparation of a protective membrane layer using polyurethane thermoplastic elastomers (*4, 5*). Recently, we demonstrated the ability to prepare selectively permeable membranes from thermoset butyl rubber compounds using the electrospinning process (*6*). These butyl rubber membranes showed improvement of moisture vapor transport over the solid butyl rubber material, while still providing resistance to liquid water and the ability to undergo high elongation (stretch).

In the rubber industry, the addition of carbon black plays a large role in the behavior of the rubber compound. The amount of carbon black loading affects viscosity, die swell, dimensional stability, as well as the finished product properties. Viscosity increases substantially with increased carbon black loading. This is caused by the additional attraction (van der Waals force) between aggregates. The relationship between carbon black and solution viscosity of rubber is given by Guth's equation (*7, 8*). Increasing carbon black typically increases the initial tensile modulus of the compound and generally decreases tensile elongation at break. Tensile strength often decreases with increasing filler content, however, in some black filled rubbers, tensile strength increases with carbon black loading to a maximum, and then decreases at higher loading (*8*). A number of equations have been developed to describe the tensile behavior of particulate-filled polymers, including the Mooney, Kerner, Hashin and Shtrikman, and Halpin and Tsai equations (*8*).

Although considerable research has been performed on the effect of carbon black on solid rubber properties, little work has been done to study the effect of carbon black loading on the electrospinning process. The addition of carbon black has been used to obtain a color change with temperature (themochromism) in electrospun fiber mats (*9*). The focus of this research was to study the effect of carbon black loading on the resulting fiber structure and tensile properties of electrospun thermoset butyl rubber membranes.

Experimental

Butyl rubber compounds with four different carbon black loading levels (0, 25, 50, and 75 parts per hundred rubber) were prepared using a standard formulation from the Vanderbilt Handbook (*10*) as shown in Table I. The compound was mixed at room temperature in a torque rheometer (Haake). The rubber compound was then dissolved in tetrahydrofuran (THF) (Aldrich) to prepare a solution for electrospinning at various viscosities from 0.05 Pa-s to 0.82 Pa-s. The viscosity of the solution was controlled by slowly adding solvent to obtain the desired viscosity. Viscosity measurements were performed using a Brookfield Viscometer.

TableI. Butyl Rubber Formulation

Material	Supplier	phr
Elastomer (Exxon Butyl 268)	R.T. Vanderbilt Company	100
Processing Aid (VANFRE AP-2)	R.T. Vanderbilt Company	2
Activator-metal oxide (Zinc oxide)	R.T. Vanderbilt Company	5
Activator-fatty acid (Stearic Acid)	R.T. Vanderbilt Company	1
Vulcanizing Agent (Sulfur)	R.T. Vanderbilt Company	2
Accelerator (Methyl Tuads)	R.T. Vanderbilt Company	1
Accelerator (Captax)	R.T. Vanderbilt Company	0.5

SOURCE: Reference 1.

The carbon black used in this study was a medium thermal (MT) black, ASTM N990 from Engineered Carbons, Inc. The average particle size is 250 – 350 nm, with a surface area of 9 m^2/g (BET N_2SA).

Polymer solutions were electrospun with a DC power source (Gramma High Voltage Research, Inc.) using voltages from 15 kV to charge the polymer solution contained in the syringe. The output of the polymer solution was controlled at 1 ml/minute using a metering pump connected to the syringe. An aluminum pan was used as the conductive target to collect the fiber. For mechanical testing, a rotational target was used in order to collect a uniform thickness membrane. The four different carbon black loadings were electrospun at four different viscosities, 0.05, 0.18, 0.61, and 0.82 Pa-S for each carbon black loading level (0, 25, 50 and 75 phr). Other processing parameters such as spinning voltage, and target distance, out put, etc., were held constant.

Scanning Electron Microscope (SEM) imaging was used to characterize fiber morphology. Electrospun samples were characterized using an AMRAY

1400 SEM, with a LaB$_6$ filament. The SEM images were taken at magnifications of 100 to 2000X at an accelerating voltage of 10 KeV. SEM images were printed at their original size. The fiber diameter was measured using ruler and converting to the corrected dimension using the SEM scale of the same image. At least 10 measurements were taken for each sample.

Transmission electron microscopy (TEM) imaging was performed on a Philips EM 400 transmission electron microscope to determine the dispersion of carbon black in the electrospun fibers. Fibers were electrospun directly onto a carbon coated copper specimen grid and dried in an air oven at 60°C for 10 minutes prior to TEM analysis.

Electrospun butyl samples for density and tensile test were subsequently cured in an air oven. The samples were heated to 60°C for 10 minutes to remove residual solvent followed by curing at 171°C for another 10 minutes. Density measurements were performed following ASTM D792(*11*). Tensile testing was performed on a universal testing machine (Instron Model 6025) with a crosshead speed of 5 cm/min. Specimens were prepared using a half scale ASTM D412 die (*12*). The crosshead displacement was used to measure the strain.

Results and Discussion

Fiber Morphology

Figure 1 shows the TEM images for the uncured butyl rubber membranes for a 40 phr loading. As seen in this Figure, the mechanical compounding procedure used to prepare the rubber stock, prior to electrospinning, appears to provide uniform distribution of carbon black throughout the fiber.

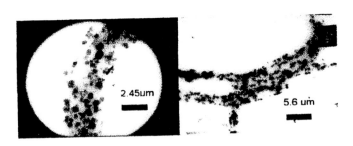

Figure 1. TEM images of electrospun butyl rubber with N990 type carbon black at 40 phr loading.

The addition of carbon black was found to play an important role in the electrospinning of butyl rubber membranes. Table II presents the results from the scanning electron microscopy (SEM) of uncured fibers as a function of solution spinning viscosity and carbon black content. For butyl rubber compounds without carbon black, it was difficult to form fibers. At low viscosity, the electrospun jet collapsed, resulting in a spray of droplets. At higher viscosity, fibers were formed, but did not maintain their shape. The resulting unfilled elastomer fibers exhibited merged fibers with many connection points to other fibers, as a result of material flow enhanced by the presence of any residual solvent after the fibers were collected.

TableII. Effect of Solution Viscosity and Carbon Black Content on Processing and Fiber Size and Shape

Carbon black (phr)	Viscosity (Pa-S)	Fiber Size (um)	Bead (size, um)	SEM Result
0	0.05			Droplets
0	0.18			Merged fibers
0	0.61			Merged fibers
0	0.82			Merged fibers
25	0.05			Droplets
25	0.18	7 - 15		Merged fibers
25	0.33	10 - 12	None	Fibers
25	0.82	16 - 22	None	Fibers
50	0.05	1 - 4	Significant	Fibers
50	0.18	3 - 6	Some	Fibers
50	0.33	6 - 8	None	Fibers
50	0.61	8 - 12	None	Fibers
50	0.82	12 - 14	None	Fibers
75	0.05	1 - 4	None	Fibers
75	0.18	2 - 5	None	Fibers
75	0.33	4 - 8	None	Fibers
75	0.61	6 - 10	None	Fibers
75	0.82	11 - 15	None	Fibers

At a carbon black loading of 25 PHR, droplets and merged fibers still occurred at low solution viscosity, however, at higher viscosity fibers were formed as shown in Figure 2. As seen in Table II, fibers were successfully made at all viscosities with carbon black contents of 50 PHR and 75 PHR. Merged fibers were the result of unvulcanized rubber molecule mobility during the fiber collection and before the curing process. Increasing the carbon black content, results in greater viscosity (7, 8) of the final fiber (less mobility) prior to curing, increased dimensional stability, and a wider range of processing conditions (viscosity).

The effect of carbon black loading in the butyl rubber compound on fiber morphology can also be seen in the density and density ratio of electrospun mats as shown in Table III. The density ratio was calculated from the electrospun mat density divided by the solid rubber density (determined as a function of carbon black loading). As seen in Table III, the density of the electrospun mat decreases as the carbon black loading increases, in contrast to a solid rubber compound where higher carbon black loading normally increases the density of the solid rubber due to the higher density of carbon black as compared to the butyl rubber. Density ratios can be used to compare the porosity of the fiber mat. A lower density ratio number indicates greater porosity of the membrane. Both density and density ratio of electrospun butyl mats decreased with increasing carbon black loading. This is due to the greater dimensional stability and decrease in fiber diameter (discussed below) with increasing carbon black content.

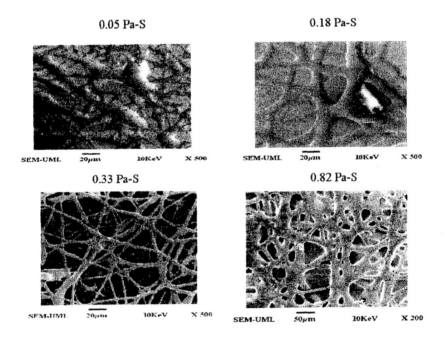

Figure 2. Fiber formation at various viscosities for butyl rubber compounds with 25 PHR carbon black loading.

TableIII. Density Properties of Butyl Rubber

Carbon black Loading (phr)	Density of spun mat (g/cm^3)	Density of solid rubber (g/cm^3)	Density ratio (Dspun/Dsolid)
0	0.97	0.97	1.00
25	0.89	1.05	0.84
50	0.80	1.12	0.72
75	0.70	1.19	0.59

Bead formation in electrospun fibers usually occurs when the fiber is spun from a low viscosity solution (13,14). The effect of carbon black loading on bead formation was studied using SEM. Bead structures were found to be significant in the fibers at the lowest viscosity (0.05 Pa-s) for the 50 PHR carbon black loaded compound. At this viscosity, the 0 phr and 25 phr carbon black loaded butyl rubber compounds did not form a stable fiber structure, but rather showed droplets and merged fibers. The development of bead formation in an electrospun mat with 50 phr carbon black loading can be seen in Figure 3. As the viscosity increased from 0.05 to 0.18 Pa-s, the distance between beads become longer, and the bead diameter larger. Bead occurrence ceased at viscosities of 0.33 Pa-s and higher. This result agrees with conclusions from other researchers (14). Increasing viscosity of the solution not only increases the fiber size, but also reduces or eliminates bead formation.

In comparison to the 50 phr carbon black loaded compound, the electrospun membranes from 75 phr carbon black loading did not exhibit the beads on string morphology at any of the viscosities evaluated as shown in Figure 4. This phenomenon can be explained in terms of the net charge density, which is inversely proportional to the resistivity. Fong and co-workers (14) reported that increasing the net charge density and the associated electrical force tends to favor the formation of smooth fibers. Adding more carbon black content in this experiment reduced the resistivity of the polymer solution, and thus eliminated the bead morphology in the fiber membrane.

The relationship between viscosity and electrospun fiber diameter has been reported by many researchers (13, 15). Similar to the electrospinning of other materials, increasing solution viscosity leads to the formation of larger diameter fibers. Figure 5 shows the plot of fiber diameter as a function of viscosity. For a given carbon black content the materials all show the expected response to solution viscosity, with the fiber diameter increasing as the viscosity increases.

For a given solution viscosity, increasing the carbon black loading leads to smaller diameter fibers. One explanation is that carbon black plays an important role in the viscosity-concentration relationship. An increase in carbon black loading leads to an increase in solution viscosity due to the incorporation of spherical particles in a liquid (7, 8). The relationship of low particle concentration to viscosity developed by Guth and Gold (7, 8) is shown in eq 1.

276

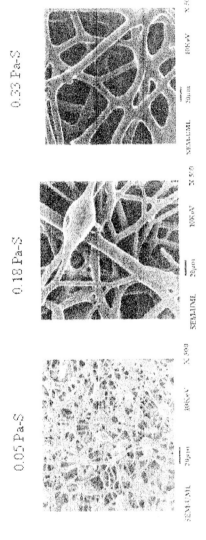

Figure 3. Bead formation in electrospun butyl rubber membranes with 50 phr carbon black loading at different viscosities.

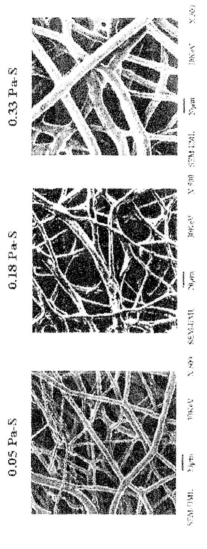

Figure 4. Fiber structures of electrospun butyl rubber membranes with 75 phr carbon black loading at different viscosities.

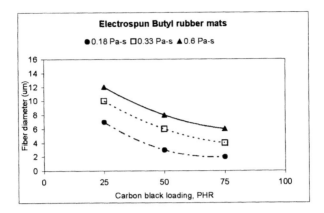

Figure 5. Fiber diameter of butyl rubber at different carbon black loadings and viscosities.

$$\eta = \eta_0 (1 + 2.5c + 14.1c^2) \tag{1}$$

Where η and η_0 are the compound and vehicle (polymer) viscosities respectively, and c is the volume concentration of the dispersed phase (in this case carbon black). From the viscosity aspect, higher carbon black loaded compounds will have less polymer content at the same solution viscosity. In this study, solution viscosity was controlled, but not the polymer concentration. Therefore, higher carbon black loaded solutions would have less polymer content compared with an unloaded compound at the same starting solution viscosity. This may explain the decrease in fiber diameter.

Conductivity of the solution from the presence of carbon black may also be a factor in the fiber diameter, as shown by other researchers there is a decrease in fiber diameter with increasing solution conductivity (16). In a previous study, the effect of carbon black loading on the electrical conductivity of electrospun mats was compared to the solid rubber sheet (17). As expected, the electrical conductivity increased with carbon black loading for both electrospun and solid rubber sheets. Comparison between the solid rubber material and the electrospun rubber showed that the electric conductivity of the microporous membranes increased to a lesser extent as compared to the solid materials as a function of carbon black content. This was explained by the decreased density of the electrospun mats as the carbon black loading increased. Although previous studies were performed using a different carbon black type, we would expect similar behavior for this system, perhaps to a lesser extent since N990 is a low structure, low conductivity black.

Tensile Properties

It is known that the physical properties of an elastomer can be affected by a change in the filler type and extent of loading (*18*). For electrospun mats one must also compensate for the change in density on the mechanical behavior. The effect of density on tensile properties of a microporous rubber membrane can be explained based on the theories of foams (*19*). Density plays the most important role in determining the tensile modulus and tensile strength of microporous material, however, density does not have a significant effect on the tensile elongation. To compensate for the density differences at different carbon black loadings, the density ratio was used to normalize these data to investigate the effect of filler loading only. Corrected values of tensile modulus and tensile strength were calculated by dividing the apparent data by the density ratio as presented in TableIV.

It has been previously demonstrated that electrospun mats show lower tensile modulus and tensile strength when compared to solid butyl rubber sheets of the same composition, even after normalization to account for differences in density (*6*). This result is likely due to the open fiber structure found in the nonwoven mat as compared to the solid rubber sheet. In the case of the nonwoven mat it would be expected that factors, such as fiber mobility and rotation, would play a large role in the mechanical properties of the structure (*20,21,22*) Farboodmanesh et al.(*23*) reported that the shear stress of fabrics depends on the fabric/composite structure and the ability to restrict the rotation of the yarns. This restriction of fiber rotation by the addition of a matrix material was found to cause changes in the mechanical behavior (*23*). Typically, the materials would show higher stress-strain response than that predicted by simple additivity of the response of the two components (*23*). Similarly, we would expect the solid butyl rubber sheet to behave like a material with restricted rotation, showing an enhanced modulus as compared to what would be expected by extrapolating the behavior of the electrospun mat to the density of the solid butyl rubber sheet. This difference in stress-strain behavior between the rubber in the nonwoven mat form versus the solid rubber sheet may be further complicated by factors such as straightening of curved or buckled fibers, which would also tend to result in lower modulus values at low elongations in the fiber form as compared to the solid material. As shown in Table IV and Figure 6 the electrospun mats show the anticipated increase in mechanical properties with carbon black loading, although they may not be directly compared to the solid rubber material.

As seen in Figure 6 (a), the addition of carbon black increased the 300% tensile modulus only slightly when the values were uncorrected for density differences. When the values were corrected for density differences, the addition of carbon black caused a significant increase in the normalized 300% tensile modulus of the electrospun butyl rubber membranes, increasing with increased carbon black content. Theoretically, adding carbon black to elastomers causes reinforcement of the polymer molecules, giving higher modulus values with increasing carbon black content for a solid rubber. Many

TableIV. Corrected Tensile Strength and Tensile Modulus of Electrospun Butyl Mats

CB Loading (phr)	Corrected Stress at Break (MPa)	Corrected 300% Modulus (MPa)
0	3.66	0.105
25	3.43	0.14
50	3.78	0.177
75	3.58	0.289

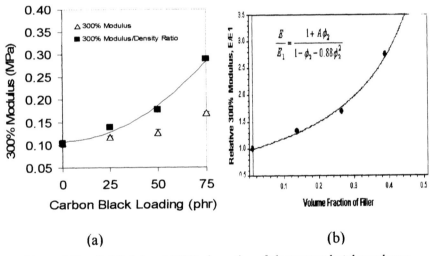

(a)   (b)

Figure 6. Tensile Modulus at 300% elongation of electrospun butyl membrane. (a) Comparison of 300 % modulus between apparent value and the value normalized by density ratio. (b) Fit of experimental normalized 300% modulus (dots) to the modified Kener or Halpin-Tsai equation (solid line).

researchers have discussed the relationship between tensile properties and the filler loading in composite materials (7, 8, 24, 25). The general equation for the filler-modulus relationship was rewriten by Nielsen and Lewis using the modified Kener or Halpin-Tsai equation (18) as shown in eq 2.

$$\frac{E}{E_1} = \frac{1 + A\,\phi_2}{1 - \phi_2 - 0.88\,\phi_2^2} \tag{2}$$

Where E and E_1 are the moduli for the filled and unfilled compounds, respectively, Φ_2 is the volume fraction of filler, and the constant A is a function of the geometry of the filler phase and Poisson's ratio of the matrix. The experimental results showed that the values of normalized 300% tensile modulus correlated well with the theoretical model as seen in Figure 6 (b).

Carbon black in rubber can decrease or increase tensile strength depending on many factors, including density, particle agglomeration, and de-wetting effects (adhesion between filler and matrix phase) (7, 8, 24, 25). In some black filled rubber, tensile strength increases with carbon black loading to a maximum, and then decreases at higher loading. Even though the agglomerates of carbon black may be strong enough to increase tensile modulus, it tends to reduce the tensile strength of the composite material. Particle agglomeration acts as a weak point or stress concentrator in the material, and thus it breaks easily when stress is applied. When density differences were not considered, the addition of carbon black decreased the apparent tensile strength at break of electrospun butyl membranes as shown in Figure 7. This result can be explained by the reduction in density with increasing carbon black content. When corrected for the density differences, there was no significant change in the corrected tensile strength at break for all carbon black loading levels. This indicates that density plays an important role in determining the tensile strength of the electrospun butyl mat.

Fillers generally decrease elongation at break because of fracture mechanisms that occur between the filler and matrix material. This is because the actual elongation of the polymer matrix is greater than the rigid filler. A simple model of tensile elongation in composite materials is shown in eq 3 (18).

$$\varepsilon_B = \varepsilon_b^0 \left(1 - \phi_2^{2/3}\right) \tag{3}$$

Where ε_b^0 is the elongation to break of the unfilled polymer. The elongation of the electrospun butyl rubber membranes decreased with increasing carbon black content as shown in Figure 8, however, more than 1000% elongation was observed in all cases. Figure 7 also presents the comparison of the theoretical curve for the relative elongation-to-break of the filled polymer as a function of filler concentration and the uncorrected experimental values for elongationto break of the electrospun butyl rubber membranes.

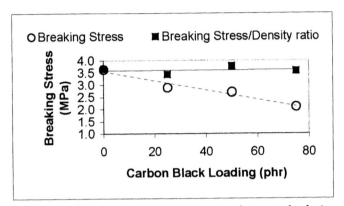

Figure 7. Tensile strength at break (apparent and corrected value) as a function of carbon black loading.

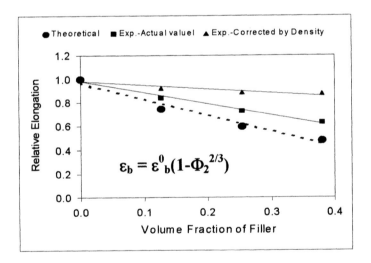

Figure 8. Comparison of experimental and theoretical elongation to break as a function of volume fraction of carbon black.

The uncorrected values of the elongation to break for the electrospun mats decreases less rapidly compared to the theoretical values, but followed the same general trend. Correction of elongation to break values for the density differences showed the results to exhibit even greater difference between the theoretical and the experimental values. These results are consistent with those of foam theory.(*19*), where it was found that the tensile elongation of a latex foam was not affected by differences in density, but rather correlated with the elongation of the solid rubber.

Conclusions

In this work we report, the effect of carbon black loading on the resulting fiber structure and mechanical properties (tensile) of a porous thermoset butyl rubber membrane produced by electrospinning. Increasing the carbon black loading provides a greater range of processing conditions (solution viscosities) for the preparation of microporous electrospun butyl fiber mats. Increased carbon black content also reduced the formation of beads in the fiber and decreased the fiber size. Density of the electrospun materials decreased with increasing carbon black content. Tensile modulus, after correcting for the difference in density, showed the expected increase in modulus with increased carbon black loading. Carbon black addition also resulted in a decrease in the tensile elongation of the butyl rubber mats as the carbon black content increased, however, more than 1000% elongation was observed in all cases. This material offers the potential to develop flexible, stretchable materials for a number of protective clothing applications.

Acknowledgements

The authors wish to acknowledge partial support from the National Science Foundation (Grant # DMI-0200498). We also wish to acknowledge the donation of materials from R.T. Vanderbilt.

References

1. Formhals, A. U.S. Patent 1,957504, 1934.
2 . Truong, Q. M.S. thesis, University of Massachusetts Lowell, Lowell, MA, 1999.

3. *Quick Selection Guide to Chemical Protective Clothing*; Forsberg, K.; Mandsdof, S.Z., Eds.; *Van Nostrand Reinhold*: New York, 1989.

4. Schreuder-Gibson, H.L.; Gibson, P.; Senecal, K.; Sennett, M.; Walker, J.; Yeomans, W.; Ziegler, D.; Tsai, P. *J. Adv. Mater.* **2002**, *34*, 44.

5. Gibson, P.; Schreuder-Gibson, H.; Rivin, D. *Colloids Surf., A* **2001**, *187-188*, 469-481.

6. Viriyabanthorn, N.; Stacer, R.G.; Sung, C.; Schreuder-Gibson H.; Gibson, P.; and Mead, J.L. Breathable Butyl Rubber Membranes formed by Electrospinning. *J. Adv. Mat.*, to be published.

7. *Rubber Material and their Compounds*; Brydson J.A., Ed.; Elsevier Applied Science: New York, 1998.

8. *Reinforcement of Elastomers*; Kraus, G., Ed.; John Wiley & Sons: New York, 1965; pp 319-390.

9. Pedicini, A.; Farris, R. J. *J. Polym. Sci.* **2004**, *42(5)*, 752-757.

10. *The Vanderbilt Rubber Handbook*; Ohm, R.F., Ed.; R.T. Vanderbilt Company: Norwalk, CT, 1990; pp 460-462.

11. *Annual Book of ASTM Standards*; Standard D792-91; American Society for Testing and Materials: Philadelphia, PA, 1998; Vol.08.01, pp 153-156.

12. *Annual Book of ASTM Standards*; Standard D412; American Society for Testing and Materials: Philadelphia, PA, 1999; Vol.09.01, pp 43-55.

13. Deitzel, J.M.; Kleinmeyer, J.; Harris, D.; Beck Tan, N.C. *Polymer*, **2001**, *42*, 261.

14. Fong, H.; Chun, I.; Reneker, D. H. *Polymer*, **1999**, *40*, 4585-4592.

15. Doshi J.; Reneker, D.H. *J. Electrostat.*, **1995**, *35*, 151-160.

16. Hohman, M., Shin, M., Rutledge, G.C., and Brenner, M. P. *Application, Phys. Fluids.*, **2001**, *13*, 1221-2236.

17. Viriyabanthorn, N. D.Eng. thesis, University of Massachusetts Lowell, Lowell, MA, 2003.

18. *Machanical Properties of Polymers and Composites*; Nielsen, L.E.; Landel R.F., Eds.; Marcel Dekker: New York, 1994; pp 377-411.

19. *Polymeric Foams*; Klempner, D.; Frisch, K.C., Eds.; Hanser, New York, 1991.

20. Chou,T.W.; *J. Meter.Sci.,*1989, *24*, 761.

21. Testa, R.B.; and Yu, L.M.; *Journal of Engineering Mechanics*, **1987**,*113*, 1631.

22. Farboodmanesh, S.; Chen, J.; Mead, J.L.; and White, K.; Yesilalan,E.; Laoulache, R.; and Warner, S.B.; paper presented at the meeting of the Rubber Division, American Chemical Society, San Francisco, CA, April 28-30, 2003, paper # 45.

23. Farboodmanesh, S.; Chen, J.; Mead, J.L.; and White, K.; Paper presented to the Rubber Division, ACS., Pittsburgh, PA. Oct 16-18 (2002).

24. *Rubber Processing*; White J.L., Ed.; Hanser, New York, 1995.

25. *Rubber Technology*; Dick J.S., Ed.; Hanser Gardener, Cincinnati, ST, 2001; pp 297-317.

Chapter 20

Investigation of the Formation of Carbon and Graphite Nanofiber from Mesophase Pitch Nanofiber Precursor

Hao Fong[1] and Darrell H. Reneker[2]

[1]Department of Chemistry and Chemical Engineering, South Dakota School of Mines and Technology, 501 East Saint Joseph Street, Rapid City, SD 57701
[2]Maurice Morton Institute of Polymer Science, University of Akron, Akron, OH 44325

Mesophase pitch nanofibers, made by a recently patented nanofiber processing technology, were stabilized, carbonized and graphitized to study the formation of novel carbon and graphite nanofibers. These pitch nanofibers were birefringent, and had diameters as small as 100 nm. Thermogravimetric analysis showed the pitch nanofibers could be stabilized in air, and the air stabilization resulted in $ca.$ 6 % weight gain, due to the absorption of oxygen. Air stabilization eliminated aromatic and aliphatic C—H structures, and created carbonyl and aryl/alkyl ether structures. After stabilization, the pitch molecules were crosslinked which prevented the nanofibers from melting during subsequent carbonization. Carbonization of the stabilized pitch nanofibers occurred in inert gas, and carbonization caused $ca.$ 23 % weight loss. During carbonization, the stabilized nanofibers underwent cyclization and evolved volatile molecules such as H_2O, H_2, HCN, N_2 and others. Carbonization generated carbon basal sheets, and the carbon basal sheets grew larger and became more oriented with the increasing of carbonization temperature. Further graphitization of the carbonized pitch nanofibers with the temperature up to 2200°C resulted in the formation of highly ordered graphitic sheets. The normals to the sheets were perpendicular to the fiber axis, which implied the graphite nanofibers were mechanically strong. The morphological, chemical and structural properties of the as-made pitch nanofibers as well as the concomitant stabilized, carbonized and graphitized nanofibers were characterized by SEM, TEM, FT-IR and XRD.

Introduction

Carbon and graphite fibers have been available since the 1960s. The high strength-to-weight ratio combined with superior stiffness has made carbon and graphite fibers the material of choice for high performance composite structures *(1)*. Commercial carbon and graphite fibers are produced from the thermal decomposition of polymeric precursors such as fibers of pitch, polyacrylonitrile (PAN) or cellulose, which are 5 microns or more in diameter *(1-7)*.

Mesophase pitch is a mostly insoluble liquid crystal pitch. Mesophase pitch fibers must be spun from a melt. Mitsubishi AR pitch, is synthetic hydrocarbon pitch with a softening point between 210°C and 260°C. Unlike coal-tar and petroleum based pitches, Mitsubishi AR pitch is derived from naphthalene and alkyl-naphthalene, and is essentially an oligomer of polynaphthalene with aromatic ladder structures. Such structure makes the mesophase pitch a liquid crystal substance at processing temperatures. Mitsubishi AR pitch has many advantages for making carbon and graphite fibers. The most important features include: (1) low spinning temperatures, AR pitch hardly deteriorates during spinning and has an excellent spinnability; (2) easy stabilization, AR pitch has a high hydrogen to carbon atomic ratio, the hydrogen can shorten the stabilization time drastically; (3) high tensile properties, carbon fibers made from AR pitch are likely to have a good graphitic orientation in the direction of fiber axis.

The recently patented nanofiber processing technology *(8)* modifies the traditional fiber producing method of air blowing, through unique spinneret design, to allow the airflow to more effectively shear the spin dopes. This technology is capable of producing mesophase pitch AR nanofibers with diameters as small as 100 nm. Additionally, the production rate of a single air jet using this technology is much higher than that of a single electrospinning jet

(9-12). The pitch nanofibers can be used as novel nano-sized precursors for preparing carbon and graphite fibers. The carbon and graphite fibers made from the pitch nanofibers have diameters in the range from tens of nanometers to hundreds of nanometers, thus they are termed "carbon and graphite nanofibers". The pitch based carbon and graphite nanofibers are low-cost compared to carbon nanotubes, and can be useful in many areas.

Previous work has demonstrated carbon and graphite nanofibers could be prepared from polymer and pitch nanofiber precursors *(9,10)*. Little has been found in the literature, however, about chemical and physical characterization during the formation (*i.e.* stabilization, carbonization and graphitization) of carbon and graphite nanofibers. In this research, the formation of carbon and graphite nanofibers from Mitsubishi AR pitch nanofibers, especially the morphological, chemical and structural changes during the preparation progress, was studied.

Experimental

Mesophase AR pitch, made by Mitsubishi Gas Chemical Company, Inc., was supplied by the US Air Force Research Laboratory at Dayton, Ohio. Mesophase AR pitch was converted to nanofibers using the recently patented nanofiber processing technology *(8)*.

Mesophase pitch nanofibers were stabilized in air using the following procedure: heating at 10°C/min from ambient temperature to 120°C, followed by isothermal treatment at 120°C for 30 min to remove the small organic molecules trapped in the pitch nanofibers; subsequent heating at 3°C/min to 300°C, and finally isothermal treatment at 300°C for 30 min to complete the stabilization. Carbonization was carried out in an inert argon environment. Two carbonization procedures were selected in the research: (1) low temperature carbonization: heating at 20°C/min from ambient temperature to 300°C; isothermal treatment at 300°C for 30 min; 2°C/min to 450°C; isothermal treatment at 450°C for 30 min; (2) High temperature carbonization: 20°C/min from ambient temperature to 300°C; isothermal treatment at 300°C for 30 min; 3°C/min to 900°C; isothermal treatment at 900°C for 30 min. High temperature carbonized pitch nanofibers were further graphitized by heating at 10°C/min from 900°C to 2200°C in an argon environment. The stabilization and carbonization were conducted using a Lindberg Hevi-duty furnace (Model 54453) made by a Division of Sola Basic Industries at Watertown, Wisconsin. The graphitization was conducted at the U.S. Air Force Research Laboratory in Dayton, Ohio, using a high temperature vacuum furnace. During heat treatments (*i.e.* stabilization, carbonization and graphitization), the fibers were processed in a relaxed state and no tension was applied to the samples.

The morphological, chemical and structural properties of the as made pitch nanofibers as well as the concomitant stabilized, carbonized and graphitized nanofibers, were investigated by polarized optical microscopy (Olympus CX31-P), scanning electron microscopy (SEM, JEOL JSM5310), transmission electron microscopy (TEM, JEM 1200XII), X-ray diffraction (XRD, Rigakurotaflex RU-200), thermogravimetric analysis (TGA, TA Instruments 2950) and Fourier transform infrared spectroscopy (FT-IR, Thermo Nicolet Avatar 360).

Results and Discussion

Microscopy Analysis

As made mesophase pitch nanofibers were examined using a polarized optical microscope. Turning the axis of pitch nanofibers to a 45° angle with respect to the crossed polarizer caused the bright nanofibers to become dark and the dark nanofibers to become bright. This birefringence indicated that the pitch molecules were oriented along the fiber direction. The pitch nanofibers were also examined by SEM. Fig. 1 shows the representative morphology. Unlike polymeric nanofibers (11,12), which are continuous nanofibers with identifiable fiber ends rarely seen under microscope, mesophase pitch nanofibers are found broken frequently (Fig. 1a). High magnification SEM images revealed that many broken pieces of very thin nanofibers (Fig. 1b) were attached to the surfaces of relatively thicker nanofibers. We believe that the nature of Mitsubishi AR pitch, which is the rod-like oligomeric polynaphthalene and lacks sufficient molecular chain entanglement, causes the pitch nanofibers to be brittle and easy to break during the nanofiber processing and handling.

Thermogravimetric Analysis (TGA)

TGA was employed to investigate the weight changes during the formation of carbon and graphite nanofibers. The characteristic weight changes observed during the air stabilization of pitch nanofibers were attributed to two causes: (1) absorption of oxygen, leading to the gain weight; (2) decomposition which produced small volatile molecules, resulting in weight loss (13). Weight changes of the pitch nanofibers during air stabilization, as a function of temperature and time were shown in Fig. 2. The stabilization procedure was described in the experimental section. The sudden drop at 120°C and sudden increase at 300°C in Fig. 2a occurred during isothermal treatment. In stabilization, at low temperature (below 180°C), there was about 0.5 % weight loss that was due to the evaporation of trapped small organic molecules in pitch nanofibers. The onset temperature for weight gain was around 180°C. The total weight gain was about 6 % after the stabilization was completed. Compared to the conventional pitch fibers *(1,4-7)*, the weight gain during the stabilization of nanofibers happened much faster. The small fiber diameter and the concomitant large specific surface area accounted for the faster stabilization rate.

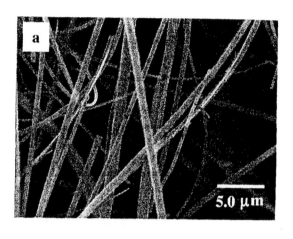

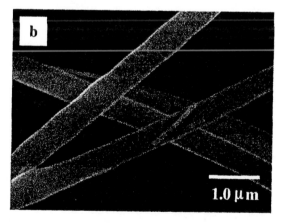

Figure 1: Representative SEM images of as-made mesophase pitch nanofibers

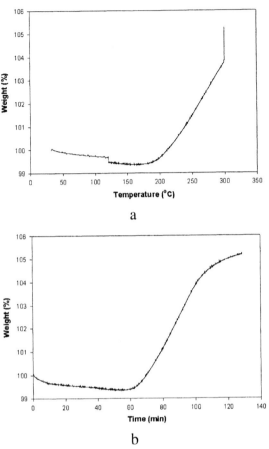

Figure 2: Weight changes of as-spun mesophase pitch fibers during stabilization in air: (a) versus temperature, (b) versus time.

Fig. 3 showed the weight loss of the stabilized pitch nanofibers during carbonization. Similar to stabilization, the weight change at 900°C in Fig. 3a occurred during isothermal treatment. As shown in Fig. 3, the weight loss continued throughout the carbonization process. However, the rates of weight loss before the temperature reached 500°C and after the temperature was above 700°C were different. It was believed that during the early stages of carbonization at 500°C or lower, the stabilized nanofibers probably underwent further cyclization in the uncyclized portion, forming ladder structure, crosslinking, and some chain-scission reaction, while evolving some small molecules of hydrogen, water, carboxylic acids, and other volatiles. When the temperature was between 500°C and 700°C, the carbon basal planes from the aromatized structure in the stabilized nanofibers began to form and increase in size *(14,15)*. As these reactions and structure rearrangements occurred within the nanofibers, consolidation and densification occurred. When the temperature was over 700°C, condensation reactions between heterocyclic rings and the resulting evolving gases of HCN, N_2 *etc.*, were dominant *(14,15)*. The carbon basal planes, as a consequence, became larger and more oriented. Fig. 3 also showed there was a small weight gain when the nanofibers were kept isothermally at 900°C. The reason is not understood.

FT-IR Analysis

FT-IR was employed to study the chemical properties of different nanofibers (*i.e.* molecular composition of both as-made pitch nanofiber and the resulting stabilized and carbonized nanofibers). In Fig. 4, "curve a" shows the FT-IR spectrum of the as-made pitch nanofibers, and "curve b" shows the spectrum of the stabilized nanofibers. The absorption band associated with the highly conjugated carbon/carbon double bonds (1600 cm^{-1}) remained almost the same before and after the air stabilization. The intensity of aromatic and aliphatic C—H structures, with absorption bands in the ranges of 2850-3100cm^{-1} and 1400-1450 cm^{-1}, was significantly reduced. After the stabilization in air, the intensities of the carbonyl band (1720 cm^{-1}), the ester band (1770 cm^{-1}), and the aryl alkyl ether band (1243 cm^{-1}) were all increased. The intensities of absorption bands near 750, 810, and 870 cm^{-1}, which were attributed to aromatic groups with one (870 cm^{-1}), two (810 cm^{-1}) or three (750 cm^{-1}) adjacent C—H groups in individual aromatic nuclei, were also reduced. The stabilization in air was a complicated process with both oxidation and oxygen induced decomposition occurring. Oxidation and decomposition both generated gaseous products, which eventually diffused out of the material. After stabilization in air, the pitch nanofibers formed crosslinked structures which prevented the fiber from changing shape during the subsequent heating to the carbonization temperature.

Curves "c" and "d" in Fig. 4 show the spectra of low and high temperature carbonized pitch nanofibers, respectively. After carbonization in argon at the relatively low temperature of 450°C, the absorption bands of functional groups disappeared, implying that carbon basal structures were formed. Further increasing the carbonization temperature did not make FT-IR spectrum much different.

292

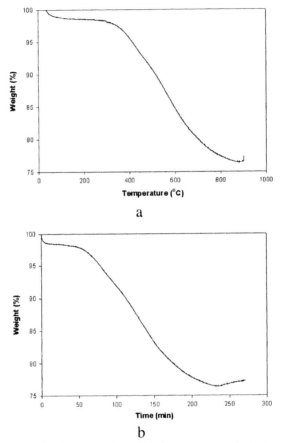

a

b

Figure 3: Weight changes of stabilized mesophase pitch fibers during carbonization: (a) versus temperature, (b) versus time

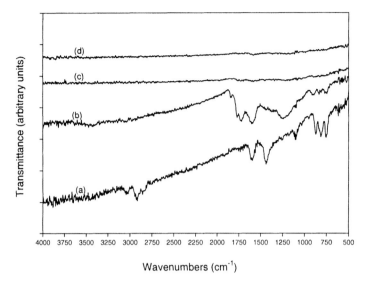

Figure 4: FT-IR spectra of different nanofibers. (a) as-made pitch nanofibers, (b) stabilized nanofibers, (c) low temperature carbonized nanofibers, (d) high temperature carbonized nanofibers

XRD Analysis

XRD was used to obtain structural parameters of different nanofibers. As shown in Fig. 5, the as-made, stabilized and carbonized pitch nanofibers had similar turbostratic structures. All the nanofibers had broad diffraction peaks around $2\theta = 25°$, which were indexed as (002) diffraction peaks. Their crystallites were similar to, but not as ordered as, those of perfect graphite. The size of these turbostratic crystallites in the direction perpendicular to the turbostratic layers could be characterized by L_c, a crystal size parameter determined from XRD data. The interplanar spacing $d_{(002)}$ and the crystal size parameter L_c were calculated by using the Bragg equation and the Scherrer equation *(16)*, respectively.

$$d_{(002)} = \frac{n\lambda}{2\sin\theta}$$

$$L_c = \frac{\lambda}{\beta\cos\theta}$$

Where λ was the wavelength of the X-ray (0.1542 nm), and β was the width of the diffraction peak measured at half its height. The results of calculations were shown in Table 1.

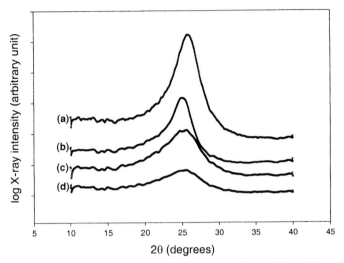

Figure 5: XRD of nanofibers: (a) as made pitch fibers, (b) stabilized pitch fibers, (c) pitch fibers carbonized at 450 °C, (d) pitch fibers carbonized at 900 °C

Table 1: Characterization of crystallites in nanofibers

	$2\theta_{(002)}$ (°)	β (Radian)	$d_{(002)}$ (nm)	L_c (nm)
As made pitch nanofibers	25.99	0.07416	0.343	2.134
Stabilized pitch nanofibers	25.00	0.06367	0.356	2.481
Low temperature (450°C) carbonized pitch nanofibers	25.38	0.10412	0.351	1.518
High temperature (900°C) carbonized pitch nanofibers	25.55	0.09738	0.349	1.624

As shown in Table 1, both the interplanar spacing ($d_{(002)}$) and the crystal size parameter (L_c) of the as made pitch nanofibers became slightly larger after the air stabilization. The oxidation and decomposition during stabilization slightly expanded the tightly stacked pitch planar structure and resulted in the slightly larger interplanar spacing. Meanwhile, since the stabilization occurred in the temperature range of 180-300°C, the crystallite of rod-like mesophase pitch molecules could also be annealed to be larger. The interplanar spacing and the crystal size parameter of the low temperature carbonized nanofibers were smaller than those of the stabilized nanofibers. This implied that the low temperature carbonization was able to consolidate and tighten the stacked sheets and thereby reduce the interplanar spacing. One the other hand, the release of small molecules, such as water, also partially destroyed the existing crystalline structure and made L_c smaller. Carbonization at high temperature made the interplanar spacing smaller and crystal size larger. This indicated that the high temperature carbonization created carbon basal crystallites, which could undergo structural rearrangement to grow larger. The interplanar spacing of the commercial pitch-based carbon fibers (17) and the high temperature carbonized pitch nanofibers were the same (0.349 nm). However, the crystal size (1.624 nm) was smaller than the reported value (3.5-4.5 nm). The reason was because the reported value was measured using the pitch-based carbon fiber carbonized at 1300°C (17). It was reported that the mechanical properties of the carbon fibers were affected by the size of crystallite (17), the carbonization temperature of 900°C might be not high enough to obtain carbon nanofibers with superior mechanical properties.

Graphitization

To further grow the crystallites and improve the structural perfection of nanofibers, high temperature carbonized pitch nanofibers were graphitized in argon at 2200°C. It is known that graphitization could result the carbon basal structures that formed during carbonization, to form graphitic sheet structures. Fig. 6 shows the TEM image and electron diffraction pattern of a segment of a graphitized pitch nanofiber. In the images, there were graphitic sheets aligned with the direction that the normals to the sheets were perpendicular to the fiber axis. The spacing between the graphitic sheets was approximately 3 to 4 nm, which was about 10 times larger than the interlayer spacing of natural graphite. The orientation of these graphitic sheets was a reasonable consequence of the high orientation of the molecular structures observed in the as made pitch nanofibers.

Scion image software (Scion Corporation, Frederick, Maryland), was used to analyze the diffraction pattern of the graphitized pitch nanofiber. Fig. 7a is a three dimensional graph of the intensity of the diffraction pattern shown in Fig. 6. Fig. 7b is a graph of the intensity along the equator of the diffraction pattern. The well-defined series of peaks, (002), (004), (006), (100), (110) clearly indicate the formation and growth of graphite crystallites. The interlayer spacings of these peaks agrees with the reported values (18,19). The asymmetrical intensity distribution in the meridian spots and layer lines, and some unexplained features such as low optical density of the film near the center of the intense (002) peaks and the central spot are not understood.

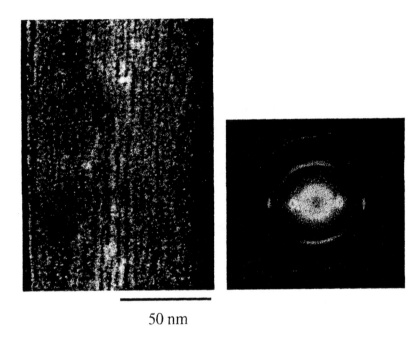

50 nm

Figure 6: TEM image and electron diffraction pattern of a representative graphited pitch nanofiber

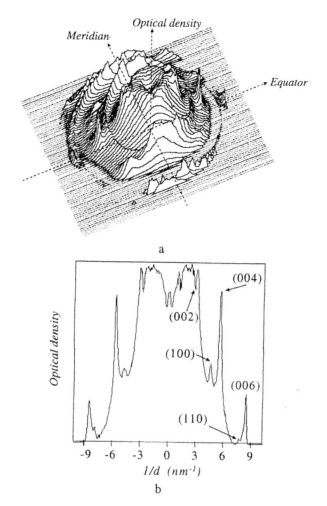

Figure 7: Digitized X-ray diffraction data from the mesophase pitch based graphite nanofiber.

Summary

Mesophase pitch nanofibers, with the diameters as small as 100 nm, were produced using the recently patented nanofiber processing technology. The pitch molecules were oriented, as evidenced by strong birefringence of the nanofibers. The formation of carbon and graphite nanofibers using these pitch nanofibers, especially the morphological, chemical and structural variations during the stabilization, carbonization and graphitization, was studied. Thermogravimetric analysis results showed that the air stabilization of pitch nanofibers began at temperatures below 200°C, and the stabilization caused about 6% weight gain. The stabilization in air eliminated aromatic and aliphatic C—H (2850-3100 and 1400-1450 cm^{-1}) structures and created carbonyl (1690-1750 cm-1) and aryl alkyl ether (1200-1275 cm^{-1}) structures. Carbonization began when the temperature reached 300°C in inert gas, and continued with the increasing of temperature. There was about 23% weight loss during the carbonization. The morphology and structures of the as made pitch nanofiber as well as the concomitant stabilized, carbonized and graphitized nanofibers were investigated by electron microscopy and X-ray diffraction. The results showed that the interplanar spacing (d_{002}) became larger after stabilization in air, and approached the value of natural graphite after carbonization. The crystal size (L_c) in the carbonized pitch nanofibers was smaller than the reported value of commercial carbon nanofibers, implying the carbonization temperature was not high enough to create carbon nanofibers with superior mechanical properties (however, these carbon nanofibers may be useful in other applications such as filtration and catalyst supporting material). Graphitization at 2200°C generated highly ordered graphitic sheets, indicating the graphite nanofibers were mechanically strong and could be used to prepare high strength and/or high modulus composites.

Acknowledgements

This work was supported by the Air Force Materials Laboratory in Dayton, Ohio, the Army Research Laboratory in Aberdeen, Maryland, and also by the Nelson Research Award.

References

1. *Technology of Carbon and Graphite Fiber Composite;* Delmonte, J. Ed. Van Nostrand Reinhold: New York, **1981**.
2. *Carbon Adsorption Handbook*; Cheremisinoff, P.N., Ellerbush, F., Eds.; Science Publishers: Ann Arbor, MI, **1978**, 810-859.
3. *Chemistry and Physics of Carbon;* Walker, P.L., Jr. Ed. Marcel Dekker: New York, **1973,** Vol. 9.
4. *Carbon Fibers;* Donnet, J.B. and Bansal, R.C., Eds.; Marcel Dekker: New York, **1984**.

5. *Strong Fibers. Handbook of Composites*; Kelly, A. and Rabotnov, Y.N. Eds. Series Amsterdam: North Holland, **1985**, Vol. 1.
6. *Carbon Fibers and Their Composites*; Fitzer, E. Ed. Springer-Verlag: New York, **1985**.
7. *Carbon fibers, filaments and composites;* Figueiredo, J.L. et al. Eds., Kluwer Academics Publishers: Netherlands, **1990**.
8. Reneker, D.H. *Process and Apparatus for the Production of Nanofibers.* US Patent 6,520,425B1, **2003**. .
9. Chun, I.; Reneker, D.H.; Fong, H. et al. *Journal of Advanced Materials*; **1999**, 31(1), 36.
10. Fong, H.; Chun, I. and Reneker, D.H. *The 24th Biennial Conference on Carbon, Charleston, SC*, **1999**, 380.
11. Reneker, D.H. and Chun, I. *Nanotechnology;* **1996**, 7, 216.
12. Fong, H. and Reneker, D.H. In *Structure Formation In Polymeric Fibers: Chapter 6, Electrospinning and Formation of Nanofibers;* Salem, D.R. & Sussman, M.V. Eds., Carl Hanser Verlag, **2001**.
13. Wapner, P.G.; Kowbel, W. and Wright, M.A. In *Metal Matrix, Carbon, and Ceramic Matrix Composites*; Buckley, J. D. Ed., NASA, Washington, DC, **1987**.
14. Fitzer, E. and Fiedler, A.K. *Polymer Preprint;* **1973**, 14, 401.
15. Watt, W.; Johnson, D.J. and Parker, E. *Proceedings of the 2nd International Plastics Conference on Carbon Fibers;* Plastics Institute, London. **1974**.
16. *X-ray Diffraction Methods in Polymer Science;* Alexander, A.E. Ed., Wiley-Interscience: New York, **1969**.
17. Mochida, I.; Yoon, S.H.; Takano, N.; Fortin.; F, Korai, Y. and Yokoawa, K. *Carbon*; **1996**, 34 (8), 941.
18. *Encyclopedia of Chemical Technology, 3rd Edition*; Herman, M.F. Ed., Wiley-Interscience: New York, **1978**., Vol. 4, 556,
19. *Carbon and Graphite Handbook*; Mantell, C.L. Ed., John Wiley & Sons, Inc.: New York, **1968**.

Chapter 21

Mechanical Behavior of Nonwoven Electrospun Fabrics and Yarns

Sian F. Fennessey, Angelo Pedicini, and Richard J. Farris[*]

Polymer Science and Engineering Department, Silvio O. Conte National Center for Polymer Research, University of Massachusetts at Amherst, Amherst, MA 01003

Continuous, electrospun nanofibers are expected to out-perform their conventional counterparts in the reinforcement of composites due to their increased surface area available for adhesion and high aspect ratio (l/d). The mechanical behavior of nonwoven, electrospun nanofibers was examined as a function of fiber morphology, alignment, and degree of molecular orientation. Fiber alignment and molecular orientation was improved with use of a high-speed, rotating collection device. The ultimate strength and modulus of nanofiber yarns have been shown to improve with addition of twist and through post drawing treatments.

Introduction

Electrospinning, a fiber spinning technique that relies on electrostatic forces to produce fibers in the nanometer to micron diameter range, has been extensively explored as a method to prepare fibers from polymer solutions or melts [1]. The main feature of the electrospinning process is that it is a simple means to prepare continuous fibers with unusually large surface to volume ratios and porous surfaces [2,3]. Due to the chaotic oscillation of the electrospinning jet, a characteristic feature of the electospinning process, randomly oriented and

300

isotropic structures in the form of nonwoven nanofiber mats or webs are often generated due to a lack of control over the forces driving fiber orientation and crystallization. Recent efforts have been made to control the spatial orientation of electrospun fibers for use with 1D device fabrication, which requires well aligned and highly ordered architectures through redesign of the collection apparatus [4]. Molecular orientation has been observed in electrospun fibers collected onto a parallel plate and a rotating drum collection apparatus, although the orientation has not been quantified [5]. Progress in understanding the electrospinning technique has allowed for recent engineering efforts in processes used to collect electrospun fibers for various applications, however, very limited work has addressed the mechanical properties of electrospun fiber mats [6].

The mechanical properties and reinforcing behavior of continuous nanofibers are expected to differ significantly from their finite length, conventional counterparts. Composites containing continuous fibers perform better than those prepared containing short fibers, particles, or whiskers since the reinforcement effect of a fiber is dependent upon its length to diameter ratio, according to the Halpin-Tsai equations [7] and Christensen's equation [8] of composite modulus prediction. Christensen's equation suggests that composites containing short fibers, particles, or whiskers are only expected to produce moderate improvements over non-reinforced polymer materials. Short fibers initially reinforce the composite, however, the reinforcing effect reaches a plateau above which short fibers have a limited ability to reinforce the matrix. As the fiber length increases, the reinforcing effect is initially increased and the level of the plateau is raised. Therefore, there is a greater possibility of increasing the modulus and strength of a composite by using a continuous fiber with a high aspect ratio (l/d) as reinforcement, rather than a high modulus short fiber.

In addition to a high aspect ratio (l/d), the homogeneous dispersion of fibers throughout the matrix, and good interfacial adhesion and load-transfer between the matrix and fiber lead to the improved strength and modulus of a composite [7a, 9]. Continuous, electrospun fibers are expected to have improved interfacial adhesion in comparison to conventional fibers due to electrospun fibers' large surface to volume ratio resulting in an increase in surface area available to adhere to a matrix material. Due to the electrospun fibers' irregular void structure among fibers of the fabric and hairiness of the yarns, electrospun fabrics and yarns are expected to take advantage of the mechanical interlocking mechanism of load transfer. The strength of a composite is based on adhesion between the filler and the matrix; if the adhesion is poor, the composite strength and modulus will be essentially that of the matrix material.

In the present work, the mechanical behavior of electrospun polyurethane and polyacrylonitrile fibers is investigated and the use of electrospun fibers for the reinforcement of thin films and nanocomposites is proposed. The effect of morphology, crystallinity, molecular orientation, and fiber alignment on the mechanical behavior of polyurethane fibers is determined. With consideration of the results found for the polyurethane system, the electrospinning process is optimized for the fabrication of fibers for reinforcement applications. Carbon precursor fibers from polyacrylonitrile are aligned with various degrees of orientation with the use of a rotating target. The mechanical properties of twisted polyacrylonitrile nanofiber yarns are determined as a function of twist angle. The effect of drawing on the modulus and ultimate strength of aligned polyacrylonitrile nanofiber yarns is determined.

Experimental

Electrospinning Set-up

The electrospinning apparatus consists of a polymer solution reservoir, a high voltage power supply, and a grounded target. Solution is loaded into a pipette and a wire electrode is immersed into the solution. The solution is electrospun between 9.5-15 kV horizontally across a gap of 15-20cm and collected onto a stationary target. When the polymer reservoir is empty, the process stops, the pipette is refilled and electrospinning is resumed.

The solution can also be loaded into a syringe that is oriented onto a dual syringe pump, and an electrode is clipped onto the needle. The needle, electrode and grounded target are all enclosed in order to reduce the effect of air currents on the trajectory of the electrospining jet. The flow rate of solution to the needle tip is maintained so that a pendant drop remains during electrospinning. The solution is electrospun between 8-16kV horizontally across a gap of 13-16cm onto a rotating target [5a, 6a, 10, 11] as described elsewhere [12].

Microscopy

Electrospun fibers were observed by field emission scanning electron microscopy (FESEM) and polarized optical microscopy. Samples were mounted onto SEM plates, sputter coated with gold, and examined using a JOEL JSM 6320FXV electron microscope operating at an accelerating voltage of 5kV. Measured fiber diameters include a 5% random error. Electrospun fibers were also examined using an Olympus BX51, polarizing optical microscope, to detect birefringence and with a 1st order red plate to determine the elongation sign of the fiber by determining the directions of vibration of the fast and slow components [13].

IR Dichroism

Dried, electrospun fiber bundles were examined using a Perkin Elmer Spectrum 2000 infrared spectrometer (FTIR) with a polarized wire-grid to measure dichroism as a function of collection take-up speed onto a rotating target. Spectra were acquired with the draw direction of the electrospun fibers positioned both parallel and perpendicular to the electric vector direction of the polarizer. Care was taken to examine the same region of fibers in both instances. Spectra were recorded over the range of 700-4000cm^{-1} with typically 64 scans. The dichroic ratio, D, and Herman's orientation function, f, were determined (Equation 1 and 3).

$$D = A_{||}/A_{\perp} \tag{1}$$

$$D_o = 2 \cot^2\alpha \tag{2}$$

$$f = [3 <\cos^2 \Phi> - 1]/ 2 = (D - 1)(D_o + 2)/ (D_o - 1)(D + 2) \tag{3}$$

where $A_{||}$ is the absorbance when the electric vector direction of the polarizer is oriented parallel to the fiber draw direction and $A_{\perp}$ is the absorbance when the electric vector is oriented perpendicular to the fiber draw direction, D_o is the dichroic ratio of an ideally oriented polymer, α is the transition moment angle, and Φ is the angle of the molecular segment relative to the fiber axis. The transition moment angle is the angle between the chain axis and the direction of the dipole moment change for the group of interest.

X-ray Diffraction

Dried, electrospun fibers were examined by wide angle X-Ray diffraction (WAXD) as a function of collection take-up speed onto a rotating target. Pinhole collimated, monochromated CuKα radiation was used and diffraction patterns of electrospun fiber bundles were collected with a GADDS detection system (Brucker) in air. Calcite was used as a reference to aid analysis. Diffraction patterns were collected after 10 hours of exposure per sample and a background was subtracted. The Herman's orientation function, f, was determined using the primary equatorial arcs (Equation 4).

$$f = [3 <\cos^2 \Phi> - 1]/2 = \int I \,|\sin \Phi| \,[3 \cos^2 \Phi - 1]/2 \, d\Phi \,/ \int I \,|\sin \Phi| \, d\Phi \quad (4)$$

where Φ is the azimuthal angle between the axis of the molecular segment and of the fiber and I is the scattering intensity of the reflection at that angle.

Twisted Yarns

Approximately 32cm x 2cm unidirectional tows of electrospun nanofibers were collected using a rotating target, linked together, and twisted using a Roberta electric spinner by Ertoel. The twisted yarns were rinsed in deionized water for 24 hours and then dried under vacuum at 100°C. The twist per centimeter (tpcm), denier, and the angle of twist of the yarn were determined.

Mechanical Testing

The mechanical behavior of dried, electrospun fibrous mats was determined according to ASTM 1708D with dumbbell shaped tensile specimens using an Instron 4411 and crosshead speed of 10 mm/min (50%/min. strain rate) in tension at room temperature. The mechanical behavior of dried, twisted yarns of electrospun nanofibers was examined using an Instron 5564 with a crosshead speed of 2mm/min (10%/min. strain rate) in tension at room temperature. Samples were mounted on paper tabs and had a 20mm gage length. The cross-sectional area was calculated from the denier and the density of PAN from the literature [14]. The initial modulus, ultimate strength, and elongation at ultimate strength was measured.

Drawing

Uniaxially aligned yarns were drawn using an Instron 5564 with a crosshead speed of speed of 2mm/min (10%/min. strain rate) in tension at 85°C in deionized water. Samples were mounted with a 20mm gage length and elongated to various draw ratios. The fibers were dried, remounted with a 20mm gage length, and the mechanical behavior was examined as previously described.

Polyurethane Nanofibers

Introduction

Although elastomeric nanofibers have no inherent utility in the field of fiber-reinforced composites, elastomeric polymers can be electrospun to produce highly porous nanofiber membranes that may prove to be useful as filter media and for medical applications [15]. Air and vapor transport properties of electrospun polyurethanes have been compared to commercial membranes, such as expanded polytetraflouroethylene (ePTFE) and Gore-Tex® membrane laminates, and the tensile properties of electrospun thermoplastic polyurethane were determined [16]. A fundamental understanding of the mechanical properties and behavior of electrospun fibers and nonwoven nanofiber fabrics is critical to membrane applications. In addition, predictive tools to estimate mechanical properties of electrospun material relative to their bulk analogs will allow for the design of further applications for electrospun materials.

In the present work, the mechanical behavior of electrospun polyurethane nonwoven fabric mats was characterized and compared to that of the bulk material. Differences in the mechanical response of the material as a function of morphology and molecular orientation was observed. The tensile strength of the examined electrospun polyurethane systems was compared to the respective bulk material strength by factoring the density of the electrospun material relative to the bulk.

Materials

A thermoplastic polyurethane elastomer Pellethane® 2103-80AE was received from an industrial source and dimethylforamide (DMF) was obtained from Sigma Aldrich Co. The polymer was used as received and DMF was dried prior to use. Solutions were prepared at room temperature with mixing.

Solution cast polyurethane films were prepared from 6wt% solution of polymer in THF and dried at room temperature for at least 72 hours prior to use.

Mechanical Behavior of Electrospun Thermoplastic Polyurethane

Electrospinning of Pellethane® 2103-80AE onto a stationary target results in the production of randomly oriented, entangled nanofibers with circular cross

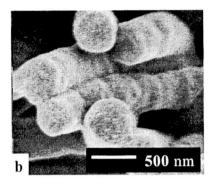

Figure 1. (a)Entangled, electrospun Pellethane® 2103-80AE collected onto a stationary target with (b) circular cross sections. Reproduced with permission from reference 6c. Copyright 2003Elsevier.

sections (Figure 1). The fibers do not have smooth surfaces, but rather the surface is rough and contains contours. The electrospun fibers are weakly birefringent as observed under cross polars, although no crystallinity is detected by wide angle X-ray diffraction (WAXD) and no melting endotherm is observed by differential scanning calorimetery (DSC) of the electrospun fibers. During the electrospinning process the spinning jet undergoes extremely large draw ratios, therefore, it would be expected that the resulting collected fibers would contain some molecular orientation. Since DMF is a high boiling solvent, it is not expected that DMF is completely evaporated during the electrospinning process. A surface, or skin, forms on the jet due to the diffusion of moisture from humidity in the environment into the fiber and solvent remains in the core of the fiber. The fibers are unconstrained after collection onto the target and tend to relax. The polymer chains are able to rearrange and any molecular orientation that may have been induced by the electrospinning process is lost.

The electrospun fabric of the polyurethane system has a film-like character due to the adhesion of the fibers to one another, forming junction points throughout the mat. The inter-fiber adhesion plays a significant role in the mechanical integrity of the electrospun polyurethane material. The cross sectional area of the electrospun mat was corrected for its density relative to the bulk material in the stress calculations (Table I). The stress-strain curve for the electrospun polyurethane is not typical of elastomeric materials, as is that for the bulk sample (Figure 2). The initial slope of the stress-strain curve for the electrospun material is lower than for the bulk due to the relatively small number of fibers bearing load at low strains. Fibers are initially unoriented, at angles to the direction of the applied load, and rotate to align with the direction of the

Table I. Density corrected, mechanical properties of Pellethane® 2103-80AE

Pellethane® 2103-80AE	Density correction factor (%)	Ultimate Strength (MPa)	Elongation at break (%)
Solution Cast Film	1.0	48.6	532
Electrospun fiber mat Isotropic	26.4	30.7	253

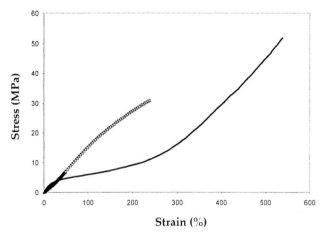

Figure 2. Mechanical behavior of a solution cast film (——) versus an electrospun fabric mat(◁) of Pellethane® 2103-80AE with density correction.

applied load with increasing strain; strain-induced alignment of fibers in the electrospun mat. With strain-induced alignment, the number of fibers bearing load increases and the slope of the stress-strain curve steepens between 0-60% strain. An analogous increase in slope, due to strain induced orientation, does not occur until 400-500% strain for the solution cast film. From ~40% strain to break at ~250% strain, the stress is greater in the electrospun than the bulk material; the bulk stress-strain curve is relatively flat in this region.

No difference in the percent crystallinity, or in thermal transitions was observed between the solution cast and electrospun material. Pellethane® 2103-80AE is completely amorphous in the bulk and electrospun form, therefore it is difficult to attribute the difference in mechanical response to any measurable changes in morphology or crystallinity. Thermoplastic polyurethanes possess substantial strength and modulus at temperatures well above their glass transition temperature T_g, where they would ordinarily be expected to be a viscous liquid.

308

The mechanical strength and stiffness of thermoplastic polyurethanes can be ascribed to the glassy state of the hard blocks acting as "fillers" for the elastomeric blocks [17]. The higher stress in the electrospun material may be a result of preexisting molecular orientation and/or more efficient cooperation among hard polyurethane segments in the electrospun fibers.

Regardless of density, the ultimate strain of the electrospun material is expected to be the same as that of the bulk. Approximately 52% reduction in the strain-to-failure for the electrospun polyurethane is observed. The surface of electrospun polyurethane fibers contains nanoscale cracks (Figure 3), which act as stress concentrators when an external load is applied. Tensile strength is ultimately a flaw limited property and the cracking phenomenon is a contributing factor to the decreased strength of the electrospun material.

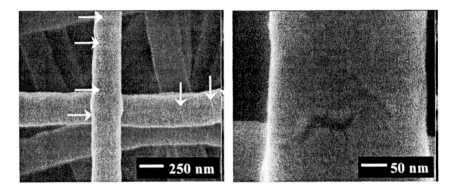

Figure 3. Surface of electrospun Pellethane® 2103-80AE fibers

Aligned Electrospun Polyurethane Fibers

The alignment of electospun fibers is induced with the use of a rotating collection target and the alignment improves as the surface velocity of the target is increased. The trajectory of a continuous electrospinning jet is affected by the rotation of the target. As the fiber collects onto the target surface, it is electrostatically attached to the surface and can be used to stretch the following tow of fiber from its spiraling path to align with the rotation direction of the target. As the rotation speed increases, the effective draw is increased resulting in better alignment of the collected fiber and less deviation between the fiber and rotation direction due to the influence of looping jet trajectory.

Figure 4 shows aligned electrospun Pellethane® 2103-80AE fibers that were collected onto a rotating target with a surface velocity of 9.8 m/s, as well as a comparison of the stress-strain curves for aligned and isotropic electrospun polyurethane mats. The molecular orientation of the electrospun fibers was

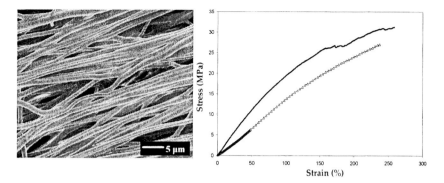

Stress (MPa)

Strain (%)

Figure 4. Left: Aligned Pellethane® 2103-80AE fibers electrospun onto a rotating target with a surface velocity of 9.8m/s. Right: Stress-strain curve of an isotropic(<) and aligned (—) electrospun Pellethane® 2103-80AE. Reproduced with permission from reference 6c. Copyright 2003 Elsevier.

determined by IR dichroism. The dichroism of the amine (-N-H) group stretching vibration, 3320cm^{-1} [18], was measured. The amine group is oriented approximately perpendicular to the draw direction of the fiber, therefore the absorbance for perpendicular polarization is greater then for parallel polarization (Figure 5). The solution cast polyurethane film and the electrospun fiber mat collected on stationary target show no dichroism. The aligned fiber mat has a small dichroic ratio, indicating low molecular orientation has been induced through the use of a rotating target.

The fiber alignment and slight increase in molecular orientation of the fibers collected onto a rotating target results in an increase of the initial slope of the stress-strain curve, and a higher stress between 0% strain and break in comparision to that of the isotropic fibers collected onto a stationary target. The mechanical behavior of electrospun fibers is influenced by the surface morphology, alignment, and molecular orientation of the fibers. By controlling these variables, the electrospinning process may be used to manufacture continuous nanofibers with high strength and modulus to reinforce thin films or nanocomposites.

Polyacrylonitrile Nanofibers

Introduction

The high strength and modulus of carbon fibers make them useful in the reinforcement of polymers, metals, carbons, and ceramics, despite the fibers'

310

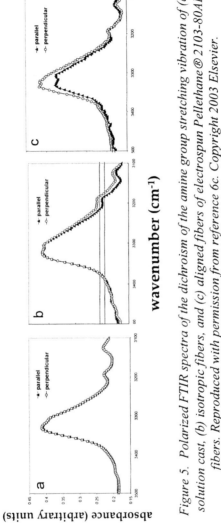

Figure 5. Polarized FTIR spectra of the dichroism of the amine group stretching vibration of (a) solution cast, (b) isotropic fibers, and (c) aligned fibers of electrospun Pellethane® 2103-80AE fibers. Reproduced with permission from reference 6c. Copyright 2003 Elsevier.

brittle nature. The market for carbon fibers is dominated by fibers made from Polyacrylonitrile due to their combination of good mechanical properties, particularly tensile strength, and reasonable cost.

Current techniques of reinforcing a matrix with carbon fiber usually involve the fabrication of carbon fiber fabrics, preforms, staple yarns, or slurries in order to ease handling and processing. Polyacrylonitrile (PAN) precursor fibers prepared by conventional techniques have diameters between 10-12µm [19] and the resultant carbon fibers have a minimum diameter between 5-7µm. The bulk of production cost incurred during carbon fiber production is due to long heating times required to stabilize and carbonize the precursor fiber, in addition to engineering costs to maintain tension on fibers during stabilization. The stabilization step converts the precursor from a linear polymer to a cyclized, highly condensed thermally stable structure. The overall reaction of cyclization of the nitrile groups is highly exothermic and depends upon the temperature, the stretch applied to the precursor fiber, and the surrounding environment. Oxidation occurs concurrently with the cyclization reactions and is a diffusion controlled process. The rate of oxidation depends upon the diameter and chemical composition of the fiber in addition to the temperature and atmosphere. A decrease in the fiber diameter of the precursor fiber would result in a decreased stabilization time [20]. If ultra-fine mechanically useful PAN fibers could be produced by electrospinning, they could be carbonized roughly 1000 times faster than conventional fibers and therefore might be a lower cost, high-output means of producing submicron carbon fibers. Also, the fibers prepared by electrospinning have diameters so small that skin-core effects caused by differential stabilization should be eliminated.

Since carbon fibers fail at critical flaws, reducing the fiber diameter lowers the probability of encountering a critical flaw in a given test length. The strength of a carbon filament increases as the diameter decreases [20]. Carbon fibers produced from electrospun precursor nanofibers would be on a similar size scale as vapor grown carbon (CCVD) filaments, which have diameters between 1µm and 100µm. CCVD filaments suffer from the same difficulties as carbon nanotubes in regards to reinforcing composite materials due to their low aspect ratio (l/d) and difficulties in dispersing the reinforcer within the composite. Carbon nanofibers have been produced from electrospun precursor fibers and characterized physically and structurally [21], although limited study of their mechanical properties has been made [4b] and no molecular orientation study of the precursor fiber prior to carbonization has been completed.

In the present work, the electrospinning process of fiber production has been examined in regards to the preparation of PAN nanofibers with the purpose of preparing carbon nanofibers for the reinforcement. Tows of unidirectional and molecularly oriented PAN nanofibers were prepared using a high speed, rotating take-up wheel. The effect of voltage and take-up speed on the alignment and molecular orientation of the generated fiber was examined. The aligned tows were twisted into yarns and the mechanical properties of the yarns were determined as a function of twist angle. The effect of post treatment on the mechanical properties of uniaxially aligned precursor yarns was determined.

Materials

Two polyacrylonitrile copolymers were received from an industrial source and N,N-dimethylformamide was obtained from Sigma Aldrich Co. The copolymers have a number average molecular weight, (M_n), and polydispersity index of $2.0E5 g/mol$ and 1.18, and $1.6E5 g/mol$ and 2.02 (determined against styrene standards by gel permeation chromatography), respectively. The polymer and solvent are dried prior to use; solutions are prepared at room temperature with mixing.

Alignment and Orientation of Polyacrylonitrile Fiber Yarns

It is generally thought that the better the degree of molecular orientation in the original PAN fiber, the better the mechanical properties, in particular the modulus, of the resultant carbon fiber [22]. Electrospinning of polyacrylonitrile from ~10-15wt% solution in DMF onto a stationary grounded target results in the production of isotropic, entangled fibers that do not birefringe under cross polars nor show a diffraction pattern by wide angle X-ray diffraction (WAXD).

Partially aligned and oriented PAN nanofibers with diameters in the range of $0.27\mu m$ to $0.29\mu m$ were prepared by electrospinning a ~15wt% PAN in DMF solution at 16kV onto a grounded, rotating collection wheel with a surface velocity of 3.5m/s to 12.3m/s. The collected fibers were observed to be birefringent under cross-polars, and exhibit positive elongation when observed under cross-polars with a 1st order gypsum red plate. The overall average diameter of fibers prepared from a ~15wt% solution electrospun at 16kV range are smaller than those prepared at similar conditions onto a stationary target, since the fibers were constrained and not allowed to shrink.

The molecular orientation of the nanofibers was examined by IR dichroism (FTIR) and wide angle x-ray diffraction (WAXD). The dichroism of the nitrile-($-C\equiv N$) group stretching vibration, $2240cm^{-1}$ [23], was measured by FTIR for ~15wt% solutions electrospun at 16kV over a distance of ~15cm onto a collection wheel rotating between 0m/s and 12.3m/s. The nitrile group is oriented approximately perpendicular to the draw direction of the fiber, and the nitrile-stretching vibration shows a strong perpendicular dichroism [23c]. The dichroic ratio decreases from approximately unity at 0m/s take-up speed to 0.75 at 9.8m/s, suggesting that the electrospun fibers are unoriented or isotropic when collected onto a stationary target and become gradually more oriented as the take-up speed is increased up to 9.84m/s. At a take-up speed above 9.8m/s, the dichroic ratio was found to increase to 0.91 and 0.88 at 11.0m/s and 12.3m/s take-up speeds, respectively. A transition moment angle of 73° [23b, 23f], the angle between the direction of the nitrile group's dipole moment change and the chain axis, was used to calculate the chain orientation factor, f. The chain orientation factor followed a similar trend to the dichroic ratio, with a maximum orientation of 0.23 for the nitrile-stetching vibration at 9.84m/s take-up speed (Figure 6).

Wide angle X-ray diffraction (WAXD) patterns were collected from bundles of fibers electrospun from a ~15wt% PAN in DMF solutions at 16kV collected onto a target rotating with a surface velocity between 0m/s and 12.3m/s. The diffraction pattern showed two equatorial peaks; a weak peak at $2\theta = 29.5°$ corresponding to a spacing of d ≈ 3.03Å from the (1120) reflection and a strong peak at $2\theta = 16°$ corresponding to a spacing of d ≈ 5.3Å from the (10 $\overline{1}$0) reflection. The equatorial peaks at $2\theta = 29.5°$ and $2\theta = 16°$ are common to the fiber diffraction pattern of PAN with hexagonal packing [24]. The arc width of the (10 $\overline{1}$0), strongest equatorial, reflection provides an indication of the degree of molecular orientation within the fibers. The Herman's orientation function increases gradually with take-up speed from 0 at 0m/s to 0.23 ±0.01 at 8.61m/s and plateaued to 0.21 ±0.07 at 9.81m/s to 12.3m/s (Figure 6). The orientation measurements from the dichroism and the WAXD are in agreement. The orientation is increased with increasing target surface velocity to a speed of approximately 8.61m/s. As the surface velocity of the target is increased further the molecular orientation of the fibers decreases slightly and plateaus [12].

Mechanical Properties of Polyacrylonitrile Fiber Yarns

Unidirectional tows of PAN nanofibers prepared from ~15wt% PAN in DMF solutions electrospun at 16kV onto a target rotating with a surface velocity of 9.8m/s were linked and twisted into yarns. Yarns of twisted electrospun PAN nanofibers were prepared with an angle of twist ranging from 1.1° to 16.8° and a denier between 326 and 618 with the average yarn denier being 446. The stress-strain behavior of the yarns was examined and the modulus, ultimate strength, and elongation at the ultimate strength were measured as a function of twist angle. The initial modulus and ultimate strength increase gradually with twist angle from 3.8 ± 1.1GPa and 91.1 ± 5.5MPa with a twist angle of 1.1° to 5.8 ± 0.4GPa and 163 ± 12MPa with a twist angle between 9.3° and 11.0°, respectively. The modulus and ultimate strength of the yarns decreased with angles of twist greater than 11.0° (Figure 7). The surface of the broken filaments were damaged or roughened due to the frictional and normal forces in action as the twisted yarns are exposed to tension, as observed by FESEM.

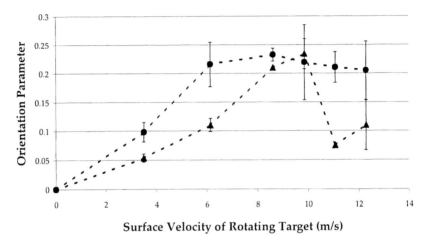

Figure 6. Orienation parameter determined by dichroism (7) and WAXD (,) as a function of the surface velocity (m/s) of the rotating target. Reproduced with permission from reference 12. Copyright 2004 Elsevier.

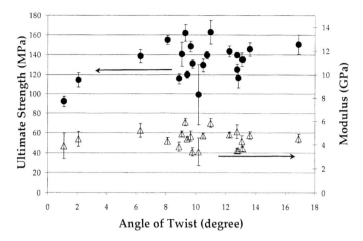

Figure 7. Mechanical properties of twisted Polyacrylonitrile nanofiber yarn as a function of twist angle (°). Reproduced with permission from reference 12. Copyright 2004 Elsevier

The stress-strain behavior of the yarns appeared to be similar to that of commercially produced PAN fibers. Commercial PAN fiber has an ultimate strength of approximately 512MPa (after post-treatment), according to the literature [25]. Commercial PAN precursor fibers are drawn prior to stabilization which decreases the fiber diameter and reduces the probability of encountering a critical flaw in a given test length. The ultimate strength of commercial PAN precursor fibers is approximately three times larger than the yarns of electrospun nanofibers; the twisted yarns were not drawn. PAN precursor fiber with a diameter of 155μm, as measured by laser diffraction, and an average denier of 80 was prepared by dry-jet solution spinning in our laboratory and the mechanical properties were measured prior to post treatment. The initial modulus and ultimate strength of the dry-jet solution spun fiber was 2.6 ±0.1GPa and 56 ±13MPa, respectively. The initial modulus and ultimate strength of the twisted electrospun PAN yarn with a twist angle of 1.1° and 11° are both approximately 1.5 times, and 2.2 and 2.9 times greater than that of the dry-jet solution spun PAN fiber prior to post-drawing, respectively [12].

Post-treatment of Polyacrylonitrile Fiber Yarns

The effect of post stretching of unidirectional electrospun PAN fiber yarns at elevated temperature on the mechanical properties was investigated.

Unidirectional tows of PAN nanofibers prepared by electrospinning onto a rotating target with a surface velocity of 9.8m/s were drawn above the glass transition temperature to 100% elongation. The initial modulus and ultimate strength increase with draw ratio from 3.0 ± 0.3 GPa and 114 ± 11 MPa at zero draw to 4.8 ± 0.5 GPa and 253 ± 47 MPa at 100% draw, respectively (Figure 8). The elongation at break decreases from 25 ± 6% at zero draw to 9.4 ± 2% at 50% draw, and then remains approximately constant with increasing draw to

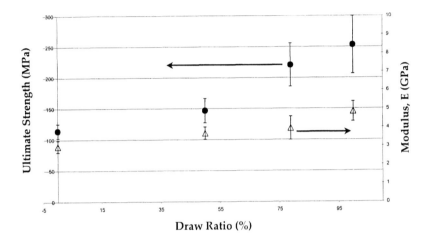

Figure 8. Mechanical properties of uniaxially aligned Polyacrylonitrile as a function of elongation (%)

100% draw. Unidirectional tows could not be drawn more than 100% at 85-90°C and provide reproducible results. The ultimate strength of unidirectional yarns drawn to 100% is approximately half that of commercial PAN precursor fibers, which are drawn to a maximum of 600% elongation.

The degree of orientation determines the mechanical properties of PAN fibers. It is also known that the mechanical properties of carbon fibers made from PAN fibers depend on the orientation imparted by the latter. The electrospinning process with rotating collection system provides a technique to prepare oriented carbon precursor fiber yarns with fiber diameters in the sub-micron range.

Conclusions

Electrospun fibers collected onto a stationary target are isotropic, with little or no molecular orientation, and entangled. The molecular orientation and alignment of electrospun fibers is controlled and altered through the use of a rotating collection device. The ultimate strength and modulus of electrospun yarns is increased with the addition of twist to a critical twist angle. Post drawing of electrospun fibers increases their axial orientation and mechanical strength. Aligned, electrospun fibers with a high degree of orientation are expected to surpass conventional fibers in the reinforcement of composites due to the increased surface area available for adhesion, their high aspect ratio (l/d), and expected property improvement with smaller diameter fibers.

References

1. Larrondo, L. and R.S. Manley. *J. Polym. Sci.* **1981**, *19*, 921.
2. Deitzel, J.M.; Harris, K.D. and N.C. Beck Tan. *Polymer* **2000**, *42*, 261-272.
4. (a) Theron, A.; Zussman, E.; Yarin, A.L. *Nanotechnology* **2001**, *12*, 384. (b) Li, D.; Xia, Y. *Nano Lett.* **2003**, *3*, 555. (c) Zussman, E.; Theron, A.; Yarin, A.L. *Appl. Phys. Lett.* **2003**, *82*, 973. (d) Li, D.; Wang, Y.; Xia, Y. *Adv. Mater.* **2004**, *16*, 361.
5. (a) Lee, S.; Yoon, J.; Suh, M. *Macromol. Res.* **2002**, *10*, 282. (b) Dersch, R.; Liu, T.; Schaper, A.K.; Greiner, A.; Wendorff, J.H. *J. Polym. Sci.: Part A* **2003**, *41*, 545. (c) Kameoka, J.; Orth, R.; Yang, Y.; Czaplewski, D.; Mathers, R.; Coates, G.W.; Craighead, H.G. *Nanotechnology.* **2003**, *14*, 1124. (d) Sundaray, B.; Subramanian, V.; Natarajan, T.S.; Xiang, R.; Chang, C.; Fann, W. *Appl. Phys. Lett.* **2004**, *84*, 1222.
6. (a) Lee, K.; Kim, H.; Ryu, Y.; Kim, K.; Choi, S. *J. Polym. Sci.: Part B* **2003**, *41*, 1256. (b) Yang, K.S.; Edie, D.D.; Lim, D.Y.; Kim, Y.M.; Choi, Y.O. *Carbon* **2003**, *41*, 2039. (c) Pedicini, A.; Farris, R.J. *Polymer* **2003**, *44*, 6857.
7. (a) Kardos, J. L. High Performance Polymers. Baer, E. and A. Moet, Eds. New York, Hanser Publishers **1991**, Ch 6. (b) Halpin, J.C. and J.L. Kardos. *Polym. Eng. & Sci.* **1976**, *16*, 344. (c) Halpin, J.C. Revised Primer on Composite Materials: Analysis. Pennsylvania, Technomic Publishing Co. **1984**, Ch. 6.
8. Christensen, R.M. Mechanics of Composite Materials. New York, John Wiley & Sons Inc. **1979**, Ch. 3.
9. Schwartz, M. Composite Materials: Properties, Nondestructive Testing, and Repair. New Jersey, Prentice Hall Inc. **1996**, Ch 1.
10. MacDiarmid, A.G.; Jones, W.E.; Norris, I.D.; Gao, J.; Johnson, A.T.; Pinto, N.J.; Hone, J.; Han, B.; Ko, F.K.; Okuzaki, H.; Llaguno, M. *Synth. Met.* **2001**, *119*, 27.

318

11. (a) Ding, B.; Kim, H.; Lee, S.; Lee, D.; Choi, K. *Fibers and Polymers* **2002**, *3*, 73. (b) Kang, Y.; Kim, H.; Ryu, Y.; Lee, D. ; Park, S. *Polymer (Korea)* **2002**, *26*, 360.

12. Fennessey, S.F.; Farris R.J. *Polymer* **2004**, *45*, 4217.

13. Chamot, E.; Mayson, C. Handbook of Chemical Microscopy Volume 1. New York, John Wiley & Sons **1938**, Ch.10.

14. (a) Brandrup, J. and E.H. Immergut. Polymer Handbook. 3rd Ed. New York, John Wiley & Sons Inc. **1989**, V/57. (b) Brandrup, J.; Immergut, E.H. and E.A. Grulke. Polymer Handbook. 4th Ed. New York, John Wiley & Sons Inc. **1999**, VII/11.

15. (a) W.L. Simm, "Apparatus for the Production of Filter by Electrostatic Fiber Spinning," U.S. Patent 3,994,258 (**1976**). (b) D. Groitzsch, E. Fahrback, "Microporous Multilayer Nonwoven Materials for Medical Applications," U.S. Patent 4,618,524 (**1986**). (c) C.H. Bamford, K.G. Al-Lamee, "Functionalisation of Polymers," U.S. Patent 5,618,887 (**1997**). (d) G.E. Martin, I.D. Cockshott, F.J.T. Fildes, "Fibrillar Lining for Prosthetic Device," U.S. Patent 4,044,404 (**1977**).

16. H. Schrueder-Gibson, P. Gibson, K. Sencal, M. Sennett, J. Walker, W. Yeomans, D. Ziegler, P.P. Tsai, *J. of Adv. Mater.* **2002**, *34*, 44.

17. S.L. Cooper; A.V. Tobolsky. *J. of Applied Polym. Sci.* 1966, *10*, 1837.

18. Estes, G.M.; Seymour, R.W.; Cooper, S.L. *Macromolecules* 1971, *4*, 452.

19. Chung, D.D.L. Carbon Fiber Composites. Boston, Butterworth-Heinemann **1994**, Ch 2.

20. Fitzer, E.; Manocha, L.M. Carbon Reinforcements and Carbon/Carbon Composites. New York, Springer-Verlag **1998**, Ch1.

21. (a) Chun, I.; Reneker, D.H.; Fong, H.; Fang, X.; Deitzel, J.; Beck Tan, N.; Kearns, K. *Adv. Mater.* **1999**, *1*, 36. (b) Kim, C.; Yang, K.S. *Appl. Phys. Lett.* **2003**, *83*, 1216.

22. (a) Bahl, O.; Mathur, R.; Kundra, K. *Fibre Sci. Tech.* **1981**, *15*, 147. (b) Chari, S.; Bahl, O.; Mathur, R. *Fibre Sci. Tech.* **1981**, *15*, 153.

23. (a) Bashir, Z.; Church, S.; Waldron, D. *Polymer* **1994**, *35*, 967. (b) Bashir, Z.; Tipping A.; Church, S. *Polymer International* **1994**, *33*, 9. (c) Bashir, Z.; Atureliya, S.; Church, S. *J. Mater. Sci.* **1993**, *28*, 2731. (d) Dalton, S.; Heatley, F.; Budd, P. *Polymer* **1999**, *40*, 5531. (e) Zbinden, R. Infared Spectroscopy of High Polymers. New York, Academic Press **1964**, Ch V. (f) Jasse B.; Koenig, J. *J. Macromol. Sci.- Rev. Macromol. Chem.* **1979**, *C17*, 61.

24. (a) Bashir, Z. *Acta Polymer* **1996**, *47*, 125. (b) Davidson, J.; Jung, H.; Hudson, S.; Percec, S. *Polymer* **2000**, *41*, 3357.

25. Moncrieff, R.W. Man-Made Fibres, 6th Ed. New York, John Wiley & Sons **1975**.

Chapter 22

Uniaxial Alignment of Electrospun Nanofibers

Dan Li, Jesse T. McCann, and Younan Xia[*]

Department of Chemistry, University of Washington, Seattle, WA 98195

The conventional setup for electrospinning has been modified to generate uniaxially aligned arrays of nanofibers over large areas. By using a collector with conductive strips separated by an insulating gap of variable width, these arrays could be directly fabricated during the electrospinning process. Directed by electrostatic forces, the charged nanofibers were stretched across the gap in a uniaxially aligned fashion. It was also versatile to stack the nanofibers into multilayered architectures with well-controlled hierarchical structures by using multiple electrode pairs patterned on an insulating substrate. Both methods have been applied to nanofibers consisting of organic polymers, carbon, ceramics, and composites. The facile formation of uniaxially aligned arrays of nanofibers with controlled positions and orientations on a solid substrate makes it possible to fabricate electrospun nanofiber-based devices and systems.

Introduction

Electrospinning represents a simple and versatile tool for manufacturing nanofibers (*1,2*). Besides polymers, this technique has recently been extended to produce carbon, ceramic, and composite nanofibers (*1-3*). Because of the bending instability associated with a spinning jet (*4,5*), electrospun fibers are often deposited on the surface of a collector (a piece of conductive substrate) as randomly oriented, nonwoven mats. Such mats are of great importance for applications in membrane separation and filtration, reinforcement of composites, texturing, sensing, scaffolding for tissue growth, enzyme immobilization and in electronic uses such as supercapacitors, actuators or photovoltaic devices (*1,2*). Many of the important applications of nanofibers, however, rely on the secondary and tertiary structures of these materials. For example, the fabrication of electronic and photonic devices often requires well-aligned and highly-ordered architectures (*6-9*). Even for fiber reinforcement and tissue engineering, uniaxial alignment of electrospun fibers is beneficial to improve their performance (*10*).

A number of approaches have been developed to fabricate electrospun fibers with well-ordered secondary structures. Alignment of electrospun fibers was observed by several groups when a cylinder with high rotating speed was used as the collector (*10-12*). Flow of air may also favor the orientation of fibers along the shearing direction (*13*). In order to improve the degree of orientation, Zussman and co-workers modified the design of a drum and used a tapered, wheel-like disk as the collector (*14*). It was found that most of the fibers could be collected on the sharp edge. The collected fibers were oriented parallel to each other along the edge. They further demonstrated that nanofiber crossbars could be readily fabricated using this collector (*15*). With the use of a similar setup, Natarajan, Xu, and coworkers have also fabricated well-aligned nanofibers (*10,16*). In addition to drums, metal or wooden frames have been explored by several research groups to collect electrospun nanofibers as relatively aligned arrays. Deitzel and co-workers have demonstrated that electrospun fibers could be aligned into parallel arrays using a multiple field technique (*17*), while Vaia and co-workers reported that they have fabricated aligned yarns of nylon-6 nanofibers by rapid oscillation of a grounded frame within the jet (*18*). Wendorff and co-workers have reported that they used a metal frame to collect parallel arrays of polyamide nanofibers (*19*). However, neither the mechanism of alignment nor a way to transfer the aligned fibers for use in the fabrication of devices has been reported.

We have recently demonstrated that uniaxially aligned electrospun nanofibers can be obtained by using a collector with an insulating gap (*20,21*). This method allows the fibers to be transferred or directly deposited onto a solid substrate for device fabrication. It is also convenient to fabricate multilayered hierarchical structures of well-aligned nanofibers by using suitable electrode configurations.

Experimental

Figure 1A illustrates the schematic setup used for our electrospinning experiments. It is similar to the conventional one except for the use of two separated conducting strips as the collector. The collector could be fabricated from two pieces of conducting silicon wafers by separating with a void gap, or two conductive gold strips deposited on a quartz or plastic substrate by evaporation of gold through a physical mask. The gap width was fixed at 7 mm for our experiments, though the width could be varied from tens of micrometers to several centimeters. Poly(vinyl pyrrolidone) (PVP, Aldrich, $M_w \approx 1,300,000$) was electrospun to demonstrate this concept. In a typical procedure, a 6 wt% solution of PVP in a mixture of ethanol and water (8:1.5 by volume) was loaded into a plastic syringe equipped with a stainless steel needle. The needle was connected to a high-voltage power supply (ES30P-5W, Gamma High Voltage Research Inc., Ormond Beach, FL) capable of generating up to 30 kV of DC voltage. The solution was continuously supplied using a syringe pump at a rate of 0.2 mL/h. The voltage used for electrospinning was 6 kV, and the collection distance was 9 cm. Optical micrographs (dark field mode) were recorded using a Leica microscope equipped with a digital camera. Scanning electron microscope (SEM) micrographs were obtained using a field-emission scanning electron microscope (Sirion, FEI, Portland, OR) operating at an accelerating voltage of 5 kV.

Results and Discussion

Uniaxial Alignment of Nanofibers

Figure 1B shows the electrostatic forces acting on a segment of charged nanofiber stretched across the gap. An electrospun fiber can be approximated as a string of positively (or negatively if a negative potential is applied to the solution) charged elements connected by a viscoelastic material. Near the point of collection, the nanofiber should experience two sets of electrostatic forces: The first set (F_1) originates from the splitting electric field, while the second one is between the charged fiber and the induced image charges on the surface of the two grounded electrodes (F_2). The electrostatic force F_1 is parallel to the electric

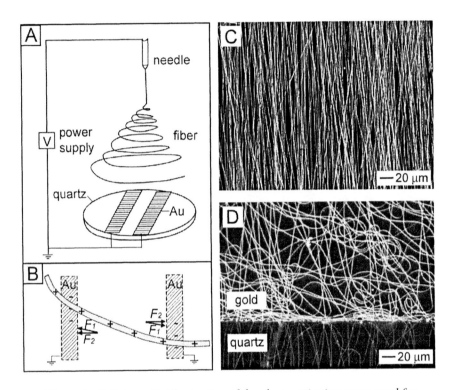

Figure 1. A) Schematic illustration of the electrospinning setup used for fabricating uniaxially aligned arrays of nanofibers. The collector was composed of a quartz wafer patterned with a pair of gold electrodes. B) Analysis of the electrostatic forces operating on a charged nanofiber spanning across the gap. F_1: the electrostatic force directly acted by the electric field; F_2: the Columbic interactions between the positive charges on the nanofiber and the negative image charges on the two grounded electrodes. C) Optical micrograph of PVP nanofibers collected between a pair of gold electrodes patterned on a quartz wafer. The collection time was 20 seconds. D) Optical micrograph of the edge of a gold electrode. The collection time was 3 seconds.

fields and therefore pulls the ends of the fiber toward the grounded electrodes. Since Columbic interactions are inversely proportional to the square of the separation between charges, the ends of the fiber closest to the grounded substrate will generate a strong electrostatic force (F_2), which will stretch the nanofiber across the gap and thereby position it perpendicular to the edges of the electrode.

Figure 1C is a dark field optical micrograph of a portion of nanofibers that were deposited on top of a quartz gap. Figure 1D is an optical micrograph that shows the edge of one of the conductive strips. These figures clearly show that the fibers were aligned as a uniaxial array across the insulating gap, while the fibers that were deposited on the conductive strip were randomly oriented.

Both the degree of uniaxial alignment and the density of the fibers on the gap were dependent on the collection time. Figure 2 shows the effect of collection time on fiber alignment and density. Fibers collected for 1 second showed some alignment, though less than 50 percent of the fibers were aligned within 10° of normal. Fibers collected for 5 seconds had better alignment, with more than 60 percent of fibers aligned within 10° of normal. Fibers collected for 30 seconds exhibited much improved alignment, with over 80 percent of the fibers aligned within 10° of normal. From these images and statistical analysis of the angles between the long axes of the fibers, it can be observed that the degree of alignment increased with increasing collection time. This phenomenon was ascribed to the repulsions between charged fibers. Unlike fibers deposited onto an electrode (where the fibers were discharged immediately upon contacting with the electrode), the fibers suspended across the insulating gap remained highly charged. These fibers had the same charge and therefore repelled each other. This type of electrostatic repulsions also tended to make these fibers align in a parallel fashion since the parallel configuration represents the lowest energy state.

When a void gap was used, the aligned fibers suspended across this gap could be easily and conveniently transferred to the surface of a solid substrate for further processing or device fabrication by moving the substrate vertically through the gap after the nanofibers had been collected. The collector with a void gap is particularly useful to manipulating single electrospun fibers (for example, for fabricating single-fiber devices). By controlling the collection time, well-separated fibers could be suspended across the gap, and single fibers could be conveniently obtained by breaking other fibers. The fabrication of arrayed crossbar junctions by this method was facile. However, the void gap was not suited for collecting fibers with diameters less than 150 nm because such thin fibers could not support their own weight and broke when they were stretched across the gap. One could solve this problem by patterning metal strips on an insulating substrate. These patterned electrodes are particularly useful for fabricating multilayered nanofiber architectures with controllable orientations.

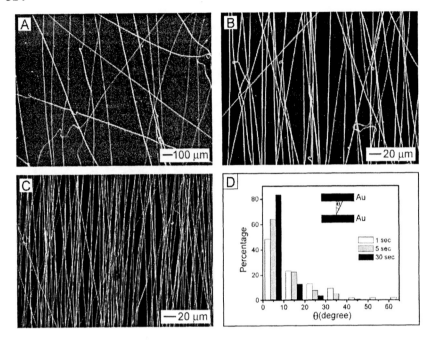

Figure 2. A-C) Typical optical micrographs of PVP nanofibers that were collected between two gold electrodes patterned on a quartz wafer. The collection times for the three samples were 1, 5, and 30 seconds, respectively. D) Distribution of angles between the long axis of a fiber and the normal to the edges of two parallel gold electrodes. The results displayed in each panel came from analysis of more than 150 fibers. Adapted in part from reference 21, © 2004 Wiley-VCH.

Multilayered Structures of Aligned Nanofibers

Uniaxially aligned arrays were only formed between the electrodes, thus their spatial orientation and position could be controlled by the location and configuration of the grounded electrodes. It was very convenient to stack the nanofibers into multilayered hierarchical structures by using multiple electrode pairs patterned on an insulating substrate. Figure 3A shows a schematic illustration of the four-electrode pattern that was deposited on a quartz wafer. When only one pair of electrodes (1-3 or 2-4) was grounded, the spun fibers were deposited solely on the grounded electrode pair and the gap between them.

By alternately grounding the electrode pairs, a double-layer mesh of fibers could be obtained, as the optical micrograph illustrates (Figure 3B). A six-electrode pattern was also used, as shown in Figure 3C. By alternately grounding the electrode pairs (1-4, 2-5, and 3-6), a tri-layered mesh was generated by the sequential deposition of three layers of uniaxially aligned nanofibers, with their long axes rotated by 60°. It is believed that architectures of greater complexity could be generated by controlling the electrode configuration and the sequence of voltage application.

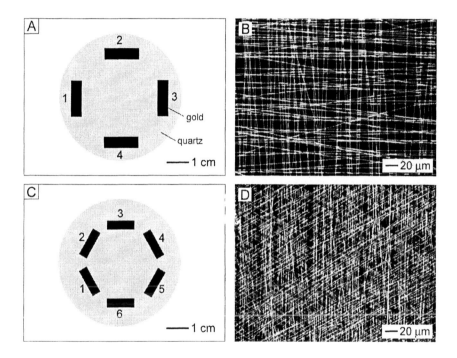

Figure 3. A, C) Schematic illustration of test patterns that were composed of four and six electrodes deposited onto quartz wafers. B,D) Optical micrograph of a mesh of PVP nanofibers collected in the center region of the gold electrodes shown in (A) and (C), respectively. During collection, the opposing electrode pairs were alternately grounded for ~5 seconds. Adapted in part from reference 21, © 2004 Wiley-VCH.

It is quite remarkable that this technique allows for the alignment and assembly of nanofibers into complex architectures concomitant with the fabrication of the fibers. By using and extending these electrode designs, multiple layers of well-aligned nanofibers with different compositions could be readily assembled to create well-ordered hierarchical architectures on a solid substrate. Such structures could find immediate use in the fabrication of both electronic and photonic devices. In addition, different arrays of nanofibers can be integrated into the same device, since a large number of electrode patterns can be readily fabricated on a substrate using conventional microfabrication methods.

Uniaxially Arrayed Fibers of Various Materials

We have successfully used the setup shown in Figure 1A to fabricate uniaxially aligned nanofibers from a variety of organic polymers including poly(ethylene oxide), polystyrene, polyacrylonitrile and polycaprolactone. Our results indicate that this approach is a generic one which can be used for many types of conventional organic polymers. We have also been able to extend the conventional electrospinning technique to fabricate functional ceramic nanofibers by spinning polymer solutions containing sol-gel precursors, followed by calcination of as-spun composite nanofibers at elevated temperatures (20-23). Hollow nanofibers have also been fabricated using a dual-capillary spinneret system (24). All of these fibers could also be collected as uniaxially aligned arrays using a collector with an insulating gap. Figure 4 shows some representative SEM images of aligned nanofibers made of different materials. In addition to pure polymer or ceramic nanofibers, organic functional molecules, biological macromolecules, nanoparticles, inorganic nanowires and carbon nanotubes could also be incorporated into solutions for electrospinning of functionalized nanofiber arrays. These aligned arrays of nanofibers with various functionalities hold much promise as building blocks for the fabrication of nanoscale devices.

Properties and Applications of Uniaxially Aligned Nanofibers

The fabrication of uniaxially aligned arrays by electrospinning allows for the exploration of a range of interesting properties and applications associated with one-dimensional nanostructures. As an example, the controlled alignment of nanofibers should result in the formation of nanostructured materials with highly anisotropic behavior. The electrical conductivities of thin films of uniaxially aligned nanofibers of both carbon and SnO_2 were found to be highly anisotropic. The ratio between the conductivities parallel and perpendicular to the long axis of the fiber was ~15 for a film with a density of ~900 nanofibers per millimeter.

Using a collector with a void gap, it was very convenient to collect single fibers for device fabrication. Compared with 1D nanostructures prepared by

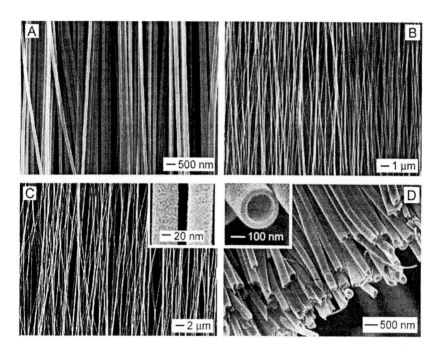

Figure 4. SEM images of uniaxially aligned nanofibers (A-C) and nanotubes (D) with various compositions: A) carbon; B) TiO₂/PVP composite; C) Sb-doped SnO₂; and D) anatase.

other chemical or physical methods, it was much easier to collect and manipulate single nanofibers. Figure 5A shows the SEM image of an individual Sb-doped nanofiber that was collected on top of a gap and subsequently transferred onto two gold electrodes separated by ~20 μm. The inset shows a nonlinear current-voltage curve measured from this nanofiber, which shows behavior similar to that of a metal oxide varistor.

Uniaxially aligned nanofibers also have the potential to be good optical polarizers. The Maxwell-Garnett model predicts that the attenuation of light is greater for the electric field polarized parallel to the long axis of an infinite cylinder than the component perpendicular to the axis *(25,26)*. Our results agree well with the Maxwell-Garnett model. Figure 5B shows Rayleigh scattering spectra of a parallel array of PVP nanofibers. It can be seen that the extinction of incident light with polarization parallel to the long axis was three times greater than the extinction of light polarized perpendicular to the long axis of the fibers.

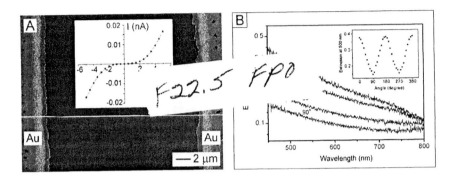

Figure 5. (A) SEM image of a single Sb-doped SnO$_2$ nanofiber stretched across two gold electrodes. The inset shows a typical I-V curve measured from the device. (B) Rayleigh scattering spectra obtained from a uniaxially aligned array of 90-nm PVP nanofibers with the polarizer oriented at various angles relative to the longitudinal axes of the nanofibers. The inset shows how the extinction at 500 nm was modulated (as a cosine wave) as a function of the angle between the nanofibers and the polarization.

Conclusion

By using collectors with an insulating gap, the electrospinning process has been modified to provide a simple and versatile method for creating uniaxially aligned arrays of nanofibers with various compositions and properties. The use of electrodes with insulating gaps between pair-wise plates allowed for the uniaxial alignment of fibers into parallel arrays via electrostatic forces. It was also possible to stack the aligned nanofibers into multilayered films with well-defined hierarchical structures by controlling the configuration for patterned electrodes. This system holds much promise for the fabrication of electrospun nanofiber-based architectures, devices, and systems.

References

1. Reneker, D. H.; Chun, I. *Nanotechnology* **1996**, 7, 216.
2. Li, D.; Xia, Y. *Adv. Mater.* **2004**, 4, in press.

3. Larsen, G.; Velarde-Ortiz, R.; Minchow, K.; Barrero, A.; Loscertales, I. G. *J. Am. Chem. Soc.* **2003**, *125*, 1154.

4. Reneker, D. H.; Yarin, A. L.; Fong, H.; Koombhongse, S. *J. Appl. Phys.* **2000**, *87*, 4531.

5. Shin, Y. M.; Hohman, M. M.; Brenner, M. P.; Rutledge, G. C. *Polymer* **2001**, *42*, 9955.

6. Kovtyukhova, N. I.; Mallouk, T. E. *Chem. Eur. J.* **2002**, *8*, 4354.

7. Huang, Y.; Duan, X.; Wei, Q.; Lieber, C. M. *Science* **2001**, *291*, 630.

8. Favier, F.; Walter, E. C.; Zach, M. P.; Benter, T.; Penner, R. M. *Science* **2001**, *293*, 2227.

9. Melosh, N. A.; Boukai, A.; Diana, F.; Gerardot, B.; Badolato, A.; Petroff, P. M.; Heath, J. R. *Science* **2003**, 300, 112.

10. Xu, C. Y.; Inai, R.; Kotaki, M.; Ramakrishna, S. *Biomaterials* **2004**, *25*, 877.

11. Kim, J.-S.; Reneker, D. H. *Polym. Eng. Sci.* **1999**, *39*, 849.

12. Kameoka, J.; Craighead, H. G. *Appl. Phys. Lett.* **2003**, *83*, 371.

13. Chun, I.; Reneker, D. H.; Fong, H.; Fang, X.; Deitzel, J.; Tan, N. B.; Kearns, K. *J. Adv. Mater.* **2003**, *31 (1)*, 37.

14. Theron, A.; Zussman, E.; Yarin, A. L. *Nanotechnology* **2001**, *12*, 384.

15. Zussman, E.; Theron, A.; Yarin, A. L. *Appl. Phys. Lett.* **2003**, *82*, 973.

16. Sundaray, B.; Subramanian, V.; Natarajan, T. S. *Appl. Phys. Lett.* **2003**, *82*, 973.

17. Deitzel, J. M.; Kleinmeyer, J. D.; Hirvonen, J. K.; Tan, N. C. B. *Polymer* **2001**, *42*, 8163.

18. Fong, H.; Liu, W. D.; Wang, C. S.; Vaia, R. A. *Polymer* **2002**, *43*, 775.

19. Dersch, R.; Liu, T.; Schaper, A. K.; Greiner, A.; Wendorff, J. H. *J. Polym. Sci. A* **2003**, *41*, 545.

20. Li, D.; Wang, Y.; Xia, Y. *Nano Lett.* **2003**, *3*, 1167.

21. Li, D.; Wang, Y.; Xia, Y. *Adv. Mater.* **2004**, *16*, 361.

22. Li, D.; Xia, Y. *Nano Lett.* **2003**, *3*, 555.

23. Li, D.; Herricks, T.; Xia, Y. *Appl. Phys. Lett.* **2003**, *83*, 4586.

24. Li, D.; Xia, Y. *Nano Lett.* **2004**, *4*, 933.

25. Rogers, J. A.; Paul, K. E.; Jackman, R. J.; Whitesides, G. M. *Appl. Phys. Lett.* **1997**, *70*, 2658.

26. Aspnes, D. E. *Thin Solid Films* **1982**, *89*, 249.

Chapter 23

Orientation Development in Electrospun Liquid–Crystalline Polymer Nanofibers

David Y. Lin[1] and David C. Martin[1–3]

[1]Macromolecular Science and Engineering Center and Departments of [2]Materials Science and Engineering and [3]Biomedical Engineering, University of Michigan, Ann Arbor, MI 48109

Banded structures in electrospun liquid-crystalline polymer, poly(hexyl isocyanate) (PHIC), are used to establish a direct relationship between fiber diameter and orientational order of polymer chains. The effect of solution concentration on the morphology of electrospun PHIC is discussed. Using polarized light optical microscopy (POM), the local molecular orientation of the polymer chains within the fibers can be measured. The degree of orientation of the polymer chains with respect to the fiber axis is found to increase with decreasing fiber diameters. Electron diffractions of PHIC are also obtained and compared with the results from POM.

Introduction

Electrospinning is the process of using electrostatic forces to distort a pendant droplet of polymer solution into a fine filament to be deposited onto a substrate (*1*). Research efforts in electrospinning in recent years have re-generated interest in this technique since its first patent was issued to Formhals in 1934 (*2*). These efforts have allowed more than 20 different types of polymer fibers to be generated by electrospinning (*3*). The ability of this process to produce extremely fine fibers (from less than 3 nm to greater than 1 μm in diameter), as well as a number of interesting fiber morphologies has made it an attractive area of research for many different applications including the manufacturing of biopolymer nanofibers (*4, 5*), protective clothing systems, separation membranes (*6*), and tissue scaffolds (*7, 8*).

Successful electrospinning requires the use of a volatile solvent. Rapid solvent evaporation during electrospinning leads to a rapid concentration change of the polymer solution jet before deposition. In many cases the jet actually undergoes the liquid-to-solid transition before being collected onto the substrate as nanofibers. Furthermore, it has been shown that the jet travels in a twisting spiral path after the onset of a "bending" instability. The onset and the extent of this bending instability are still not very well understood. Therefore, the orientational order, both local and long-range, of the electrospun nanofibers can be developed during different stages of electrospinning and may involve different process parameters in different polymer/solvent systems.

Liquid crystalline polymers (LCPs) can be readily aligned in shear fields, magnetic fields, and electric fields. The oriented microstructure can be trapped in place from solution to create high strength fibers for lightweight structural applications (*9*). Studies of the orientation order induced during solution processing can therefore provide insights about the mechanisms of electrospinning. In this study the orientational order of an electrospun LCP, poly(hexyl isocyanate) (PHIC), is examined. PHIC is readily processed from volatile organic solvents at room temperature, and can be highly crystalline in the solid state. It therefore provides a convenient system for detailed investigations. However PHIC is not very thermally stable, and therefore detailed electron optical studies are experimentally challenging.

Background

Importance of Orientation in Electrospun Nanofibers

Controlling the orientation of polymer materials has significant implications in commercially important applications. First, orientation provides the main route in which the mechanical properties of polymers can be improved. Processing methods including drawing at elevated temperatures to spinning from liquid-crystalline solutions or gels have been researched extensively because of their ability to induce segmental orientation of polymer chains. By controlling these segmental orientation during processing while preventing relaxation of the chains when the external field is removed polymers can be processed into ordered materials having excellent high-performance properties (*10, 11*).

In addition to improving mechanical properties, orientation can also be manipulated to provide desired microstructures or textures by using an externally applied field. This field may be chemical, mechanical, magnetic, or electrical in origin. An example of this is the use of surface aligned liquid crystals (LCs) in LC display applications. The most widely used alignment film is mechanically rubbed polyimide. One explanation for the alignment mechanism is that polyimide molecules acquire anisotropic orientation during the rubbing process and the anisotropic molecular force between the polyimide and LCs causes the surface alignment of LCs. In this application, both polymer main chain orientation and microgrooves obtained by rubbing contribute to the LC alignment (*9, 12-16*).

Previous Results on Molecular Alignment of Electrospun Nanofibers

The use of a volatile solvent is an important requirement of the electrospinning process. In order for the polymeric jet to be collected as fibers, significant solvent evaporation has to occur before deposition. In addition, many polymers can only be dissolved in a limited number of volatile organic solvents. This causes the polymeric jet to undergo a rapid liquid-to-solid transition during electrospinning where the chain conformation and crystal structure are locked in place in less than a second. Furthermore, many polymers have been electrospun consistently into nanofibers with diameters of 100 nm or less. The small diameter imposes a size constraint on the crystallization process such that the crystals can grow to long dimensions only along the fiber axis (*11*). The implication is that the size confinement of the small diameter fibers will result in a higher degree of polymer chain orientation in the direction of the fiber axis.

The high shear stress the electrospun jet experiences during electrospinning and the extremely fast crystallization time would suggest that electrospinning will cause polymer chains to be highly oriented in the direction of the fiber axis. However, characterization of this orientational order is difficult because of the low crystallinity of electrospun fibers as a result of very fast crystallization kinetics. Stephens et al. have shown that the electrospinning process does not destroy the chemical architecture of nylon-6, but causes some alteration in the conformation to change the resulting crystal structure from the α-form, which is found in most commercially available nylon-6, to the γ-form, which has been observed in nylon-6 melt spun at high take-up speeds (17). Molecular orientation of as-spun fibers have only been characterized by birefringence under optical microscope and weak, diffuse patterns under x-ray or electron diffraction (18). Wendorff et al. found that orientation seems to be almost absent in as-spun polylactic acid (PLA) fibers and to be locally strong yet inhomogeneous in electrospun polyamide-6 fibers (19). However, to our knowledge the degree of orientation has not been characterized in a systematic manner in as-spun nanofibers from electrospinning. Reneker et al. has shown that as-spun PPTA nanofibers have diffuse diffraction spots from electron diffraction. The sharpness of the diffraction pattern can be significantly improved by an annealing procedure to show orientation along the fibers axis (11). Chu et al. found that as-spun poly(glycolide-co-lactide) (GA/LA: 90/10) membranes exhibited a low degree of crystallinity. With annealing at elevated temperatures without drawing, the degree of crystallinity of their samples improved significantly but no overall orientation was observed. However, if the samples were annealed with drawing, the degree of orientation was found to increase with the elongation ratio (20).

Banded textures from Sheared Liquid Crystalline Polymers

Many polymeric materials contain chemical structures that can be oriented under mechanical deformation or flow, and electric and magnetic fields. When these structures contain anisotropic polarizabilities, the index of refraction of the sample will be orientation-dependent, resulting in an observable birefringence in the specimen. Measurements of birefringence can be used to characterize the preferential molecular alignment of the sample. However, quantitative conversion of the measured birefringence to molecular orientation requires an accurate estimate of the intrinsic birefringence of perfectly aligned molecules. Furthermore, light scattering in thick crystalline samples as well as the need to determine the form component make this technique more readily applicable to transparent non-crystalline polymers, crystalline fibers, thin films, and complex fluids. In many cases more reliable but less convenient methods such as WAXS, i.r. dichroism, NMR, and electron diffraction techniques can be used to provide additional data to accurately determine the molecular orientation (*21-24*).

Liquid crystalline polymers (LCPs), both lyotropic and thermotropic, orient readily under shear or elongational flow. Following thin-film shear, uniaxial fiber drawing, injection molding, or elongational flow, LCPs often exhibit the so-called "banded texture" under cross polars. This texture is composed of equally spaced extinction bands oriented perpendicular to the direction of flow and in the plane parallel to the shearing surfaces (*25*).

Characterization of Banded Textures by Microscopy

A common technique for measuring birefringence is polarized light optical microscopy. In this setup, the sample is placed between a polarizer and an analyzer. Polarized light from the first polarizer impinges on the sample and scattering-induced changes in the polarization of the light are detected after the scattered light passes through the analyzer. Using this technique, the direction in which the light is scattered can be measured as a function of the orientation of the light's polarization. One characteristic of optical anisotropy, given the sample is thin enough, is the presence of extinctions when either of the crossed polars is aligned with the axis of molecular orientation (*26-28*). The degree of orientation can be characterized by measuring the angle between the extinction direction and the direction of the aligned field.

Experimental

The PHIC was synthesized by a living anionic polymerization with the help of F. E. Filisko and T. Menna. The molecular weight and polydispersity of the synthesized PHIC are 76000 and 3.2, respectively. Polymer solutions of various concentrations between 5 and 25 wt.% of PHIC in chloroform were prepared for electrospinning. Chloroform was purchased from Fisher Scientific.

PHIC solutions were electrospun at 8-9 kV and 2-5 cm tip-to-target distance. The field was generated by a Hipotronics R10B HV DC power supply. The solution was drawn into a 3 ml plastic syringe from Becton, Dickinson, and Co. and a #23GP needle tip from EFD, Inc. was attached to the syringe. Using a syringe pump model KDS 100 from KD Scientific, feed rate of the polymer solution to form the pendant drop at the tip of the syringe needle was controlled at 0.2 ml/hr. The substrates used for collecting electrospun fibers include ~10 cm^2 Al foil (OM and SEM) and C coated mica (TEM).

Optical microscopy was done using a Nikon Optiphot with crossed polarizers and a full wave red filter. SEM was conducted on a Philips FEG operating at 5 kV, and TEM was performed on a JEOL 4000 EX operating under low dose conditions at 400 kV.

Results and Discussion

In solution, PHIC is thought to adopt a helical conformation. Chen et al. reported the most detailed analysis to date on the structure of PHIC containing rod-coil block copolymers (29). Based on their electron diffraction results they proposed an eightfold helix structure arranged on a two-chain, pseudo-hexagonal lattice (a = b = 1.51 nm, c = 1.56 nm). This 8_3 helical conformation is similar to that proposed by Shmueli et al. for crystalline PBIC (9). Martin showed the ability to develop orientation in PHIC thin films using both DC and AC electric fields, where the degree of induced orientation is a function of the strength and geometry of the electric field (9). In addition, he found that the polymer molecules were uniformly aligned parallel to the field.

Concentration Effect

The effects of various processing parameters on the morphology of electrospun nanofibers have been extensively studies for a number of polymer solutions (30-45). In general, electrospinning at low concentrations is more similar to electrospraying, producing tiny droplets with sizes range from tens to couple hundred micrometers. The reason of this break-up of the polymer jet is the inability of low viscoelastic forces to prevent the over-stretching of a charged jet during its flight to the collector. At very high concentrations, solvent evaporation will cause the polymer drop at the tip of the syringe to solidify to prevent continuous electrospinning. The morphology of electrospun fibers changes from beaded to smooth fibers within these two limits. Several groups have observed that the average diameter of the electrospun fibers increases with increasing concentration in a power law relationship in this working range. Besides introducing an external variable such as salt, polyelectrolyte, or a second solvent, it was found that concentration has the most significant role in determining the morphology and size of electrospun nanofibers.

The structure of the electrospun PHIC was found to exhibit tremendous changes as the concentration of the polymer solution was varied. Figure 1 shows the morphological and orientational changes observed in this study using OM with crossed polarizers. At 5 wt.% concentration, PHIC was electrospun into large beads (drops) of ~100μm in diameter with orientation starting to develop around the edges of the beads. As concentration increased, a mixture of large beads and fibers was generated. Further increase in concentration decreased the amount of beads present and produced continuous fibers with larger diameters. Orientation of both the beads and fibers can also be observed in these figures, with blue regions indicating +45° alignment and orange regions indicating -45° alignment. The thinnest fibers show uniform birefringence consistent with an average orientation that is uniform at these length scales (~1 μm).

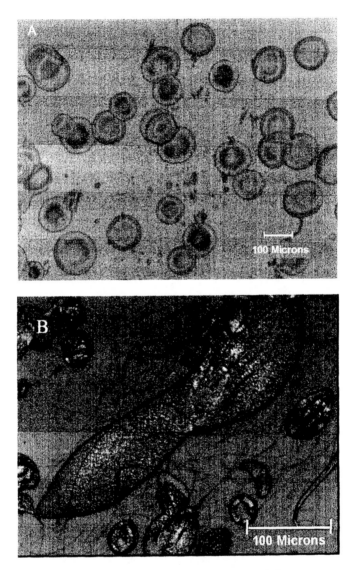

Figure 1. Concentration effect on microstructures of electrospun PHIC nanofibers at voltage of 9kV, feed rate of 0.2 ml/hr, and concentration of (A) 5 wt.%; (B) 14.5 wt.% and (C) 26.9 wt.%.

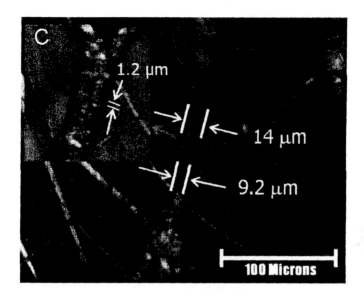

Figure 1. *Continued.*

Banded Structure

Figure 1C also shows local organization of PHIC chains into banded structures that has been observed previously in sheared solutions and drawn PHIC thin films (*29*). Viney and Putnam found that in lyotropic systems of hydroxypropyl cellulose the rate of band evolution after shear can be characterized by the time taken for the average band spacing to reach a minimum and depends on a combination of prior shear rate, specimen thickness, solvent type and concentration (*25*). Figure 2 shows a schematic model of these banded structures with two parameters that are used in this study to quantitatively characterize the orientational order of electrospun PHIC fibers.

We found that the band width, W_b, systematically increases with fiber diameter as shown in Figure 3. In fibers with very large diameters, oriented regions can be found randomly dispersed within the fibers, but do not organize into bands. These orientated regions look like those that are found in stretched beads as seen in Fig. 1B and 1C. As fiber diameter decreases, banded structures with large W_b start to appear. As fiber diameter decreases from $14\mu m$ to $5\mu m$, W_b does not change significantly. However, the morphology changes from disjointed regions in the band to continuos, smooth bands. Further decreases in fiber diameter below 5μm are associated with further decreases in W_b. Viney and Putnam proposed that band evolution rates are governed by a balance between the effects of solvent on microstructural mobility and the disclination density which stores the energy to drive post-shear microstructural change, such as band formation (*25*). Our results then suggest that as fiber diameter decreases, there are less disclination defects available to provide the energy for band formation.

For fibers with diameters smaller than 2 μm, it is difficult to observe any banded structure using OM. Therefore, we employed TEM to examine these fibers. Figure 4 shows ED pattern of electrospun PHIC fibers, compared with ED pattern of an aligned PHIC thin film. The ED pattern of electrospun PHIC fibers contains more diffuse arcs and spots compare with the sharp spots in the ED pattern of the aligned PHIC thin film. This observation indicates that electrospun PHIC filaments have lower orientational order and crystallinity than electric-field aligned PHIC thin films. This is consistent with what Zong et al. observed in electrospun poly(L-lactic acid) nanofibers (*30*). The apparent reason for this lower crystallinity is the rapid solvent evaporation during electrospinning, which prevents polymer chains from sufficiently relaxing into their equilibrium conformations. The presence of the strong 3rd, 5th, and 8th layer lines shows that molecular chains in the electrospun PHIC fibers also adopt an $8_3/8_5$ helical conformation. Finally, our ED results show that the local director field of PHIC chains is well-aligned with the fiber axis.

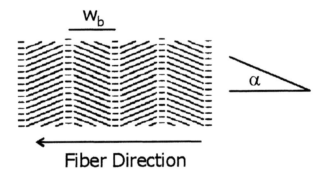

Figure 2. Schematic diagram of PHIC banded structure from electrospinning. The angle between the local director field and fiber axis is denoted α.. The width of the bands is denoted W_b.

Effect of Fiber Diameter on Band Width

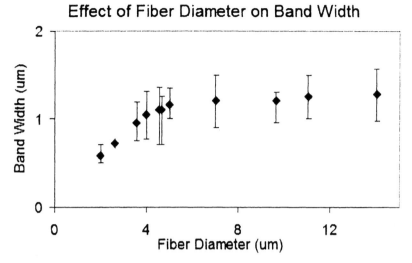

Figure 3. Relationship between fiber diameter, and band width (W_b) of electrospun PHIC fibers.

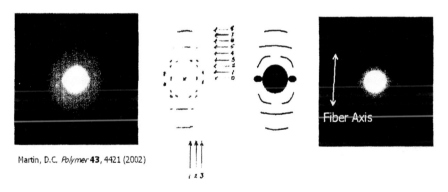

Martin, D.C. *Polymer* **43**, 4421 (2002)

Figure 4. ED of electrospun PHIC (right), compared with ED from aligned, crystalline PHIC thin film (left). (Reproduced with permission from reference 9. Copyright 2002 Elsevier.)

Figure 5 shows that there is a reasonably linear relationship between fiber diameter and α observed in the region between 2 μm and 13 μm diameter fibers. By extrapolating our data, this relationship suggests that there will be essentially complete alignment (α = 0) for fibers less than 2 μm in diameter (no bands).

340

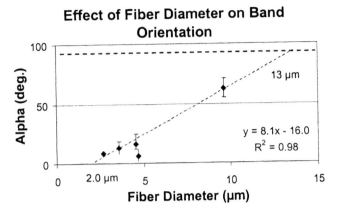

Figure 5. Relationship between fiber diameter, and α of electrospun PHIC fibers.

Conclusions

We have shown that the orientation development during electrospinning of PHIC correlates with fiber diameter. Since fiber diameter can be controlled by varying processing parameters and polymer solution properties, our results indicate a route to control the orientation of electrospun nanofibers by controlling fiber diameters. One of the most effective parameters in changing fiber diameters is polymer solution concentration. We have shown that by changing PHIC/Chloroform concentration from 26.9 wt.% to 14.5 wt.%, fiber diameter can be decreased more than one order of magnitude, although this decrease in fiber diameter is accompanied by an increase in the number of beads.

We have also shown that the as-spun electrospun PHIC fibers have lower crystallinity than aligned PHIC thin films, but the PHIC chains are highly oriented with the fiber axis, adopting an $8_3/8_5$ helical conformation.

Acknowledgements

This study was funded in part by the National Institute of Health, the National Science Foundation Integrated Graduate Education and Research and Training program at the University of Michigan and by Foster-Miller, Inc. SEM and TEM were done at the University of Michigan Electron Microbeam Analysis Laboratory.

References

1. Doshi, J.; Reneker, D. H. *J. Electrostat.* **1995**, *35*, 151-160.
2. Formhals, A. U.S. Patent 1,975,504, 1934.
3. Reneker, D. H.; Yarin, A. L.; Fong, H.; Koombhongse, S. *J. Applied Phys.* **2000**, *87*, 4531-4547.
4. Buchko, C. J. Ph.D. thesis, University of Michigan, Ann Arbor, MI, 1997.
5. Larrondo, L.; Manley, R. S. *J. J. Poly. Sci. B-Poly. Phy.* **1981**, *19*, 909-920.
6. Gibson, P.; Schreuder-Gibson, H.; Rivin, D. *Colloids and Surfaces a-Physicochemical and Engineering Aspects* **2001**, *187*, 469-481.
7. Matthews, J. A.; Wnek, G. E.; Simpson, D. G.; Bowlin, G. L. *Biomacromolecules* **2002**, *3*, 232-238.
8. Li, W. J.; Laurencin, C. T.; Caterson, E. J.; Tuan, R. S.; Ko, F. K. *J. Biomed. Mater. Res.* **2002**, *60*, 613-621.
9. Martin, D. C. *Polymer* **2002**, *43*, 4421-4436.
10. Zhao, W. Y.; Kloczkowski, A.; Mark, J. E.; Erman, B.; Bahar, I. *Macromolecules* **1996**, *29*, 2805-2812.
11. *Structure Formation in Polymeric Fibers* Salem, D. R., Eds.; Hanser Gardner Publications, Inc.: Munich, 2001.
12. Mori, N.; Morimoto, M.; Nakamura, K. *Advanced Materials* **1999**, *11*, 1049-1051.
13. Yan, L. F.; Zhu, Q. S.; Ikeda, T. *Polymer International* **2003**, *52*, 265-268.
14. Yan, L. F.; Zhu, Q. S.; Ikeda, T. *Journal of Applied Polymer Science* **2001**, *82*, 2770-2774.
15. Brostow, W.; Hibner, K.; Walasek, J. *Macromolecular Theory and Simulations* **2001**, *10*, 304-313.
16. Andresen, E. M.; Mitchell, G. R. *Europhysics Letters* **1998**, *43*, 296-301.
17. Stephens, J. S.; Chase, D. B.; Rabolt, J. F. *Macromolecules* **2004**, *37*, 877-881.
18. Fong, H.; Reneker, D. H. *J. Poly. Sci. B-Poly. Phys.* **1999**, *37*, 3488-3493.
19. Dersch, R.; Liu, T. Q.; Schaper, A. K.; Greiner, A.; Wendorff, J. H. *J. Poly. Sci. a-Poly. Chem.* **2003**, *41*, 545-553.
20. Zong, X. H.; Ran, S. F.; Fang, D. F.; Hsiao, B. S.; Chu, B. *Polymer* **2003**, *44*, 4959-4967.
21. Viney, C.; Donald, A. M.; Windle, A. H. *Polymer* **1985**, *26*, 870-878.
22. Viney, C.; Mitchell, G. R.; Windle, A. H. *Poly. Comm.* **1983**, *24*, 145-146.
23. Gervat, L.; Mackley, M. R.; Nicholson, T. M.; Windle, A. H. *Phil. Tran. Royal Soc. London Series a- Math. Phys. Eng. Sci.* **1995**, *350*, 1-27.
24. Romo-Uribe, A.; Windle, A. H. *Proc. Royal Soc. London Series a-Math. Phys. Eng. Sci.* **1999**, *455*, 1175-1201.

342

25. C. Viney, W. S. Putnam, *Polymer* **1995**, *36*, 1731-1741.
26. C. Viney, *Poly. Eng. Sci.* **1986**, *26*, 1021-1032.
27. Donald, A. M.; Viney, C.; Windle, A. H. *Phil. Mag. B-Phys. Condensed Matter Stat. Mech. Elec. Opt. and Magnetic Prop.* **1985**, *52*, 925-941.
28. Mitchell, G. R.; Windle, A. H. *Polymer* **1983**, *24*, 1513-1520.
29. Chen, J. T.; Thomas, E. L.; Ober, C. K.; Hwang, S. S. *Macromolecules* **1995**, *28*, 1688-1697.
30. Zong, X. H.; Kim, K.; Fang, D.; Ran, S.; Hsiao, S. H.; Chu, B. *Polymer* **2002**, *43*, 4403-4412.
31. Mit-uppatham, C.; Nithitanakul, M.; Supaphol, P. *Macromolecular Chemistry and Physics* **2004**, *205*, 2327-2338.
32. Deitzel, J. M.; Kleinmeyer, J.; Harris, D.; Beck Tan, N. C. *Polymer* **2001**, *42*, 261-272.
33. Nair, L. S.; Bhattacharyya, S.; Bender, J. D.; Greish, Y. E.; Brown, P. W.; Allcock, H. R.; Laurencin, C. T. *Biomacromolecules* **2004**, *5*, 2212-2220.
34. Katti, D. S.; Robinson, K. W.; Ko, F. K.; Laurencin, C. T. *J. Biomed. Mat. Res. B-App. Biomater.* **2004**, *70B*, 286-296.
35. Sukigara, S.; Gandhi, M.; Ayutsede, J.; Micklus, M.; Ko, F. *Polymer* **2003**, *44*, 5721-5727.
36. Buchko, C. J.; Chen, L. C.; Shen, Y.; Martin, D. C. *Polymer* **1999**, *40*, 7397-7407.
37. Son, W. K.; Youk, J. H.; Lee, T. S.; Park, W. H. *Polymer* **2004**, *45*, 2959-2966.
38. Lin, T.; Wang, H.; Wang, H.; Wang, X. *Nanotechnology* **2004**, *15*, 1375-1381.
39. Stankus, J. J.; Guan, J.J.; Wagner, W.R. *Journal of Biomed. Mat. Res. A* **2004**, *70A*, 603-614.
40. Mit-uppatham, C.; Nithitanakul, M.; Supaphol, P. *Macromol. Symp.* **2004**, *216*, 293-299.
41. Liu, H. Q.; Hsieh, Y.L. *J. Poly. Sci B-Poly. Phys.* **2002**, *40*, 2119-2129.
42. Koski, A.; Yim, K.; Shivkumar, S. *Materials Letters* **2004**, *58*, 493-497.
43. Lee, K. H.; Kim, H. Y.; La, Y. M.; Lee, D. R.; Sung, N. H. *J. Poly. Sci. B-Poly. Phys.* **2002**, *40*, 2259-2268.
44. McKee, M. G.; Wilkes, G. L.; Colby, R. H.; Long, T. E. *Macromolecules* **2004**, *37*, 1760-1767.
45. Kenawy, E. R.; Layman, J. M.; Watkins, J. R.; Bowlin, G. L.; Matthews, J. A.; Simpson, D. G.; Wnek, G. E. *Biomaterials* **2003**, *24*, 907-913.

Chapter 24

Morphology and Activity of Biological Fabrics Prepared by Electrospray Deposition Method

A. Taniokia[1,*], H. Matsumoto[1], I. Uematsu[1], K. Morota[1], M. Minagawa[1], Y. Yamagata[2], and K. Inoue[3]

[1]Department of Organic and Polymeric Materials, Tokyo Institute of Technology, Tokyo 152–8552, Japan
[2]Materials Fabrication Laboratory, The Institute of Physical and Chemical Research (RIKEN), Saitama 351–0198, Japan
[3]Fuence, Tokyo 150–0012, Japan

Protein nanofabrics were prepared by the electrospray deposition (ESD) method from the aqueous solutions of α-lactalbumin (α-LA), invertaze, immunoglobulin (IgG), and α-LA with poly(vinyl alcohol) (PVA), and their surface morphologies and biological activities were characterized. The surface morphologies of the deposited films were observed using scanning electron microscopy (SEM) and atomic force microscopy (AFM). The SEM and AFM images showed that the film surfaces had a fine porous structure, in which the pore diameters ranged from 40 to 600 nm. The biological activities were tested by the mechano-chemical method, a microarray-based enzyme linked immunosorbent assay (ELISA), and fluorescence immunoassay (FIA) format. It was demonstrated that the activities of the deposited protein fabrics were preserved during the ESD. The results revealed that the ESD method was useful for producing fine porous protein nanofabrics with biological activities. The porous protein fabric opens a new direction in the application of biomaterials. To improve the sensitivity of the protein nanofabric, it should be attempted to control the nano-scaled structure and fine porous morphology of the fabric surface. ESD technique also can be extended to the protein fabrics combined with polymeric nanofibers.

343

Electrospray deposition (ESD) is a versatile method for forming nano-scaled superstructures, or *nanofabrics*. This method is applicable for solute molecules, which have wide range of molecular weights (e.g., low weight molecules, synthetic polymers, proteins, and DNA) (1-10). Electrospray-deposited nanofabrics, have recently attracted much attention for applications such as biosensors and biochips (e.g., protein/DNA-microarray and microfluidic device) (5, 11), antifouling or biocompatible coatings for medical devices, high-performance filter media (12), biomaterial scaffolds for tissue engineering (13), structural color materials, and water proof materials. The ESD method consists of four steps: (i) application of strong electric field, (ii) overcoming of electrostatic forces in a critical value to the surface tension of the solution, (iii) spraying of charged droplets from the tip of the capillary, and (iv) collecting the dried droplets to form a dry nanofabric (14-16). One of the major advantages for the ESD method is that dry protein fabrics are deposited (8). The dry protein nanofabrics are suitable for preserving their functional properties (17). In the ESD process, the morphologies of the deposited nanofabrics are also easily controlled by changing the applied voltage, and viscosity, surface tension, and conductivity of solution.

We fabricated α-lactalbumin (α-LA), invertaze, immunoglobulin (IgG), and α-LA with poly(vinyl alcohol) (PVA) using the ESD technique in order to observe the fabricated structures for α-LA, invertaze, and α-LA with PVA; and examine the effects of ESD on activities of α-LA and IgG (18,19). α-LA is a compact globular protein that binds the calcium ion (Ca^{2+}) in a 1:1 ratio at a specific binding site and consists of an α-helical domain and a β-sheet domains (20). We prepared cross-linked α-LA nanofabrics and performed mechano-chemical measurements based on the conformational changes in α-LA caused by Ca^{2+}-binding in order to characterize their biological activity (21). A protein multi-microarray with several kinds of anti-IgG antibodies was also fabricated using the ESD technique. The feasibility of simultaneous and specified detection methods for numerous IgGs as antigens using a microarray-based enzyme linked immunosorbent assay (ELISA) and fluorescence immunoassay (FIA) format is also demonstrated.

The objectives of this paper are to show how (i) to prepare protein nano fabrics, (ii) to characterize their surface morphologies, and (iii) to examine their biological activities by using the outstanding proteins.

Materials

α-Lactalbumin (α-LA, Type III, calcium depleted, from Bovine Milk), poly(L-lysine) (PLL,), glutaraldehyde (25% aqueous solution) and invertaze were purchased from Sigma. 2-[4-(2-Hydroxylethyl)-1-piperazinyl] ethanesulfonic acid (HEPES) was obtained from Dojindo Lab. Calcium dichloride, sodium chloride, sodium hydroxide, and poly(vinyl alcohol) (PVA) were from Wako Pure Chemical. Antigens (mouse, human, bovine, chicken, rabbit, and guinea pig IgGs) and their corresponding anti-IgG antibodies were purchased from the Sigma and Jackson Immuno Research. ECL substrate was purchased from Amersham.

Electrospray Deposition (ESD) Method

Electrospraying was carried out using an electrospray deposition device (ESD200S, Fuence, Japan) which is shown in Figure 1. The deposition of protein was performed under the control of a local electric field, which attracts charged electrosprayed products to specified substrate areas. By masking a conducting substrate with a dielectric mask with holes, which is covered with Teflon-coated

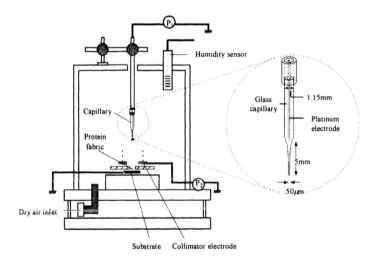

Figure 1. Schematic diagram of the ESD device. P1: Power supply for ESD, P2: Power supply for collimator (Reproduced with permission from reference 18. Copyright 2004 Elsevier.)

copper film, a local electric field was formed between the capillary and the substrate. Protein and synthetic polymer were then deposited on the substrate through the holes in the mask. To improve the efficiency of the spraying, a collimating voltage was applied to the thin copper film between the surface of the mask and the Teflon film. Biological and synthetic polymer solutions are sprayed from a glass capillary with an internal platinum microelectrode. The applied voltage between the capillary and the substrate was 3-4 kV. Deposition was performed in an acrylic chamber filled with dry air (the humidity is 15 %

346

and temperature is about 25°C). The deposited protein fabrics were then cross-linked in a vapor of glutaraldehyde aqueous solution.

Surface Morphologies of Protein Nanofabrics(18)

The surface morphologies of the deposited protein nanofabrics were observed using a field emission scanning electron microscope (FE-SEM, S-800, Hitachi) and a scanning probe microscope (SPM, AutoProbe M5, TM Microscopes) in the atomic force microscope (AFM) mode. All SEM observations were carried out at 6 kV without sample coating.

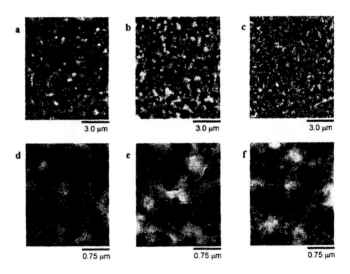

Figure 2. Surface SEM images of the electrospray deposited nanofibers after postdeposition crosslinking: (a) LA-1 (×10k), (b) LA-2 (×10k), (c) LA-3 (×10k), (d) LA-1 (×40k), (e) LA-2 (×40k), and (f) LA-3 (×40k) (Reproduced with permission from reference 18. Copyright 2004 Elsevier.)

To examine the surface morphologies of α-LA fabrics using the ESD method, we grew them from three solution concentrations of 0.4 mg/mL (LA-1), 0.73 mg/mL (LA-2) and 1.8 mg/mL (LA-3). Figure 2 and Figure 3 show the SEM images and 3D-AFM images of α-LA fabrics after postdeposition cross-linking, respectively. The SEM images showed that the fabric surfaces had a fine porous structure, which have pore sizes ranging from 40 to 600 nm. Nanofabrics from the higher concentrated protein solution had smaller pore sizes. The AFM images also demonstrated that there were remarkable irregularities in the fabric surfaces. For the LA-3 fabric, which was depoited from a highly concentrated solution, there are undulations with a peak-to-valley height of 25 nm. The protein solution with a lower concentration for the ESD method produced smoother fabrics (For the LA-1 fabric, peak-to-valley height is about 4 nm). Large changes in the morphologies of the fabrics could not be obesrved

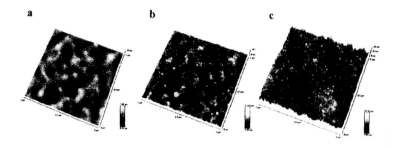

a   b   c

Figure 3. AFM images of electrospray deposited thin films after postdeposition cross-linking: (a) LA-1, (b) LA-2, and (c) LA-3 (Reproduced with permission from reference 18. Copyright 2004 Elsevier.)

before and after the postdeposition cross-linking. The control of the concentration enables ESD to tailor the surface morphology of the nanofabrics

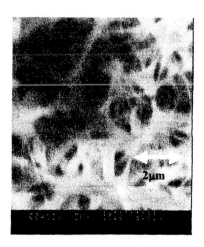

Figure 4. Surface SEM images of PVA nanofiber combined with a-LA

for specific applications; the porous and rough protein fabrics formed by ESD could potentially increase their detection capability by increasing their active protein area exposed to the analyte. Figure 4 shows the surface SEM image of the fabric from α-LA/PVA blend solution. The surface morphology has only fibrous structure. This indicated that α-LA can be sprayed with synthetic polymers and deposited one was captured on PVA nanofibers.

Figure 5 shows the SEM images of invertaze fabrics, and the fabric surfaces have also a fine porous structure, which have pore sizes ranging from 40 to 600 nm. Nanofabrics from the higher concentrated protein solution had smaller pore sizes.

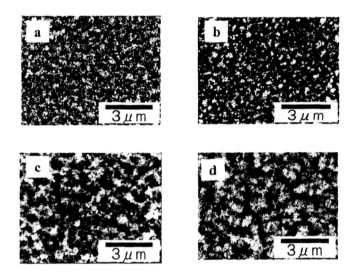

Figure 5. Surface SEM images of the electrospray deposited nanofabrics of invertaze at the solution concentration of (a) 0.5g/L, (b) 1.25g/L, (c) 2.5g/L and (d) 5.0g/l, where the spraying time is 10 min

Biological Activities of Protein Nanofabrics (18,19)

Ca^{2+}-induced conformational changes in the α-LA nanofabrics were observed by mechano-chemical method using a ligand-binding detector (MC-1, Fuence) (18). This technique is useful in the primary screening of target proteins having biological activities (21-23). Figure 6 shows a schematic diagram of the apparatus. A protein film was hooked between micro tungsten tips connected to a micro force sensor and a piezoelectric actuator and then soaked in a flow chamber supplied with a buffer solution. After equilibration in the HEPES buffer solution (10 mM HEPES + 100 mM NaCl, pH 7.6), the isometric tension changes induced by Ca^{2+}-binding were measured. The elastic constants of the

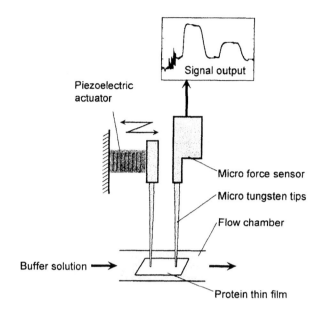

Figure 6. Schematic diagram of the ligand-binding detector based on the mechano-chemical measurements (Reproduced with permission from reference 18. Copyright 2004 Elsevier.)

fabrics were also measured at the low frequency generated by the piezoelectric actuator. To estimate the interaction between the protein and Ca^{2+}, isometric tensions in the buffer solutions with different Ca^{2+} concentrations were recorded. Figure 7 shows the Ca^{2+}-dependence of the isometric tension of the α-LA thin film (LA-3). The observed isometric tension of α-LA thin film reversibly responded to changes in the Ca^{2+} concentration in buffer solution. These changes would be due to the conformational changes in α-LA molecules induced by Ca^{2+}-binding (20), which indicated that the cross-linked α-LA film preserved the biological activity. In Figure 7, the data of poly(L-lysine) (PLL) thin film, which is electrosprayed from the 0.1 w/v aqueous solution and then cross-linked in glutaraldehyde vapor for 10 min at 30°C, is also shown for comparison. The tensile force signal of PLL thin film without Ca^{2+}-binding site showed no definite change on the Ca^{2+} concentration in buffer solution.

350

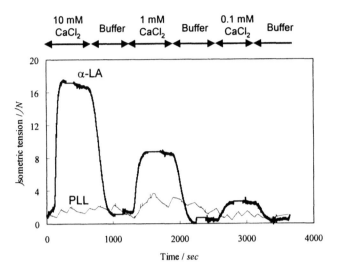

Figure 7. Interactions of α-LA and PLL thin films with calcium ion in the buffer
solution. Buffer: HEPES 10 mM + NaCl 100 mM, pH 7.4
(Reproduced with permission from reference 18. Copyright 2004 Elsevier.)

The sensitivity of a microarray-based enzyme linked immunosorbent assay (ELISA) using anti-mouse IgG microarray plates fabricated with ESD method, were also examined (19). Following the blocking procedure, the plates incubated with different concentrations of HRP-conjugated mouse IgG. The levels of antigen as low as 1ng/ml could be detected using the ECL method and X-ray film. The sensitivity of fluorescence immunoassay (FIA) using the microarray was also examined. Anti-mouse IgG was deposited and immobilized onto an indium-tin-oxide (ITO) coated slide glass. After the blocking process, the plate was placed on skimmed milk solution (2%) with different concentrations of mouse IgG conjugated with FITC as the antigen. After washing, the fluorescent signals of the antigen and antibody complexes were detected by a digital CCD camera. As shown in Figure 8 the signals were quantitatively visualized and levels of mouse IgG lower than 1ng/ml could be detected. Sensitive and simultaneous detection of various antigens could be performed by enzyme immunosorbent assay or fluorescence immunoassay. It was proved that ESD technique was also available for the medical diagnostic system.

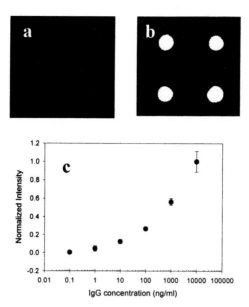

Figure 8. The sensitivity of FIA in the detection of FITC conjugated mouse IgG: (a) 0.1ng/ml, (b) 100ng/ml, (c) standard curve (normalized fluorescent intensity vs. mouse IgG concentration)

Conclusions

It was demonstrated that the activities of the deposited protein fabrics were preserved during the ESD. The results revealed that the ESD method was useful for producing fine porous protein nanofabrics with biological activities. The porous protein fabric opens a new direction in the application of biomaterials. To improve the sensitivity of the protein nanofabric, it should be attempted to control the nano-scaled structure and fine porous morphology of the fabric surface. ESD technique also can be extended to the protein nanofabrics combined with polymeric nanofibers.

References

1. Chen, C., Kelder, E. M., van der Put, P. J. J. M., Schoonman, J., *J. Mater. Chem.* **1996**, 6, 765
2. Hoyer, B., Sorensen, G., Jensen, N., Nielsen, D. B., Larsen, B, *Anal. Chem.* **1996**, 68, 3840

3. Morozov, V. N., Morozova, T. Ya, and Kallenbach, N. R., *Int. J. Mass Spectrom.* **1998**, 178, 143
4. Morozov, V. N., and Morozova, T. Ya, *Anal. Chem.* **1999**, 71, 1415
5. Morozov, V. N., and Morozova, T. Ya, *Anal. Chem.* **1999**, 71, 3110
6. Buchko, C. J., Chen, L. I., Shen, Y., and Martin, D. C., *Polymer*, **1999**, 40, 7397
7. Natalya, V., Morozova, T. Ya, Ataullakhanov, F. I., and Morozov, V. N., *Anal. Chem.* **73**, 6047 (2001)
8. Morozov, V. N., and Morozova, T. Ya, *Anal. Chem.* **2002**, 74, 927
9. Morota, K., Tanioka, A., Yamagata, Y., and Inoue, K., *Kobunshi Ronbunshu*, **2002**, 59, 706
10. Morota, K., Tanioka, A., Yamagata, Y., and Inoue, K., *Kobunshi Ronbunshu*, **2002**, 59, 710
11. Gibson, P., Schreuder-Gibson, H., and Rivin, D., *Colloids Surfaces A*, **2001**, 187-188, 469
12. Kane, R. S., Takayama, S., Ostuni, E., Ingber, D. E., and Whitesides, G. M., *Biomaterials*, **1999**, 20, 2363
13. Matthews, J. A., Simpson, D. G., Wnek, G. E., and Bowlin, G. L., *Biomacromolecules*, **2002**, 3, 232
14. Reneker, D. H. and Chun, I., *Nanotechnology*, **1996**, 7, 216
15. Fong, H., Chun, I., and Reneker, D. H., *Polymer*, **1999**, 40, 4585
16. Deitzel, J. M., Kleinmeyer, J., Harris, D., and Beck Tan, N. C., *Polymer*, **2001**, 42, 261
17. Liu, W. R., Langer, R., and Klibanov, A. M., *Biotechnol. Bioeng.*, **1991**, 37, 177
18. Uematsu, I., Matsumoto, H., Morota, K., Minagawa, Taniokia, A., M., Yamagata, Y., and Inoue, K., *J. Colloid & Interface Sci.*, **2004**, 269, 336
19. Lee, B., Kim J., Ishimoto, K., Yamagata, Y., Tanioka, A., and Nagamine, T., *J. Chem. Eng., Japn.*, **2003**, 36, 1370
20. Permyakov, E. A. and Berliner, L. J., *FEBS Letters*, **2000**, 473, 269
21. Bukatina, A. E., Morozov, V. N., Gusev, N. B., and Sieck, G. C., *FEBS Letters*, **2002**, 524, 107
22. Morozov, V. N. and Morozova, T. Ya., *Anal. Biochem.*, **1992**, 201, 68
23. Yamagata, Y., Morozov, V. N., Inoue, K., Kim, J., Ohmori, H., and Higuchi, T., in the Conference Abstracts of The 7th World Congress on Biosensors, Kyoto, Japan, 15-17 May, **2002**

Chapter 25

Optical Properties of Transparent Resins with Electrospun Polymer Nanofibers

C. Krauthauser[1,2], J. M. Deitzel[1], D. O'Brien[3], and J. Hrycushko[1]

[1]Center for Composite Materials, University of Delaware, Newark, DE 19716
[2]U.S. Army Research Laboratory, Armor Mechanics Branch, Aberdeen Proving Ground, MD 21005
[3]U.S. Army Research Laboratory, Multifunctional Materials Branch, Aberdeen Proving Ground, MD 21005

Model composites have been made using electrospun Nylon 6,6 nanofibers with diameters of ~100 nm. Resins used include both transparent epoxy and vinyl ester resins. Optical measurements using a white light source have been made to evaluate the degree of light transmission, haze, and clarity for the composites. A simple model describing the transmission of light through the composite specimen has been developed. This model relates fiber volume fraction, fiber diameter, indices of refraction of both fiber and matrix, and sample thickness to the percentage of light transmitted. The model demonstrates that submicron fiber diameters are necessary in order to maintain a high degree of transparency. Predictions of the model are in good agreement with experimental data. The work shows that processing issues such as resin wet-out of the nanofiber fabric and mitigation of void formation are key factors to obtaining clear nanofiber composites.

Introduction

There is a need for light-weight, impact resistant transparent materials for use in a variety of applications, which could include vehicle windows and windshields, hand held protective shields, face shields, and protective eyewear. Solutions using glass and layered polymer/glass composites have the disadvantage of adding a significant weight penalty to a given application and are generally restricted to use in vehicle and building applications. Some research [1] has looked at using glass fiber to reinforce resins with a matching index of refraction. In addition to the same issues of weight associated with the laminate composites, the optical clarity of these materials is generally dependent on temperature due to difference in CTE between the resin and glass fiber as well as thermal dependence of indices of refraction.

Applications requiring light weight, such as face shields and protective eyewear, usually employ a transparent polymer resin like vinyl ester, epoxy(thermoset), or polycarbonate (PC) and poly(methyl methacrylate)(PMMA)(thermoplastic). However, the impact resistance and structural properties of these materials are limited by their relatively low mechanical properties in comparison to high performance materials like Kevlar and carbon fiber reinforced composites. Reinforcement of these transparent resins with high perfomance materials is problematic due to the need to maintain optical clarity. Recent efforts by Hsieh [2] have looked at using nanoparticulate fillers and microlayering processing techniques to increase the impact performance of PC, and PC/PMMA blends. (needs a couple more examples of nano reinforcement)

We propose a novel approach to increasing the impact properties of transparent polymer resins like thermosetting vinyl ester and epoxy resins, and thermoplastic resins like PC and PMMA by reinforcing these resins with high performance polymer nanofibers. These nanoscale fibers can be produced readily using the process of electrospinning [3-7]. The electrospinning process uses an electrostatically driven jet of polymer fluid (solution or melt) to form fibers with diameters ranging from 50-500 nm. These fibers are most often collected in the form of a non-woven mat of randomly oriented fibers.

The advantages of using nanofibers in transparent composites are many. First, the small diameter (~200nm; see Figure 2a and 2b) of the fibers are below the characteristic wavelengths of visible light (λ=400-700 nm), and therefore, nanofibers dispersed in a transparent medium should not unduly scatter light in the visible range [3]. Second, nanofiber textiles have orders of magnitude greater specific surface area [4] than conventional fabrics, due to the small fiber diameter. The greater surface area will provide more interaction between the resin and reinforcing fiber, improving mechanical properties and potentially increasing the amount of energy dissipated during an impact event due to sliding friction associated with fiber pullout. The wide variety of polymer materials that can be electrospun [5] provide the engineer with great flexibility in designing a transparent composite for specific applications(ie. high performance fibers for structural applications, elastomeric fibers for toughening and impact resistance, conductive fibers for EM shielding etc.). Finally, in addition to an improvement of mechanical properties, nanofibers can provide a continuous network connection a variety of sensors for health monitoring of the composite.

In the work presented here, a simple theoretical model has been developed to predict the optical transmission properties of a transparent material reinforced with electrospun fabric. This first order model relates the variables of index of refraction, fiber volume fraction, and fiber diameter to the optical transparency (transmission of light in the 400-700nm region) of the composite system. The transmission of light in the optical region has been measured for transparent composites made from electrospun Nylon 6,6 and are found to be in good agreement with predictions made by the model.

Experimental

Electrospun fiber mats

Nylon 6,6 sub-micron fiber textiles have been fabricated using the electrospinning process. Nylon 6,6 was spun from a solution of forming acid at a concentration of 20% by wt. The electrospinning voltage was 8 kV and the distance between syringe needle and collection plate was ~ 15 cm. Estane^{tm} fibers were spun from a binary solvent of Tetrahydrofuran (THF,75%)/ Dimethyl Formamide(DMF, 25%) at a concentration of 5% by weight. The spinning voltage for the Estane^{tm} solution was 7 kV, and the distance from syringe needle to collection plate was ~ 12 cm. Each type of electrospun mat was examined using field emission scanning electron microscopy and the average fiber diameter was measured from the micrographs.

Electrospun fiber composites

Thin film Nylon–6,6 electrospun fiber composites were made using both Epon 828 epoxy and Derkane Vinyl ester resin. In order to eliminate air bubbles the thermoset resin was held under vacuum to ~10 psi for 30 minutes, prior to infusion. After the resin was degassed, the electrospun fabric was slowly lowered into the resin in order to get complete infiltration. The resin infused fabric was then placed between two glass slides that had been treated with a thin layer of mold release, and compressed under a pressure of ~25 psi. The composite was allowed to gel at room temperature. The samples were then postcured for one hour at 130 ° C. Optical properties of the electrospun fiber composite were characterized with respect to total transmission, haze and optical clarity using a BYK Haze-Gard Plus apparatus (ASTM D 1003). The illumination source used in this technique was white light.

A major question to be confronted is what are the factors controlling transmission, and thus, controlling optical transmission. There are many factors controlling optical transmission of light through a media, such as orientation of inclusions, characteristic sizes of the inclusions, volume fraction of the inclusions, and so forth. These issues will be discussed in further detail in the next section.

Theory: A look at fiber size and transparency

There are many parameters, such as transmittance, haze, clarity, that can be used as a measure of the level of transparency that characterizes a particular specimen. Transmittance is traditionally defined as the ratio of the intensity of the transmitted light through a specimen to the intensity of the incident light on the specimen. The ASTM D 1003 defines haze as that percentage of transmitted light which in passing through the specimen deviates from the incident beam by more than 2.5° on average, whereas clarity is evaluated in an angle smaller than 2.5°. In this section, in an attempt to present a simple theory for understanding the effects on transparency of infusing nano-scale fibers into transparent resins, our focus will be exclusively on determining transmittance; the issues of haze and clarity will be treated elsewhere.

There are many factors that can affect optical transmittance, which include fiber diameter and volume fraction of the fibers in the resin. Optical transmittance, as given above, is expressed as [1] :

$$T_{\%} = \frac{I_t}{I_0} \qquad (1)$$

As is known, the scattering of electromagnetic signals by any material is related to the optical heterogeneity of that material. In addition to scattering, the material could also absorb some of the energy of the electromagnetic signal. These two pieces combined affect the optical clarity of the incident light, and make up the attenuation that diminishes the optical clarity, thus

Attenuation = Scattering + Absorption

When incident light of intensity I_0 traverses a slab of heterogeneous material over an optical path d_o, the transmitted intensity can be given by (Beer's Law)

$$I_t = I_0 \exp(-Ad_o) \qquad (2)$$

assuming that effects from multiple scattering can be ignored, and that phase and wavelength of the light after the scattering is unchanged (coherent scattering). Here, A is the attenuation coefficient, and is given by

$$A = \sum_i n_i \left(\sigma_{a,i} + \sigma_{s,i} \right) \qquad (3)$$

where n_i is the particle density for the i-th material, $\sigma_{a,i}$ and $\sigma_{s,i}$ are the absorption and scattering cross-sections, respectively, for the i-th material. In this paper, we will assume that the inclusions are, to good approximation, non-absorbing. For non-absorbing particles, only the scattering term is important, and thus $A = \sum_i n_i \sigma_{s,i}$. Furthermore, it is assumed for this paper that the non-woven mat can be approximated as a collection of non-interacting, randomly oriented, cylindrical fibers with high aspect ratio. By non-interacting, it is meant that there is no weave associated with the collection, and furthermore, the individual fibers are, on average, sufficiently far from neighboring fibers (greater than 4-5 fiber diameters).

Let $V_{f,i}$ be the volume fraction of the i-th material, and v_i be the volume of a single particle of material i, then it is clear that $n_i = V_{f,i} / v_i$, and thus

$$A = \sum_i \frac{V_{f,i} \sigma_{s,i}}{v_i} \qquad (4)$$

Letting $\sigma_{g,i}$ to be the so-called geometrical cross section, we define the scattering efficiency, Q, by

$$Q_i = \frac{\sigma_{s,i}}{\sigma_{g,i}} \tag{5}$$

and thus

$$A = \sum_i \frac{V_{f,i} Q_i \sigma_{g,i}}{v_i} \tag{6}$$

Here it becomes necessary to specify the types and extension of the materials embedded in the transparent resin. To a good approximation, there are two basic types of inclusions in the transparent resin: infinitely long cylinders and spheres. Each has to be taken into account when determining the attenuation factor, thus there will be 2 basic terms:

$$A = A_c + A_s \tag{7}$$

For cylindrical strands, it is clear that $\sigma_g = d_f L$, where d_f is the fiber strand diameter and L is the strand length, and $v = \pi d_f^2 L / 4$. Thus,

$$A_c = \frac{4 V_{f,c} Q_c}{\pi d_f} \tag{8}$$

For normally unpolarized incident light, Q_c can be expressed as:

$$Q_c(n_f, n_M, d_f, \lambda) = \frac{1}{x} \sum_{n=0}^{\infty} \varepsilon_n \left(\left\| B_{nl} \right\|^2 + \left\| B_{nll} \right\|^2 \right) \tag{9}$$

where

$$B_{nl} = \frac{J_n(mx) J_n'(x) - m J_n'(mx) J_n(x)}{J_n(mx) H_n'(x) - m J_n'(mx) H_n(x)}$$

$$B_{nII} = \frac{mJ_n(mx)J'_n(x) - J'_n(mx)J_n(x)}{mJ_n(mx)H'_n(x) - J'_n(mx)H_n(x)}$$

Here, $\varepsilon_n = 2$ when $n > 0$ and $\varepsilon_0 = 1$, $x = \pi d_f / \lambda$ and λ is the wavelength in the matrix material, $m = n_f / n_M$ is the relative refractive index of the polymer fiber material (n_f) and the matrix material (n_M), J_n is a Bessel function of the first kind, H_n is a Hankel function of the second kind, and the primes denote differentiation of the functions with respect to their arguments.

It is useful to have some bounds on what can be expected in terms of the transmission ratio for the case of cylindrical fibers perfectly infused (no gaps, voids, bubbles, etc.) by a transparent resin. The electrospun fibers that will be considered experimentally have a characteristic diameter of $d_f \sim 100 - 300 \, \text{nm}$, the volume fraction of the fibers in the matrix material goes as $V_f \sim 0.10$, the thicknesses of the resin infused samples would go as $d_0 \sim 0.05 - 0.10 \, \text{nm}$, and the wavelength of light going through the matrix material goes as $\lambda \sim 400 - 700 \, \text{nm}$. In terms of indices of refraction, the fiber material was chosen to have an index of refraction reasonably close to that of the resin material, thus

$$n_{nylon6,6} = 1.53$$
$$n_{polyurethane} = 1.5 - 1.6$$
$$n_{epoxy} = 1.55 - 1.60$$
$$n_{polyester} = 1.52 - 1.54$$

Within this range of values, $x, mx \sim 0.45 - 2.36$. For our purposes, it is practical to choose $n_{vinyester} = 1.52$, thus $m = 1.00658$. If we consider the above samples, allowing only the fibers to be the scatterers, and taking 200 nm as the average fiber diameter, and 550 nm as the average wavelength of light, the predicted transmissions are 99.31%, 99.34%, and 99.48% for the samples in Figures 6A, B, C respectively. When compared to the 95%-98% transmission as measured, there is clearly a not insignificant role being played by the inclusions and voids that have resulted from processing.

One can go through a similar analysis of considering the voids and inclusions. We consider these voids and inclusions as essentially spherical balls made up mostly of air, with a characteristic diameter of approximately 5 microns, and volume fraction of approximately 0.1%; the index of refraction, n_i, would be 1.0. For these spherical inclusions, we derive through a fashion similar to that done above the attenuation factor. Clearly, $\sigma_g = \pi d_s^2 / 4$ where d_s is the diameter of the sphere and $v_s = \pi d_s^3 / 6$, thus

$$A_s = \frac{3V_{f,s} Q_s}{2d_s} \tag{10}$$

The scattering efficiency for non-absorbing spheres is given by

$$Q_s = \frac{2}{x^2} \sum_{n=1}^{\infty} (2n+1) \left[\|b_{nI}\|^2 + \|b_{nII}\|^2 \right] \tag{11}$$

where

$$b_{nI} = \frac{\psi_n'(mx)\psi_n(x) - m\psi_n(mx)\psi_n'(x)}{\psi_n'(mx)\zeta_n(x) - m\psi_n(mx)\zeta_n'(x)}$$

$$b_{nII} = \frac{m\psi_n'(mx)\psi_n(x) - \psi_n(mx)\psi_n'(x)}{m\psi_n'(mx)\zeta_n(x) - \psi_n(mx)\zeta_n'(x)}$$

and

$$\psi_n(z) = \left(\frac{\pi z}{2} \right)^{\frac{1}{2}} J_{n+\frac{1}{2}}(z)$$

$$\zeta_n(z) = \left(\frac{\pi z}{2} \right)^{\frac{1}{2}} H_{n+\frac{1}{2}}(z)$$

again, J_n is a Bessel function of the first kind, H_n is a Hankel function of the second kind, with $x = \pi d_s / \lambda$ and λ is the wavelength of light in the matrix media. If we assume the inclusions have an average diameter of approximately 5 microns, a volume fraction of approximately 0.1%, and the average wavelength is 550 nm, the transmission from the combination of the non-woven mat and "air" inclusions for the samples given in Figures 6A, 6B, and 6C are 93.51%, 94.39%, and 95.95%, respectively. Since it was assumed that the volume fraction of the "air" inclusions was constant, the theoretical predictions seem to lag behind the experimental measurements.

It is worthwhile to note briefly and in a general way the effects of the different parameters of the model (volume fraction, the ratio of the fiber and matrix indices of refraction, fiber diameters, etc.). In general, for toughening and durability issues of reasonably thick films of transparent resins, to maintain a relatively high level of transmittance, it is important to keep the fiber diameters at the submicron level, as the following graph illustrates (Figure 1).

Of singular importance is the ratio between the fiber and matrix indices of refraction. The scattering efficiency is very sensitive to this ratio, and if the ratio significantly departs from 1 (>1.5), there are profound effects on the scattering efficiency, and thus the transmission, without a concomitant offset in either the volume fraction or fiber diameter or both. To maintain good transmission, it is, in some sense, a balancing act between these three parameters. This is very important when one wishes to consider a fiber material with an index of refraction that departs significantly from the index of refraction of the resin material.

Results and Discussion

Electrospun fiber mats as a suitable textile for making durable transparent composites

Fiber mats were electrospun from a 20%(by wt.) solution of Nylon 6,6 in formic acid. Figure 2a shows an SEM micrograph of a typical fiber mat, while 2b shows a histogram of the distribution of fiber diameters. From this micrograph we see that the electrospun fibers are uniform in shape and do not exhibit any of the radical deviations in fiber morphology that have been reported elsewhere [4]. Additionally, it is clear that the electrospun fiber mats are continuous and that the fiber aspect ratio is essentially infinite to a good approximation. Figure 2b shows the distribution of fiber diameters measured from several SEM micrographs taken from different areas of a piece of electrospun nylon fabric. The distribution is log normal, and the majority fiber diameters are below 150nm (Figure 2a) and all are below 300 nm, which is well below the wavelength of visible light (400-700 nm). These results show that the properties of the Nylon 6,6 electrospun fiber mat are consistent with the assumptions made in the theoretical model discussed above.

362

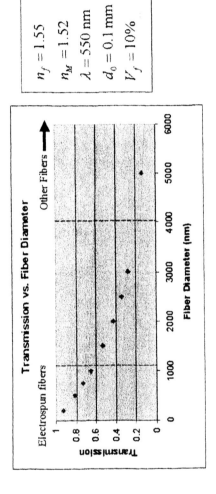

Figure 1. Predicted transmission as a function of fiber diameter. Indices of refraction for nylon 6,6 (n_f) and polyester resin (n_M) were obtained from the Polymer Handbook, 4th edition.

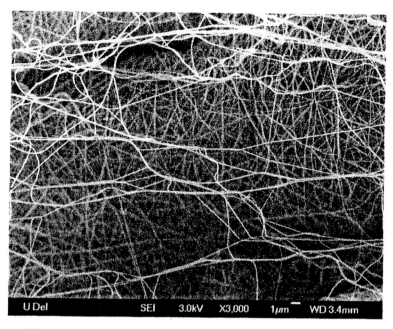

Figure 2a – SEM picture of electrospun Nylon 6,6 non-woven mat

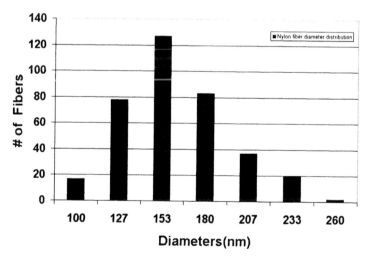

Figure 2b – Histogram of diameters for Nylon Fibers

Nanofiber composite fabrication

Initial attempts in making a nanofiber composite focused on the infusion of non-woven electro-spun mats of Nylon 6,6 with a commercially available amine based epoxy resin, Epon 828. An example of the electrospun nylon fabric is shown in Figure 3. As can be seen, the fabric is opaque in air, and has a definite texture that is the result from the collection of the electrospun fibers on a metal screen. Pieces of fabric were then immersed in a small amount of resin and allowed to gel at room temperature (Figure 4). A qualitative visual inspection of the sample in Figure 4 illustrates the relatively high degree of visual clarity that is maintained. However, when the samples were backlit, two observations were made that were not obvious when the samples were viewed in reflected light. First, a faint grid pattern, corresponding to the texture of the fabric was clearly evident. This is not surprising since it has been observed [4] that electrospun fibers will fall preferentially on the conductive portions of a patterned collection target, resulting in a regular variation in fiber volume fraction. Secondly, a noticeable degree of haze was detected in the region of the fiber mat in the composite. Inspection of the samples using optical microscopy (Figure 5), revealed the presence of voids ranging from 1 to a few tens of microns in diameter throughout the sample.

These observations illustrate two key challenges in the manufacture of transparent composites with electrospun fibers, the need for a uniform distribution of fibers throughout the matrix and the importance of complete wetting of the electrospun fabric with the matrix resin. To address these concerns, several modifications were made.

Our subsequent efforts were devoted to the infusion of vinyl ester into electrospun non-woven mats of Nylon 6,6. The migration to vinyl ester was due primarily to the greatly reduced viscosity over the epoxy resin. Initially, the volume fraction of the non-woven mat in the resin was very small (<0.1%), however, the optical clarity of the non-woven mat/resin mixture was reduced, and furthermore, the contours of the non-woven mat were clearly visible. There was no pre-wetting of the non-woven mat in the situation of using the vinyl ester, in that none of the ingredients that are used to make the vinyl ester lent themselves for use for the purposes of wetting. Thus, there are micro-voids surrounding the non-woven mats due, perhaps, to trapped air that was not released before final curing of the vinyl ester. With this in mind, as well as a desire to increase the fiber volume fraction, an alternative manufacturing procedure was necessary. Thus, the next generation of Nylon-Vinyl Ester composite was made by applying pressure during the cure cycle to force out the air bubbles, as well as increase the fiber volume-fraction. This provided composite samples with volume fractions on the order of 10%, as well as nearly eliminating the number of macro bubbles and voids, and significantly reducing the amount of micro air bubbles accumulating around the non-woven mat during manufacture (see Figure 6A-C).

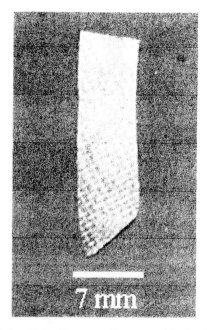

Figure 3. Nylon Nanofiber mat (See page 14 of color inserts.)

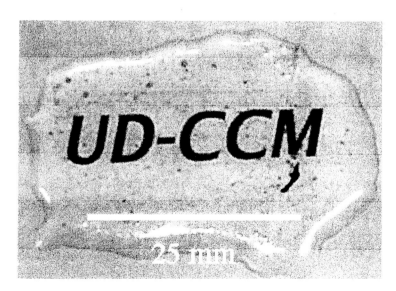

Figure 4. Nanofiber composite (See page 14 of color inserts.)

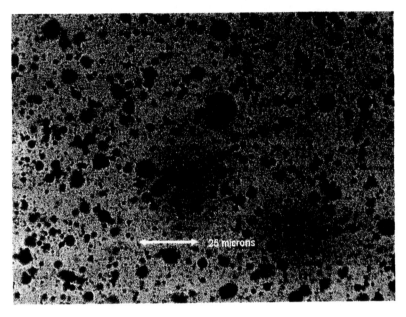

*Figure 5. Optical micrograph of voids in the electrospun fiber composite
(See page 15 of color inserts.)*

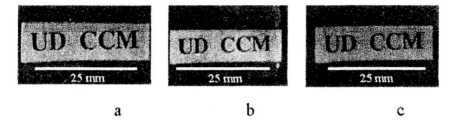

a   b   c

*Figure 6. Nylon/Vinyl ester nanofiber composites films ~9% fiber volume
fraction (See page 15 of color inserts.)*

Figure 6A-C show Nylon-Vinyl Ester composite sample with fiber volume-
fractions of approximately 9%. Three sets of samples were originally made for
the purposes of statistical consistency, but instead have provided samples of
varied thickness due to processing optimization issues. While it is clear from
Figures 6A, B, and C (which correspond to specimens 1, 2, and 3, respectively in
Table 1) that there is some diminution in optical clarity, principally due to dust
and other foreign materials being absorbed during processing, nonetheless there
is nearly complete visual transparency of the non-woven mat infused with the
resin. Optical transmission studies done on the samples in Figure 6 would bear
this out, as seen in Table 1.

Table 1: Optical characteristics of transparent samples

Specimen	Thickness (mm)	Transmission (%)	Haze (%)	Clarity (%)
neat epoxy	0.030	93.6±0.2	1.5±0.2	99.7±0.0
#1	0.100	88.8±0.2	24.4±0.8	88.6±0.8
#2	0.085	90.1±0.4	18.7±0.9	88.8±1.8
#3	0.060	91.6±0.3	18.5±0.1	88.8±0.3

The third column in Table 1 indicates the transmission of 4 samples: the
three mentioned above, and a control, labeled as "neat epoxy". The control was
simply a .03-mm thick film of the vinyl ester without being infused into a non-
woven mat. For the purposes of the remaining part of the paper, it is more useful
to consider the transmission relative to the control. Since the transmission is the
ratio of measured intensity of light passing through a transparent media to the
measured intensity prior to entering the transparent media, taking the ratio of the
transmission of the composite to the transmission of the "neat" control provides

a measure of the transmission through the non-woven mat. Thus, the transmissions for samples 1, 2, and 3 would be 94.9%, 96.3%, and 97.9%, respectively, as compared with the theoretical predictions of 93.5%, 94.3%, and 96.0%, respectively, as given above. For the samples discussed here, there would be some error associated with the fact that the samples are not all the same thickness, however, this point is mitigated by the fact that the transmissions for all samples are high. The last two columns of Table 1, dealing with haze and clarity, and which have been defined above, are not discussed in this work, but are presented for the sake of completion.

Conclusion

As noted, there is a general need to improve the performance and durability of transparent media for use in a variety of applications, most notably in vehicle windows and windshields, face and eye shields, and so on. Previous methods have had notable disadvantages, not the least of which being significant weight penalties. We have proposed here a novel approach that is intended to improve the mechanical properties of transparent polymer resins. This approach incorporates electrospun polymer nano-fibers into the transparent resins as a means of reinforcing them without adversely affecting the transparent properties of these resins.

Several generations of composites fabricated from electrospun non-woven mats infused with transparent resins have been processed. Each generation arose out of a need to adjust a processing issue that affected the transparency of the composite. In most cases the processing issue dealt with the problem of the necessity of a full wet-out of the non-woven mat in order to maintain the transparency of the matrix material. Other issues involved the formation of inclusions and voids that occurred post-wetting. Still other issues involved ways of increasing fiber volume fraction in the transparent composites.

Once the processing issues were dealt with for the most part, without overall loss of optical clarity, we investigated optical clarity by means of transmission studies on the theoretical as well as experimental levels. It is important to determine, both qualitatively and quantitatively, what is the level of transmission. On the theoretical level, we assumed that the nanoscale fibers were sufficiently separated to assume that each fiber acted as an independent scatterer, without any absorption. For the purposes of experiment, we performed out transmission experiments on electrospun nylon 6,6 infused with vinyl ester. It turns out that the indices of refraction of both are nearly the same, and the fiber diameters are on the order of 200 nm. Even with a fiber volume fraction of 10%, the fiber-resin composite maintained a high degree of transparency (95%-97% transmission). Theoretical predictions of the nano-scale fibers, acting as scatterers predicted 99+% transmission, but when voids and extraneous inclusions were taken into account, theoretical predictions fell to 93%-95%, thus providing bounds, as it were, on the level of transmission possible under the present assumptions of our model. Thus it seems reasonable that, given the good correlation between theory and experiment, that our assumptions are a good approximation. Given the sensitivity of the theory to the parameters of the ratio of fiber-matrix indices of refraction, fiber diameter, and fiber volume fraction, it stands to reason that when a greater disparity exists between the indices of refraction for the fiber and the matrix, greater car will have to be given to monitoring the fiber diameters as well as the fiber volume fraction. Clearly, further work is needed to determine on a practical level what level of disparity is possible without rendering the transparency of the composite as completely unsustainable.

References

1. H.C. van de Hulst; "Light Scattering by Small Particles," Dover Publications: New York, 1981.
2. Kerns, J.; Hsieh, A.; Hiltner, A.; Baer, E. *Journal of Applied Polymer Science*, **2000**, *77*, 1545-1557.
3. Bergshoef, M.M.; Vancso, G.J. *Advanced Materials*, **1999**, *11*, 1362-1365.
4. Deitzel, J.M.; Kleinmeyer, J.; Harris, D.; Beck Tan, N.C. *Polymer*, **2001**, *42*, 231-272.
5. Huang, Z.M.; Zhang, Y.Z.; Kotaki, M.; Ramakrishna, S. *Composites Science and Technology*, **2003**, *63*, 2223-2253.
6. Yarin, A.L.; Koombhongse, S.; Reneker, D.H. *Journal of Applied Physics*, **2001**, *90*, 4836-4846.
7. Fridrikh, S.V.; Yu, J.H.; Brenner, M.P.; Rutledge, G.C. *Physical Review Letters*, **2003**, *90*, 144502-1-144502-4.

Chapter 26

Some Structural Observations of Self-Assembled, Fibrillar Gels Composed of Two-Directional Bolaform Arborols

Jirun Sun[1], Keunok Han Yu[2], Paul S. Russo[1,*], John Pople[3], Alyssa Henry[1], Bethany Lyles[1], Robin S. McCarley[1], Gregory R. Baker[4], and George R. Newkome[5]

[1]Department of Chemistry and Macromolecular Studies Group, Louisiana State University, Baton Rouge, LA 70803
[2]Department of Chemistry, Kunsan National University, Kunsan City 573–360, South Korea
[3]Stanford Synchrotron Radiation Laboratory, Menlo Park, CA 94025
[4]Department of Chemistry, University of South Florida, Tampa, FL 33620
[5]Departments of Chemistry and Polymer Science, The University of Akron, Akron, OH 44325–4717
*Corresponding author: chruss@LSU.edu

Arborols are dumbbell shaped molecules (bolaform amphiphiles) in which a hydrophobic spacer separates two hydrophilic end groups. They are a valuable model for naturally occurring fibers, such as actin or amyloid. Applications to materials science can be envisioned. On cooling from warm aqueous or methanolic solutions, arborols spontaneously assemble into long fibers. When the solutions are above a certain concentration that depends on the hydrophilic / hydrophobic balance, this leads to thermally reversible gels stabilized by a mechanism that is poorly understood. With the help of wide-angle X-ray scattering, details of the arborol fiber and gel structure were obtained on wet gels. The characteristic dimensions of the fibers vary in a sensible fashion with the molecular specifics. Solvent character appears to affect the average domain length of arborols stacked into fibers. Fluorescently labeled arborols were prepared. The label does not prevent incorporation into the fibrillar structure, rendering fibril bundles visible in wet gels. Bundles are visible in concentrated gels, but not in less concentrated sols. These results are consistent with observations of dried arborols using atomic force microscopy and with previously published freeze-fracture electron microscopy and small-angle X-ray scattering experiments on dried gels.

Hydrogels are key ingredients in a number of biological systems, such as the vitreous humor of the eye and the synovial fluid that lubricates skeletal joints. They are also applied increasingly in industry—e.g., in foods, deodorants, cosmetics, and chromatography (1). The preparation and properties of hydrogels have become active areas of investigation, consistent with their various applications (2).

Several systems have been found in which hydrogels are formed by self assembly of small molecules (3-8). Arborols are bolaform amphiphiles (9-11) that can lead to hydrogels by formation of extended fibers, similar to organogelators such as bis-urea derivatives (12). The factors that turn a mesh of long fibers into a true gel are not fully understood. First synthesized by Newkome et al. (13), arborols dissolve in warm water or water/alcohol mixtures and gel on cooling (11). The two-directional $[m]$-n-$[m]$ arborols consist of two hydrophilic regions, each with m terminal hydroxyl groups, connected by a linear alkyl chain with the formula C_nH_{2n}. An example appears in Figure 1.

The extended, fibrillar nature of arborols is reminiscent of several biological systems that form networks of extended fibers from proteins, such as F-actin (14) and fibrinogen (15). Arborols are even more efficient in the sense that the starting molecule has a molar mass of only ~1 kD. A closer biological comparison may be β-amyloid, the 4kD peptide comprising most of the plaques found in the brains of patients afflicted with Alzheimer's disease (16). Methods for studying β-amyloid and other fibrous biogels can be developed at lower cost through arborol studies. Arborols provide a simple, synthetic system whose viscoelastic properties could be controlled by using temperature to adjust fiber length. Arborols also provide a convenient platform for the exploration of fibrillar self-assembly and possibly the opportunity to make controlled rigid rod systems. For example, designed inhibitors could limit the rod growth to create a self-assembled system that spontaneously forms a lyotropic liquid crystalline phase, with associated optical and viscoelastic properties, that would last only so long as it's not heated. Such a system would combine the speed of small molecule liquid crystals with the material efficiency of comparatively dilute polymer lyotropic liquid crystals.

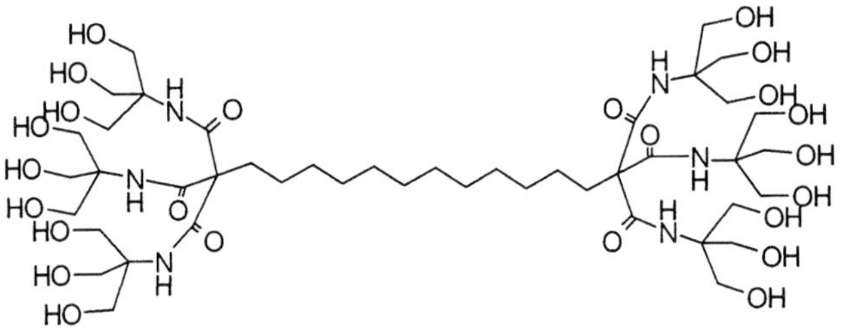

Figure 1. Structure of [9]-12-[9] arborol

The study of arborols is still at its beginning. Fibers, or perhaps bundles of fibers, were first visualized at low resolution by fluorescence microscopy, taking advantage of a dye that fluoresces only in a hydrophobic environment such as the fiber core (*17*). Freeze-fracture transmission electron microscopy confirmed the presence of fibers with lengths on the micrometer scale (*18*). Long associations of the fibers were evident, suggesting fibrils, but jamming of the fibers during the freeze step is a worrisome artifact in freeze fracture microscopy. Small-angle X-ray scattering (SAXS) on dried gels further supports the existence of bundled fibrillar structures (*18*). The model shown in Figure 2 was used to interpret the SAXS results on dried gels. The dumbbell structure of the arborols is modeled as two spheres, radius r, attached at their center by a line, length b. To simulate fibers, the dumbbells are stacked orthogonally with a separation, a. When modeling the dried gels, the distance, a, between neighboring alkane chains was found to be ≈ 0.5 nm. The distance, b, between two hydrophilic parts of one arborol was typically 1.4 nm for most of the dried arborol gels. By modeling multiple fibers, it was found that certain features of the SAXS pattern on dried gels were consistent with multiple fibrils. It was not clear whether multiple fibrils, or bundles, were present in wet gels. Engelhardt, *et al.* studied the assembly and gelation of a dilute [9]-10-[9] solution by static and dynamic light scattering (*19*). The broad light scattering transitions, compared to the sharper DSC transitions, suggested that weak interconnections between fiber bundles account for gelation.

Here, we report investigations of wet gels using wide angle X-ray scattering (WAXS), atomic force microscopy (AFM) in contact mode, polarized optical microscopy (POM), and fluorescence microscopy after covalent attachment of a dye label. The results are discussed with a view towards establishing whether or not molecular specifics, such as spacer length or number of hydrophilic groups, are reflected in the wet gels, what variations may occur as a result of solvent character, and whether gel-stabilizing bundles exist in the wet gels.

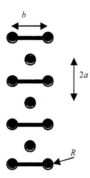

Figure 2. Model for single arborol fibers(18).

Experimental

Molecular Characterization: Arborol purity was tested by proton and ^{13}C NMR, IR and MALDI-TOF mass spectrometry. In most cases, the NMR and FTIR signals met expectations (*13*). In FTIR, a new preparation of [9]-12-[9] showed a small peak at 1735 cm[-1] which suggests the existence of an ester group, thus incomplete amidization in the last step of synthesis. However, the ^{13}C and proton NMR spectra of this sample match expectations for the pure material. Ironically, this is the only arborol sample that gives an interpretable MALDI-TOF spectrum. Its observed mass (1009) is close to that of pure [9]-12-[9] (1080) less the difference between one tris(hydroxymethyl)amino-methane group (120) and the ethoxy group (45) it was supposed to displace: 1080 - 120 + 45 = 1005. HPLC was carried out according to the same methods described in ref. 20, and the results indicated that the purity of this sample was 93%. The new [9]-12-[9] forms a gel and behaves normally in all regards.

Solution preparation. All arborols were dried in a vacuum oven at 40 °C overnight before use. The solutions were prepared by dissolving arborols in dust-free, de-ionized water from a Barnstead Nanopure water purification system at about 80 °C.

WAXS: Wide-angle X-ray scattering was carried out at Stanford Synchrotron Radiation Laboratory on beamline I-4. The distance between detector and sample was 700 mm, and the diameter of the beam stop was 6 mm. At the selected wavelength of 1.488 Å, the scattering vector magnitudes ranged from 0.48 to 15.9 nm^{-1}. The arborols were dissolved in water at about 80 °C and then loaded into X-ray capillary cells (Charles Supper) of diameter 1.0, 1.5 or 2.0 mm, centrifuged to the bottom and sealed by flame.

Seven samples were prepared with five arborols ([6]-7-[6], [6]-8-[6], [6]-10-[6], [6]-11-[6] and [6]-13-[6]). The arborols were dissolved in Nanopure water or a mixture of methanol and water (volume ratio = 1:1) respectively to make a 5% solution. The samples prepared from arborols [6]-7-[6] and [6]-8-[6] did not form gels, and the solubility of [6]-13-[6] in water was not high enough to make a 5% solution. A Mettler FP80 microscopy oven provided convenient and rapid temperature control. The background scattering was measured from Nanopure water at 25 °C.

Microscopy: Polarized light microscopy studies were made on an Olympus-BH outfitted with a Kodak DC290 digital camera. The [9]-12-[9] arborol was dissolved in water at about 80 °C. One end of a rectangular microslide (0.1 x 4.0 mm, Vitro Com Inc.) was flame sealed. The open end was inserted into molten [9]-12-[9] arborol solution, and sample was drawn into the microslide when trapped air contracted on cooling. The solution was centrifuged to the sealed end of the microslide before gelation could take place. To evenly distribute the gel, the sealed microslides were heated again and then centrifuged.

Fluorescence Microscopy and Labeling of Arborol: The experiments were carried out on a Leitz Metallux 3 epi-illumination microscope fitted with a K2BIO confocal adapter, a Nipkow disk design (*21*). Images were acquired with a Dage 66 SIT (silicon-intensified target) camera. An FITC filter cube was selected. Arborols were labeled by 5-(4, 6-dichlorotriazinyl)-aminofluorescein (5-DTAF), which is an effective dye for labeling hydroxyl groups (*22-24*). Arborol [9]-12-[9] (24 μmol, 25.6 mg) was dissolved in Nanopure water to a concentration of 3 mg/ml, heated to dissolve the mixture, cooled and allowed to sit for about 20 minutes until the gelation was complete. Dye (49 μmol 5-DTAF, 24.3 mg) was dissolved in a pH = 10 sodium hydroxide solution. The solution was poured on the top of the arborol gel and allowed to set for about 2 hours; two layers were formed. The upper one contained unreacted 5-DTAF, and the bottom layer contained fluorescently labeled arborol gel. The upper layer was withdrawn by a pipette. The gel was first washed with pH = 10 NaOH solution, and then rinsed with several portions of Nanopure water. The gel was dried by blowing dry nitrogen overnight. The success of labeling was confirmed by MALDI (although the values of Arborol and labeled Arborol were not correct, the difference was, as discussed in Results and Discussion). No more than 11% of the hydroxyl groups can be labeled by this procedure; the actual labeling is less as indicated by the unreacted dye removed during purification.

Preparing samples for the Fluorescence Microscope: Unlabeled [9]-12-[9] arborol (25 mg) and its dyed counterpart (5 mg) were placed in a vial and dissolved in Nanopure water to a concentration of 3 mg/ml. The mixture was heated to dissolve the arborols, then cooled at 25 °C to produce a strong gel. The gel was put on a cavitated microscope slide and covered by a glass cover slip, observed under the fluorescence microscope when it was wet and checked again after several hours when it was dry.

AFM: The atomic force microscope experiments were carried out on a Digital Instruments Nanoscope III multimode SPM in contact mode. The tip was a silicon nitride probe from Digital Instruments, type NP-S. The solution was applied on a freshly cleaved mica surface five minutes before putting in the microscope. Four concentrations (2%, 0.2%, 0.1% and 0.05 %) of [9]-12-[9] were examined.

MALDI: MALDI-MS was carried out on a Bruker ProFLEX III MALDI-TOF mass spectrometer. A small amount of arborol was placed in a sample cell and covered with 200 μL of 50 % water/ethanol. The cell was suspended in a water bath and heated until the arborol dissolved. The sample was mixed with an α-cyano-4-hydroxycinnamic acid (CCA) matrix using 30% methyl cyanide (MeCN), 70% water, and 0.1% trifluoroacetic acid (TFA) as an ionizing agent.

FTIR: Samples were prepared by placing a small amount of the arborols in a mortar with slightly wet potassium bromide (KBr). The contents were ground into a fine powder and set in a die to form a translucent pellet.

Results and Discussion

Figure 3 shows that the WAXS patterns are nominally symmetrical, which indicates that there is no preferential alignment of the fibrillar structures. The effect of varying the length of the hydrophobic spacer was studied using [6]-7-[6], [6]-10-[6] and [6]-11-[6] arborols in water (Figure 4) and [6]-8-[6], [6]-10-[6], [6]-11-[6] and [6]-13-[6] arborols in 5% methanol/water mixtures (Figure 5). Results are grouped in Table 1. A frequent observation is a peak just beyond the beamstop at about 2.45 nm^{-1}. This feature recalls the peaks seen in *dried* fibrils at about 3 nm^{-1} in our earlier study, which were identified as side-by-side alignment of fibrils, although twist along the fibril axis cannot yet be excluded. If the peaks do represent side-by-side alignment, it is sensible for them to have shifted towards lower scattering angles in the hydrated fibrils. Other aspects of the WAXS patterns also are reasonable, although they are near the limits of resolution of the experiment. The [6]-11-[6] arborol has a longer spacer chain than [6]-10-[6] and the measured *b* value is about 0.09 nm bigger for the former molecule. Arborol [6]-8-[6], which has the shortest spacer chain, is observed to have the largest *a* value (step length along the fibril).

Conversely, arborol [6]-13-[6], which has the longest spacer chain, exhibits the smallest *a* value, but it shares that characteristic with arborol [6]-11-[6]. This suggests that a longer spacer chain permits closer approach of the molecules, but the effect tops out at eleven methylene groups. Arborol [6]-7-[6] does not form a gel at 5% in water. Its WAXS plot does not have the characteristic peak around 0.5 nm, which suggests that the dumbbells are no longer stacked orthogonally; however, the existence of peak at 1.47 nm indicates that there are some ordered arrangements.

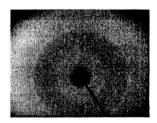

Figure 3. WAXS pattern for [6]-10-[6] in water.

Table 1 Summary of WAXS results

Sample Number		1	2	3	4	5	6	7
Arborol		[6]-7-[6]	[6]-8-[6]	[6]-10-[6]	[6]-10-[6]	[6]-11-[6]	[6]-11-[6]	[6]-13-[6]
Solvent		H_2O	MeOH + H_2O	H_2O	MeOH + H_2O	H_2O	MeOH + H_2O	MeOH + H_2O
Gel?		N	N	Y	Y	Y	Y	Y
Peaks found (nm)	$a^{\#}$		0.58 ± 0.02	0.54 ± 0.01	0.55 ± 0.01	0.52 ± .01	0.52 ± 0.01	0.52 ± 0.01
	$b^{\#}$			1.17 ± 0.02	1.17 ± 0.02	1.26 ± 0.03	1.20 ± 0.02	1.15 ± 0.10
	Others	1.47 ± 0.07	n/r*	2.45 ± 0.15	2.45 ± 0.12	2.45 ± 0.13	2.45 ± 0.15	2.89 ± 0.20

Note: #see text for definition of the dimensions *a* and *b*. *n/r = not resolved from beamstop. MeOH + H_2O mixtures were at 1:1 volume ratio.

The effects of solvent on the [6]-10-[6] and [6]-11-[6] arborols were studied using water and a mixture of methanol and water (1:1 volume ratio). The WAXS spacings of [6]-10-[6] and [6]-11-[6] are indicated directly in Figure 6 and Figure 7, respectively.

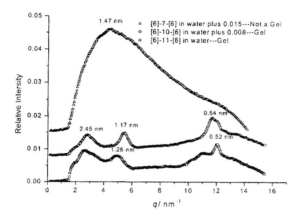

Figure 4. WAXS for [6]-n-[6] arborols with different spacer lengths in water. Curves are offset vertically for clarity.

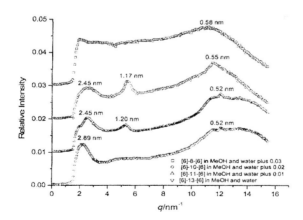

Figure 5. WAXS for [6]-n-[6] arborols with different spacer lengths in MeOH and water. Curves are offset vertically for clarity.

The spacings are similar in both solvents, but the peak indicating the distance along the fibril (parameter a) is broadened by addition of methanol, which suggests a reduction in the contiguous length of the arborol stacks comprising the fibrils.

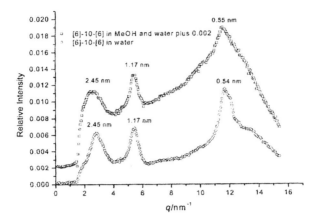

Figure 6. The Effect of Solvents on [6]-10-[6]. Curves are offset vertically for clarity.

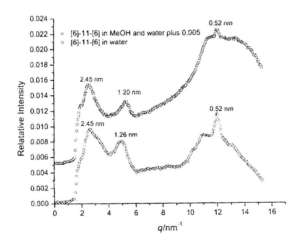

Figure 7. The Effects of Solvents on [6]-11-[6]. Curves are offset vertically for clarity.

Centrifuging the sample after melting the gel led to an uneven distribution of arborols in the slice (Figure 8A). Under crossed polarizers in an optical microscope, the arborol gel appears to be a collection of fibers aligned at 45 or 135 degrees of angle (Figure 8B). The intensity of light changes periodically as the sample is rotated on the stage. Without the polarizers, the fibers appear to be randomly distributed. These observations indicate that the fibers possess birefringence, but polarized light microscopy does not clearly elucidate the structure.

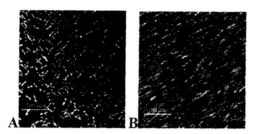

Figure 8. A 3 wt% [9]-12-[9] arborol gel, as viewed between crossed polar in an optical microscope. A: Unevenly distributed part of sample introduced by centrifuging; B: Evenly distributed part.

Fluorescence microscopy usually provides an unambiguous contrast mechanism: one sees the labeled object (at the risk of altering the structure by the presence of the dye). The possible reaction mechanism for DTAF (Figure 9) labeling of arborols as described in Experimental Section involves nucleophilic aromatic substitution by an addition-elimination pathway (*23*).

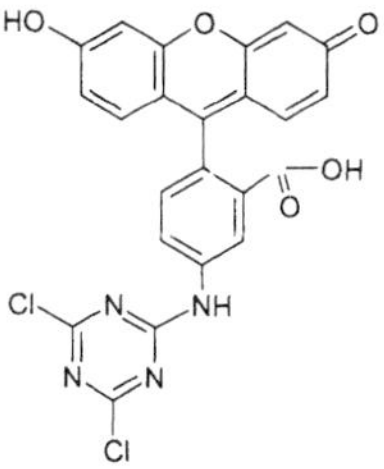

Figure 9. The Structure of 5-DTAF

As the selected two-directional arborol has nine hydroxyl groups at each end, it should be easy to dye, and the MALDI results confirmed this suggestion. The MALDI spectrum contains peaks around 1470 D, which is in the range of molecular weight for one 5-DTAF labeled [9]-12-[9]. As discussed in Experimental, the molecular weight of [9]-12-[9] that we used is 1009 by MALDI and the molecular weight of 5-DTAF is 495. If one 5-DTAF is attached to a hydroxyl group, HCl will be released. So the molecular weight of labeled [9]-12-[9] is 1468, very close to the observed 1470. While this is encouraging, other species are surely present—some with no dyes attached and perhaps others with more than one dye attached. Another potential problem is that the DTAF label is not much smaller than the arborol molecule itself. To assess the behavior of the (partially) dyed arborols, the DSC melting point of a 3.0 wt% [9]-12-[9] unlabeled gel was compared to a 3.0 wt% sample containing 2.7 wt% unlabeled arborols and 0.3 wt% from the labeled preparation. There was no difference in the melting point with or without label. Detailed studies of the enthalpies of melting as a function of labeled content may prove revealing, but fluorescence microscopy confirms that the labeled arborols do participate in fibril formation.

When the gel was wet, the fluorescence contrast of the network to its surroundings was weak, but the fibrous textures did emit strongly enough to be caught by confocal microscopy. Figure 10A is the result of processing by Image J software (National Institutes of Health) of three pictures, each of these an average of 50 video frames from the same spot in the microscope. Two well-focused pictures, in which fibers were barely visible, were summed. From this result was subtracted another picture taken slightly out of focus to reveal the structures shown. The contrast improved after the solvent was evaporated (Figure 10 B and C). These figures suggest that very long fibers exist, confirming the impressions from freeze-fracture electron microscopy. The gel is *not* destroyed after the arborol is labeled, but it is not clear yet whether it is affected in more subtle ways that might be revealed by additional SAXS, DSC and AFM studies.

Figure 10. Fluorescence microscopy of [9]-12-[9]. A. Confocal image taken from wet gel. B. Taken after the gel was dried. C. One specific spot in the dried gel.

The atomic force microscope (AFM) can be used to check the topographical, elastic, and frictional properties of hydrogels (*25-27*). Four (2, 0.2, 0.1, and 0.05%) concentrations of [9]-12-[9] were tried. The AFM images, Figure 11, were obtained in contact mode in air at 25 °C. Solutions, 0.1% and 0.05% [9]-12-[9], did not produce clear AFM images. For 0.2% [9]-12-[9], the sample is not a gel. One can still see fibers, but not large bundles, in the upper part of Figure 11. The appearance calls to mind a network, but the fibers probably just collapsed as the solution dried to produce that illusion. A 2% [9]-12-[9] preparation does, however, form a gel on cooling, and one can see large fiber bundles in the lower part of Figure 11. This is strong evidence that the bundles strengthen the gel, as suggested from the SAXS data on dried gels.[28] The freeze fracture EM images, with the usual risks of artifact, also suggest that the strengthening occurs as fiber bundles share individual fibers to make effective crosslinks.

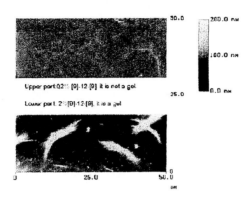

Figure 11. AFM image for 0.2% and 2% [9]-12-[9] in contact mode.

Conclusion

This is the first time that WAXS has been used to check the properties of wet arborol gels. The structures found by previous studies of dried samples seem to exist in the wet gels. SAXS studies on wet gels may reveal additional parameters of the fibrils, such as thickness and how it changes during melting, addition of fiber inhibitors, etc. For arborols with 6 hydroxy groups at each end, the hydrophobic chain must include more than ten (but less than 14) methylene groups to induce gelation (*20*). If the chain length is shorter, the arborol fails to form large bundles of fibers, which are the key to formation of a strong gel. AFM gives visual information on the function of bundles. At high concentration (2%) for [9]-12-[9], there are plenty of large bundles to support a gel. At low concentration (0.2%) mostly thin fibers are observed and the system does not gel. Bundles effectively crosslink the long fibers. POM confirms the birefringence of arborol gels, but the textures of the samples are not yet fully interpretable. Fluorescence microscopy on both dry and wet samples confirms the rod like structure. Covalent attachment of DTAF dye does not prevent incorporation of arborols into fibrillar structures.

Some questions remain. It is not understood how the fibers respond to forces such as shaking, shearing or sonication. It is not known if a broken gel will heal through slow rearrangements of freely diffusing arborol molecules or small protofibrils. Molecules that inhibit the fibril formation could lead to control over the length, beyond thermal control, resulting in a self-assembling lyotropic liquid crystalline system or solutions with tunable viscosity. Such inhibitors can be imagined to take the form of half an arborol, and they may have unusual surface behavior. Finally, rheological and microrheological assessment of these fibrillar networks is overdue.

Acknowledgement

We acknowledge the support of the National Science Foundation through award DMR-00-75810 (PSR), DMR-01-96231 (GRN). We thank Jed Aucoin and Jowell Bolivar for help with AFM. Jirun Sun thanks Professor George Newkome and his group for hospitality during a visit to the University of Akron.

References

1. Abdallah, D. J.; Weiss, R. G. *Langmuir* **2000**, *16*, 352-355.
2. Molyneux, P. *Chemistry and Technology of Water-Soluble Polymers*; Plenum: New York, 1983; pp. 1-20.
3. Bhattacharya, S.; Acharya, S. N. G. *Chemistry of Materials* **1999**, *11*, 3121-3132.
4. Franceschi, S.; de Viguerie, N.; Riviere, M.; Lattes, A. *New Journal of Chemistry* **1999**, *23*, 447-452.
5. Jokic, M.; Makarevic, J.; Zinic, M. *Journal of the Chemical Society-Chemical Communications* **1995**, 1723-1724.
6. Kogiso, M.; Ohnishi, S.; Yase, K.; Masuda, M.; Shimizu, T. *Langmuir* **1998**, *14*, 4978-4986.
7. Kogiso, M.; Hanada, T.; Yase, K.; Shimizu, T. *Chemical Communications* **1998**, 1791-1792.
8. Oda, R.; Huc, I.; Candau, S. J. *Angewandte Chemie-International Edition* **1998**, *37*, 2689-2691.
9. Fuhrhop, J. H.; Wang, T. *Chemical Reviews (Washington, DC, United States)* **2004**, *104*, 2901-2937.
10. Newkome, G. R.; Moorefield, C. N.; Vogtle, F. *Dendrimers and Dendrons: Concepts, Synthesis, Applications*; Wiley-VCH: Weinheim, Germany, 2001.
11. Escamillia, G. H. *Advances in Dendritic Macromolecules*; JAI press: Greenwich, Conn, 1995; pp. 157-190.
12. Estroff, L. A.; Hamilton, A. D. *Angewandte Chemie-International Edition* **2000**, *39*, 3447-3450.

13. Newkome, G. R.; Baker, G. R.; Saunders, M. J.; Russo, P. S.; Gupta, V. K.; Yao, Z. Q.; Miller, J. E.; Bouillion, K. *Journal of the Chemical Society-Chemical Communications* **1986**, 752-753.

14. Wong, G. C. L.; Tang, J. X.; Lin, A.; Li, Y. L.; Janmey, P. A.; Safinya, C. R. *Science* **2000**, *288*, 2035-2039.

15. Madrazo, J.; Brown, J. H.; Litvinovich, S.; Dominguez, R.; Yakovlev, S.; Medved, L.; Cohen, C. *Proceedings of the National Academy of Sciences of the United States of America* **2001**, *98*, 11967-11972.

16. Benzinger, T. L. S.; Gregory, D. M.; Burkoth, T. S.; Miller-Auer, H.; Lynn, D. G.; Botto, R. E.; Meredith, S. C. *Biochemistry* **2000**, *39*, 3491-3499.

17. Newkome, G. R.; Baker, G. R.; Arai, S.; Saunders, M. J.; Russo, P. S.; Theriot, K. J.; Moorefield, C. N.; Rogers, L. E.; Miller, J. E.; Lieux, T. R.; Murray, M. E.; Phillips, B.; Pascal, L. *Journal of the American Chemical Society* **1990**, *112*, 8458-8465.

18. Yu, K. H.; Russo, P. S.; Younger, L.; Henk, W. G.; Hua, D. W.; Newkome, G. R.; Baker, G. *Journal of Polymer Science Part B-Polymer Physics* **1997**, *35*, 2787-2793.

19. Engelhardt, T. P.; Belkoura, L.; Woermann, D. *Berichte der Bunsen-Gesellschaft-Physical Chemistry Chemical Physics* **1996**, *100*, 1064-1072.

20. Newkome, G. R.; Baker, G. R.; Saunders, M. J.; Russo, P. S.; Gupta, V. K.; Yao, Z. Q.; Miller, J. E.; Bouillion, K. *Journal of the Chemical Society-Chemical Communications* **1986**, 752-753.

21. Yin, S.; Lu, G.; Zhang, J.; Yu, F. T. S.; Mait, J. N. *Applied Optics* **1995**, *34*, 5695-5698.

22. Helbert, W.; Chanzy, H.; Husum, T. L.; Schulein, M.; Ernst, S. *Biomacromolecules* **2003**, *4*, 481-487.

23. Ahmed, F.; Alexandridis, P.; Neelamegham, S. *Langmuir* **2001**, *17*, 537-546.

24. Benzinger, T. L. S.; Gregory, D. M.; Burkoth, T. S.; Miller-Auer, H.; Lynn, D. G.; Botto, R. E.; Meredith, S. C. *Biochemistry* **2000**, *39*, 3491-3499.

25. Matzelle, T. R.; Geuskens, G.; Kruse, N. *Macromolecules* **2003**, *36*, 2926-2931.

26. Kurokawa, T.; Gong, J. P.; Osada, Y. *Macromolecules* **2002**, *35*, 8161-8166.

27. Nakayama, Y.; Nakamata, K.; Hirano, Y.; Goto, K.; Matsuda, T. *Langmuir* **1998**, *14*, 3909-3915.

Chapter 27

Nanofilms and the Emerging Nanotechnology

Ejembi J. Onah

Department of Chemistry and Chemical Biology, Cornell University, Ithaca, NY 14853

Nanofilms or nanofiber of fluorine containing polymers because of their unique properties including: Low permittivity, low friction, thermal stability, high resistance to chemical attack especially oxidation hold a central position in the emerging nanotechnology. Important areas where fluorine containing polymer films hold a lot of promise are: Biomedical application, resist technology in photo and radiation lithographies as well as insulators or dielectrics in an emerging"nanoelectronics". Fluorine containing polymer specifically fluoropolyimide, fluoropolyamic acid, and fluoropolyacrylates have been synthesized. The polymers are soluble in wide solvents. Their Langmuir-Blodgett (LB) monolayers (polymer as film on the water surface or single molecules arranged on the substrate) and their ultrathin films (monolayers deposited on the substrate as thin films in nanometer range) have been fabricated. Techniques like AFM, surface plasmon resonance (SPR), and dielectric spectroscopy have been used to obtain the surface morphology, film thickness and dielectric constant. The dielectric constant of 1.5 is the lowest so far reported according to my knowledge.

The concept of nanofilms or nanofibers is used to describe films where the characteristic dimensions are less than about 1,000 nanometers. These films find applications in many areas in the emerging nanotechnology[1] especially for biomedical applications or nanoelectronics as dielectrics or insulator films.

As postulated by Von Neuman; the emerging nanolectronics will produce computers that are made to execute more complicated work. To achieve this role at a greater speed and less cost is the challenge to nanoelectronics in the emerging nanotechnology. To do this the computer has to be made faster, complex and smaller; the miniaturization idea by Richard Feyman. These ideas gave birth to Nanotechnology by Drexler[2-4].

Where is the position of nanofilms or nanofiber in all these?

Microelectronics now transforming to nanoelectronics has a dominant influence in our lives. This industrial revolution was originally driven by the need for very small and lightweight electronic circuits for military and aerospace applications. With the development of, first, the transistor and later, the integrated circuit (IC), microelectronics has now grown into a multibillion dollar industry, and its application is ubiquitous. Device miniaturization has brought us from small-scale (SSI), medium-scale (MSI), and very large-scale integration (VLSI) with 10^5 or more components per chip.[5] The next stage may be very very large-scale integration (VVLSI).

In the transformation from microelectronics to nanoelectronics, research has been going intensively in all areas of microelectronics to provide for ever higher performance (speed) and density at very low cost. Active foci include the scaling of devices and the search for novel materials and processing technologies for interconnects and packaging in lithography and dielectrics. [5-35] The development of new insulators for interconnects and packaging is one approach to increase the speed. [5] Compatibility with thin-film fabrication techniques, low-dielectric losses, good adhesion to a variety of substrate materials, and thermal, mechanical and chemical stability have made polymer materials attractive choices as the interlayer dielectric (ILD) and intermetal dielectric (IMD). In its simplest form, the multilevel interconnect structure comprises a metal pattern on a substrate, a polymer layer on top with metalized vials, and a second metal pattern on top of the polymer.

As operating frequencies of electronic devices enter the gigahertz range and as the dimensions of electronic devices approach the submicron level called nanotechnology, dielectric media with low dielectric constants (<3) become increasingly more important for the reduction of signal coupling among transmission lines. Therefore, there is a clear need to understand the chemistry of materials used in these applications to evaluate better their effects on the electrical and physical performance of devices.

For a material to be suitable as an interconnect dielectric in addition to a low dielectric constant, it should be able to withstand temperatures higher than the 470°C necessary for the subsequent heating steps without the evolution of volatile by-products. Other important considerations are compatibility with other materials, long-term thermal, chemical, and electrochemical stability, ease of fabrication and low cost. SiO_2, which has a dielectric constant of 3.5-4.0, has been used as interlayer dielectric in the industry. Further breakthroughs in high-performance chips hinge critically on the development of new insulators with dielectric constants much lower than that of SiO_2. It is generally believed that to achieve such a low dielectric constant organic polymeric materials must be considered instead of the traditional inorganic materials.

Fluoropolymers especially fluoropolyimides because of their unique features such as: low permittivity, low friction and high resistance to chemical attack especially oxidation have a central position in this current drive towards the development of low dielectric organic polymeric materials to replace traditional inorganic insulators.[6,7,27,35]

Maier etal [27] and Houghman [35] have had extensive reviews on different polymers that have potential of being used as ILD and IMD. They include poly(benzoxazole)s, poly(phenylquinoxaline)s, polynorbornene, siLK, poly(silsesquioxane), etc.These polymers may have high thermal stability and can be used as IMD and ILD, but many of them are very rigid and cannot be easily processed. [27] In addition their dielectric constants are between 2 and above. Where lower dielectric constants are required, they have difficulty of being applied.

Aromatic polyimides have excellent thermal stability in addition to their good electrical properties, light weight, flexibility and easy processability. Polyimides, since the invention of integrated circuits (ICS) have been applied to insulation materials in electronic devices. This is because of their relatively low fabrication cost and high performance.

However, conventional polyimides have disadvantages as dielectrics for use in microelectronics like: high coefficient of thermal expansion (CTE) compared to Si and SiO_2.[5] The difference in CTE between polyimides and other materials produces peeling, bending, and cracking in electronic devices. Low thermal-expansion polyimides can be achieved by a linear polymer and molecular construction with only rigid groups such as phenyl rings and imide rings .[9,10] Water absorption is other problem with insulation materials, because it causes corrosion of metal wiring and instability of electrical properties such as the dielectric constant. Conventional polyimides have relatively high water absorption because of the presence of polar imide rings.

However, the water absorption decreases when fluorine is introduced into polyimide molecules because of their hydrophobic nature. [5-11,35] Hougham [35]has an extensive review of fluorinated polyimide that has been synthesized.

Apart from new low-dielectric materials, an accurate method to deposit the dielectric as a uniform film is also required. Owing to the way fabrication technology in the microelectronics industry has developed and because larger silicon wafers are being used (currently, the technical difficulty of controlling thin-film uniformity for larger wafers has been a challenging area for spin coating), a methodology with the ability to deposit polymers with appropriate properties as conformal films like the Langmuir-Blodgett (LB) technique [6, 7, 17-19] and chemical deposition [5, 20] would be useful. Through multilayer formation, thickness of micrometer range can be reached.

However, it is generally considered difficult to fabricate LB films of fluoropolymers or fluorine containing polymers directly because of their rigidity and hydrophobicity. [21] As a result, there are only few reports in the literature on monolayers and LB films of fluoropolymer or fluorine containing polymers by direct fabrication. [21-23] Direct fabrication is the method whereby synthesised polymers are spread in water from their solution and deposited on the substrate by its movement vertically or horizontally. Indirect fabrication is a method whereby a precursor like polyamic acid is synthesised and deposited through the LB technique. The imidization process then takes place on the substrate by heating. Direct fabrication of LB films of fluoropolyimides has not yet been published. The indirect method has been used on conventional polyimide. [24-26]

Gas phase deposition is also used to fabricate films of polymers. The polymers are deposited from solvent called chemical vapour deposition (CVD). It has the advantage of being solvent free. However, there is problem of internal stress of polymers, especially if they are deposited below the glass transition temperature. This can be avoided if the polymers are synthesized on the substrate at lower temperature in the presence of a macroinitiator. This is a new enhanced method of gas phase polymerization developed in Institute of Polymer Research (IPF), Dresden, Germany.

The main objective of the work to be described here is to design and synthesize organic dielectric materials, and develop new techniques to deposit these materials as thin films with low dielectric constants, for use as ILDs and IMDs. In this work concentration will be giving to synthesis of fluoropolymers (imide, amide, esters) and their film fabrication using the LB technique. The fluoropolyimide will be used for LB technique to form LB monolayer and films. [21-23] The rigidity of such polymers sometimes make them to be fabricated as films through the LB technique. The acrylate because they can undergo chain polymerization in the presence of an initiator is going to be used for gas phase polymerization. This work will try to fabricate films from these polymers and select those that can easily form LB films for further characterization to be used in microelectronics or nanoelectronics. The development of low-dielectric-constant materials as ILDs is crucial to achieve low power consumption, reduce signal delay, and minimize interconnect cross-talk for high-performance VLSI devices.

Materials and methods.

4,4'-Hexafluoroisopropylidenediphthalic anhydride (6FDA) and 4,4'-hexafluoroisopropylidenedianiline were purchased from Aldrich Chemical Co. as 98% pure. N-Methyl-2-pyrolidinone (NMP) and β-picoline were purchased from Fluka as 99.8% pure. Tetrafluoropropyl methacrylate was purchased from ABCR as 98% pure. 4,4'-Azo-bis(isobutylnitrile) (AIBN) was purchased from Fluka as more than 98% pure. All reagents were used without further purification. The water used for the monolayer experiments was purified using a Milli-Q Plus system (18.2 MΩ). Hot concentrated chromic acid was used for preparing the substrate. All the substrates were made hydrophobic in the presence of 1,1,1,3,3,3-hexamethyldisilazane (HMDS) (Aldrich, 98%). Substrates used included silicon wafer, gold, glass, quartz, and mica (Good Fellow). They were made hydrophobic by exposure to vapors of HMDS. Spreading solvents were THF (Fluka, 99.5%) and chloroform (Fluka, 99.5%). The gold substrates for film deposition were prepared as metal films on glass (76 mm × 26 mm × 1 mm) from 99.9% pure granulated gold. Nuclear magnetic resonance (NMR) spectroscopy was performed on a Bruker DRX 500 spectrometer operating at 500.13 MHz for ^{1}H NMR and 125.77 MHz for ^{19}F NMR, using tetramethylsilane as internal standard for ^{1}H NMR and trichlorofluoromethane as internal standard for ^{19}F NMR. Infrared spectroscopy was performed using a Bruker IFS 66 V/S. All readings for the IR measurement were done on a film on ATR crystal or diamond. The size exclusion chromatography (SEC) measurements were carried out with modular chromatographic equipment, a single-column Hibar PS 40 (Merck) containing a refractive index detector at ambient temperature. Thermogravimetry was done on a TGA 7 (Perkin-Elmer).

Synthesis of Polyfluoroamic acid (PA-1).

4, 4'-hexafluoroisopropylidene diphthalic anhydride (6FDA) (1.12 g, 2.52 mmol) and 4, 4'-hexafluoroisopropylidene dianiline (0.842 g, 2.52 mmol 1.0eq) were added into a flask. NMP (30.0 mL) was added and the mixture stirred overnight at room temperature under N_2. At this stage a little highly viscous polyamic acid was withdrawn for further study. IR (cm^{-1}, film): 3357-2500 (OH), 2888 (C-H), 1663 (C=O), 1520 (NH), 1391 (C=C), 1293 (C-F).

Synthesis of Polyfluoroimide (PI-1)

The polyamic acid was converted into polyimide after stirring overnight in the presence of acetic anhydride (3 mL) and ß-picoline (0.3 mL). It was then precipitated out from methanol to yield a white solid quantitatively.

^{1}H-NMR (d$_8$-THF): δ (ppm): 8.0 (d, 2H), 7.9 (d, 2H), 7.8 (s, 2H), 7.6-7.4 (m, 8H).

^{19}F-NMR (d$_8$-THF): δ (ppm): -62.52 (s, 6F), -62.69 (s, 6F).

IR (cm^{-1}, film): 2975 (C-H), 1731 (C=O), 1540 (C=C), 1370 (C-N), 1194 (C-F), 748 (C-N). Other fluoropolymers PAI-(1-4) were synthesized and characterized as reported elsewhere.[5]

Gas phase polymerization

The gas phase polymerization of tetrafluoropropyl methacrylate (TFPM) was carried out in the presence of macroinitiator, poly (octadecene-co-maleic anhydride) modified with tert.-butylhydroperoxide using the poly(octadecene-co-maleic anhydride bought from Polysciences Inc. (Mw 30, 000 – 50, 000 g/mol). The macroinitiator was coated onto different hydrophobic substrates (quartz, gold, glass, etc) by the LB technique. Tetrafluoropropyl methacrylate (TFPM) 2.18 g, 0.01 mol, 1.55 mL was added into a thick glass and the reaction allowed proceeding under pressure at 80 °C.

^{1}H-NMR (CDCl$_3$)): δ (ppm):6.13 (t, H), 4.35 (m, 2H), 1.90 (d, 2H), 1.20 (dd, 3H).

Monolayer Formation, Deposition and thin film measurement

A computer-controlled KSV 3000 system (KSV Instruments Finland) LB trough was used in a dust-free box (microelectronic room) with the temperature controlled to 20 ± 1 ° C. After the LB trough was thoroughly cleaned, it was filled with 18.2 mℓ cm deionized water. The surface was cleaned again by moving barriers toward the dipping well and sucking away any surface-active agents with a capillary glass tube connected to suction. The final level of the subphase was about 1.0 mm above the rim of the trough. Since the polymers could not dissolve in a well-known spreading solvent such as chloroform, a mixture of solvents was used. The polymer (1.0 mg) were dissolved in 100 μL of TFA and made up to 1.0 mL with perfluoro(methylcyclohexane). A mixture (100 μL, 1.0 mg/mL) was used as the spreading solution on the subphase surface, and the solvent was evaporated for about 20-30 min. The LB films were prepared by the vertical deposition method. After the surface pressure reached 10.0 mN/m, about 30 min was spent to establish equilibrium of the monolayer. From the π-A isotherm, the spread monolayer is found to be a solid condensed film at this surface pressure. The monolayers were transferred onto an appropriate substrate (gold and hydrophobic quartz, 76 mm × 26 mm × 1 mm; mica, 10 mm × 10 mm × 1 mm). Usually the first layer was deposited at a speed of 0.2 mm/min and subsequent layers at a speed ranging from 0.5 mm/min to 1.0 mm/min. Thin films were measured by surface plasmon resonance (SPR)

Atomic Force Microscopy

The films were deposited on fresh mica substrate (10 mm × 10 mm × 1 mm) by the LB technique already described. The film was then imaged at ambient temperature using an atomic force microscope (Nanoscope Digital Instruments, Inc.) having a pyramidal Si_3Na tip with radius of curvature approximately 20 nm.

Dielectric Spectroscopy

The dielectric measurement was carried out on LB films and spin-coated films with a Novocontrol dielectric interface and a Solartron impedance analyzer, SI 1260. This analyzer is cleaned by heating in chromic acid and washed with a Millipore water system. After it has been thoroughly rinsed with water, it is blown dry in a stream of N_2. An electrode of aluminum stripes is evaporated on the substrate using a Leybold metal coater, as already described. The LB films prepared on hydrophobic glass (26 mm × 24 mm × 1 mm) substrate were used.

Results and discussion

The thermal properties of the fluoropolymers were evaluated by TGA. The temperature at 1 % weight loss, examined by TGA analysis showed value of 580 °C for PI-1. The thermal stability of the PI-1 having fluorine containing diamines is higher than that obtained earlier by Hamciuc et al [36] for a related polymer having non-fluorinated diamine derivatives. This author reported decomposition temperatures (1 % weight loss) in the range of 422 °C compared to 580 °C reported in this paper for PI-1. This indicates that higher fluorination increases thermal stability. The molecular weights of the polymers obtained from size exclusion chromatography (SEC) were 17 100 (M_n), 30 600 (M_w), and 1.79 (M_w/M_n).

Poly(tetrafluoropropyl methacrylate) (PTFPM) was synthesized by both radical homopolymerization and chemical vapor deposition (CVD) or gas-phase polymerization [5] as shown in Scheme 1. The radical homopolymerization was carried out in the presence of AIBN as initiator in a dry homopolymerization tube. It was a model experiment for the gas-phase polymerization. The colorless product obtained was soluble in common laboratory solvents such as chloroform, methanol, DMAc, THF, dioxane, DMF, etc. The polymer was analyzed by ^{1}H NMR, and IR confirms the structure of the polymer. The ^{1}H NMR is shown in Figure 4.The ^{1}H NMR spectrum in deuterated chloroform for PTFPM exhibited a triplet peak at 6.1 ppm (1H), multiplets at 4.4 (2H) and 2.00 ppm (2H), and two singlet peaks at 0.8-1.1 ppm (3H). The multiplets at 2.00 ppm are due to the unique tacticity of the polymer.

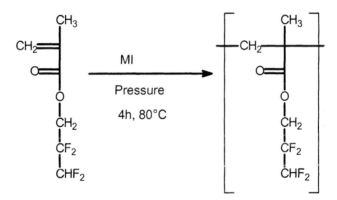

Scheme 1: Gas phase polymerization of tetrafluoropropyl methacrylate

The gas-phase polymerization was carried out in the presence of a macroinitiator (MI), poly(octadecene-co-maleic anhydride) modified with tertbutyl hydroxide. The macroinitiator was brought onto the substrate as a monolayer by the LB technique.The film thickness measurement by ellipsometry shows that the film thickness increases with reaction time in the presence and in the absence of a macroinitiator. There is a greater increase in thickness in the presence of an initiator, which indicates that the macroinitiator speeds up the reaction, as expected. Structurally, there was no difference between the products obtained from the two processes, except that the product obtained from the gas-phase polymerization is solvent-free and pure. This is a new enhanced method of gas-phase polymerization first reported for the polymer.

The isotherms of the polymers were   0.13, 0.21 and 0.23 nm² as area per repeat unit   The collapse points were 50, 62 and 60 mN/m for PTFPM, PA-1 and PI-1. For the polymers, the a.r.u. at the collapse point was observed to be smaller than expected from their molecular structure. This indicates that the structure of the monolayers of these polymers is different from that one would expect from the model of well ordered monolayers of low molecular weight amphiphiles. One can assume that the repeat units are superimposed on one another or that they exist as coiled structures. Compression-expansion experiments on the monolayers show that there are no hystereses for pressures below the collapse point. This shows that during compression no irreversible aggregation occurs.

Multilayers were formed with the polymers. Relatively low dipping speeds were necessary for successful transfer. This behaviour can be attributed to the rigid polymer chains resulting in a much higher viscosity of the polymer film in comparison to those of low molecular weight amphiphiles. Transfer ratios (T.R.) of 0.35-0.5 were obtained depending on the polymer and the substrate. These values are far from ideality (T.R. =1). Possible reasons may be a reorganization of the monolayers during transfer. Such reorganization like molecules arranging themselves in a manner different from that observed from the water surface results in erroneous records for area of monolayer actually removed from the subphase, thereby affecting transfer ratios. T.R. is the area of monolayer removed from subphase divided by area of substrate immersed in water. For a T.R. = 1, the area of monolayer removed from subphase = area of substrate immersed in water. If there is reorganization on the substrate, this value deviates from ideality of 1. However, the transfer ratios were reproducible for repeated monolayer depositions to prepare LB multilayers. As an example, Figure 1 shows the UV experiment of 24 monolayers of the PI-1. Due to the aromatic π−systems the polymers show remarkable absorption up to 330 nm. For the different number of monolayers, a linear dependence of the absorbance at 216 nm on the layer number was observed. Small deviations are within the experimental error range. These results indicate a reproducible monolayer transfer and LB layers of good quality too.

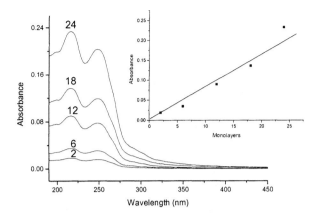

Figure 1. The representative UV (PI-1) of monolayers of the polymer
and changes in absorbance with film thickness at 216 nm.
(Reproduced with permission from reference 10.
Copyright 2003 American Chemical Society.)

Angle dependent reflectivity measurements (SPR) were used to determine the
thickness of one monolayer on a 50 nm gold film. The film thickness per
monolayer (single molecule) is 1.0 nm, 1.4 nm and 2.45 nm for PTFPM, PA-1
and PI-1 respectively. Just like UV, there is a linear relation-ship of monolayer
transfer and absorbance, indicating a good transfer and quality films. The
aromatic nature of the imides makes them thicker.

Film thickness for gas phase polymerization varies from 3.28-10.06 nm in
the presence of an initiator and 0.12-0.84 nm in the absence of an initiator as
established by ellipsometry (table 1). The thickness is by a multiple of 27 greater
in the presence of an initiator compared to the thickness in the absence of an
initiator at 5 h reaction time. These results are as expected, that is the presence
of initiator increases the rate of reaction and yield of products.A representative
AFM image of a monolayer on mica for PI-1 is exhibited in the Figure 2. At a
scan size of 5 nm a mean roughness of less than 1.0 was obtained, indicating a
highly ordered film.

The real part of permittivity (dielectric constant) for PI-1, as exhibited in
Figure 3, shows decrease at 392.5 K as the frequency is increased from 0 to 10^7
Hz at various temperatures. For the PI-1 LB film thickness of 52 nm, the real
part of permittivity is kept constant at 1.5 at the temperatures of 295 and 305 K.
This is the lowest permittivity so far obtained to our knowledge.[5] The real part
of permittivity for PTFPM (Figure 13b) increases with increasing frequency at
the given temperatures (273-373 K) and frequency (0-3 MHz). Controlled

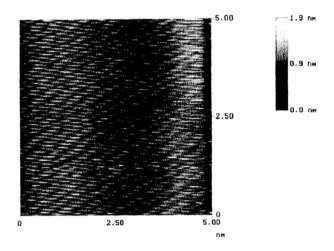

Figure 2. AFM (5 nm x 5 nm) of a monolayer of PI-1 on mica.

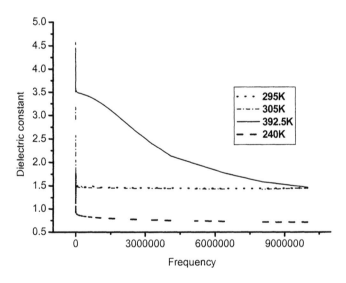

Figure 3. Dielectric spectroscopy of PI-1 (52 nm) at various temperatures.

Table 1. Ellipsometry film thickness measurement at different reaction times on Gold Substrate with or without MI.

Time(h)	Thickness/nm(MI)	Thickness/nm(without MI)
5	3.28	0.12
15	3.45	0.17
25	10.06	0.84

polarizations can be obtained in the experiment, leading to different dielectric constants. Decreasing polarization leads to decreasing dielectric constant and vice versa.

Conclusions

The fluorine containing polymers form very stable monolayers at the air-water interface as can be seen by the high collapse pressure and no hysteresis of the isotherms. Multilayers of the polymer films with varying thickness can be formed onto various substrates like gold, mica and quartz with monolayer (single molecule) thickness in the nanometer range. The films are very homogenous on the substrate as seen by AFM with mean area roughness of less than one. The dielectric constant of 1.5 is reported for PI-1 which is the lowest so far reported according to my knowledge In the future, these fluoropolymeric materials will have potentials as insulator films or dielectrics in nanoelectronics in the emerging nanotechnology. They would enable us fabricate an entire new generation of microelectronics that are cleaner, faster, stronger, lighter, and can execute a complicated tasks.

Acknowledgment

This research work was funded by a grant from the Ministry for Science and Culture, Dresden, Saxony State, Germany.

References

1 Merkle, R.C. Molecular Nanotechnology in Frontiers of Supercomputing II: A National Reassessment, edited by Ames, K. R. 1992, University of California Press

2 Feynman, R. There's Plenty of Room at the Bottom, a talk by at an annual meeting of the American Physical Society given on December 29, 1959. Published in Caltech's Engineering and Science, February 1960.

3 Drexler, K. E. Nanomachinery: Atomically precise gears and bearings, , in IEEE Micro Robots and Teleoperators Workshop, Hyannis, Cape Cod, November 1987.

4 Drexler, K. E. Engines of Creation, Doubleday, 1986.

5 Onah E.J. Polymers with low dielectric constants: Synthesis and Fabrication of their utrathin films, University Press/web publishers,Germany, 2003, Chapter 1

6 Onah, E.J.; Oertel, U.; Froeck, C.; Kratzmüller, T.; Steinert, V.; Janke, A.; Voigt, D. ; Lunkwitz,K. Macromol. Mater. Eng. **2002**, 287, 412.

7 Sheirs, J. Modern Fluoropolymers; John Wiley Publishers, 1997, Chapter 1.

8 Satou,H .; Suzuki, H .; Makino, D . in Polyimides, Wilson, D . Stenzenberger, H . D .; Hergenrother P . M . (eds .), Blackie, Glasgow and London 1990, 227.

9 Matsuura, T.; Yamada,N .; Nishi, S .; Hasuda, Y .; Macromolecules **1993**, 26, 419.

10 Onah, E.J. Chem. Mat. **2003**, 15, 4104.

11 Mercer, F . W .; Goodman, T . D . High Performance Polym. **1991**, 3, 297-310.

12 St. Clair, A. K.; St. Clair, T. L.; Winfree, W. P. Polymer Mater. Sci. Eng. **1988** , 59, 28.

13 Kame, K. M.; Wells, L. A.; Cassidy, P. E. High Perform. Polymer **1991**, 3, 191 .

14 Snow ,A. W.; Grifith , J. R .; Soalen , R. L.; Greathouse , J. A.; Lodge , J. K. Polym . Mater . Sci . Eng . **1992** , 66 , 466.

15 Babb , A .; Ezzell ,B . R .; Clement , K . S .; Richney ,W . R .; Kennedy , A . P . Polym . Preprints **1993** , 34 , 413 .

16 Babb , D . A .; Ezzell , B . R .; Clement ,K . S .; Richney ,W . R .; Kennedy , A . P . J . Polym . Sci . : Polym . Chem . Ed . **1993** , 31 , 3465.

17 Laschewsky , H . Ringsdorf , G . Schmidt , Thin Solid Films 1985 , **134** , 153 .

18 Tredgold, R . A .; Winter ,C .S .Thin Solid Films **1983** , 99 , 81.

19 Hasegawa , T .; Nishijo , J .; Watanabe, M . Langmuir **2000** , 16 , 7325.

20 Saraf, R . F .; Dimitrakopoulos , C .; Toney ,M . F .; Kowalczyk , S .P . Langmuir **1996** , 12 , 2802.

21 Parada-Rodriguez , J . M .; Kaku , M .; Sogah , D . Y . Macromolecules 1994 , **27** , 1571.

22 Schneider ,J.; Erdelen,C.; Ringsdorf, H.; Rabolt, J.F. Macromolecules **1989** , 22 , 3475.

23 Sekiya , A.; Ishida ,H.; Tamura ,M.; Watanabe , M. Chem . Lett . **1987** , 1593.

24 Suzuki , M .; Kakimoto , M .; Konishi ,T.; Imai, Y.; Iwamoto, M.; Hino , T. Chem. Lett . **1986** , 39

25 Kakimoto, M .; Suzuki, M.; Konishi, T.; Imai,Y.; Iwamoto, M.; Hino, T. Chem. Lett. **1986** , 823.

26 Nishikata,Y.; Kakimoto , M .; Imai ,Y. J. Chem . Soc. Commun . **1988** , 1040.

27 Treichel, H.; Withers, B.; Ruhl, G.; Ansman, P. ; Würl, R.; Müller, C.; Dietlmeier, D.; Maier G. in Handbook of low and high dielectric constant materials and their applications ed. H. S. Nalwa Acad. Press, San Diego, 1999, Chapter 1.

28 Fryd, M. Polyimides, ed. Mittal, K. L. Plenum Press, NY 1984, 337.

29 Bessonov, M. I.; Koton, M. M.; Kudryavtsev, V. V.; Laius, L. A. Polyimides; thermally stable Polymers, Consultants Burea, NY, A division of Plenum Pub. Corp. NY, 10013.

30 Sun, J.; Sze, W.; Rosenmeyer, T.; Wu, A. Mat. Res. Soc. Symp. **1997**, 443, 89.

31 Resnick, P. R. Polym. Prepar. **1990**, 31, 312.

32 Singh,R.; Sharanpani, R. DUMIC Conference 1996, 78.

33 Onah,E. J.; Oertel, U.; Nagel, J.; Lunkwitz, K.; Leo, K.; Sun, Y.; Fritz, T. in press.

34 Cassidy, P. E.; Fitch, J. W. in Modern Fluoropolymers; high perf. Polymers for diverse application, Sheirs, J. (ed.), John Wiley & Sons Ltd, England 1997, Chapter 8.

35 Houghman, G. in Fluoropolymers 2-Properties, Hougham, G.; Cassidy, P. E.; Johns, K.; Davidson,T. Kluwer Academic/Plenum Publishers, New York 1999, Chapter 13

36 Hamciuc, C.; Brunma, M.; Mercer,F.W.; Kopnick, T.; Shultz, B. Macromol. Mater. Eng. **2000**, 276/277, 38

Indexes

Author Index

Subject Index